LACTIC ACID BACTERIA
A Functional Approach

Editors

Marcela Albuquerque Cavalcanti de Albuquerque
Department of Food and Human Nutrition
School of Pharmaceutical Sciences
University of São Paulo
São Paulo, SP, Brazil

Food Research Center - FoRC
University of São Paulo
São Paulo, SP, Brazil

Alejandra de Moreno de LeBlanc
Centro de Referencia para Lactobacilos (CERELA-CONICET)
San Miguel de Tucumán
Argentina

Jean Guy LeBlanc
Centro de Referencia para Lactobacilos (CERELA-CONICET)
San Miguel de Tucumán
Argentina

Raquel Bedani
Department of Biochemical and Pharmaceutical Technology
University of São Paulo
São Paulo, SP, Brazil

CRC Press
Taylor & Francis Group
Boca Raton London New York

CRC Press is an imprint of the
Taylor & Francis Group, an **informa** business
A SCIENCE PUBLISHERS BOOK

Cover credit: Fabiana Zan

CRC Press
Taylor & Francis Group
6000 Broken Sound Parkway NW, Suite 300
Boca Raton, FL 33487-2742

Version Date: 20191210

International Standard Book Number-13: 978-1-138-39163-5 (Hardback)

Library of Congress Cataloging-in-Publication Data

Names: Albuquerque, Marcela Albuquerque Cavalcanti de, 1984- editor. | LeBlanc, Alejandra de Moreno de, editor. | LeBlanc, Jean Guy (Jean Guy Joseph), editor. | Bedani, Raquel, editor.
Title: Lactic acid bacteria : a functional approach / editors, Marcela Albuquerque Cavalcanti de Albuquerque, Alejandra de Moreno de LeBlanc, Jean Guy LeBlanc, and Raquel Bedani.
Other titles: Lactic acid bacteria (Albuquerque)
Description: Boca Raton : CRC Press, Taylor & Francis Group, [2020] | Includes bibliographical references and index. | Summary: "The book deals with advances made in the functionalities of lactic acid bacteria (LAB) such as, their effect on vitamin D receptor expression, impact on neurodegeneratives pathologies, production of B-vitamins for food bio-enrichment, production of bacteriocins to improve gut microbiota dysbiosis, production of metabolites from polyphenols and their effects on human health, effect on reducing the immunoreaction of food allergens, as biological system using time-temperature to improve food safety, and the use of probiotic in animal feed. The book also reviews the use of LAB and probiotics technologies to develop new functional foods and functional pharmaceutical"-- Provided by publisher.
Identifiers: LCCN 2019056212 | ISBN 9781138391635 (hardback)
Subjects: MESH: Lactobacillales--physiology | Probiotics--therapeutic use | Nutritive Value | Lactobacillales--metabolism | Functional Food--microbiology
Classification: LCC QR82.L3 | NLM QU 145.5 | DDC 579.3/7--dc23
LC record available at https://lccn.loc.gov/2019056212

Visit the Taylor & Francis Web site at
http://www.taylorandfrancis.com

and the CRC Press Web site at
http://www.crcpress.com

Printed and bound by CPI Group (UK) Ltd, Croydon, CR0 4YY

Preface

Lactic Acid Bacteria have been used for centuries in the food and beverage industries, first for food preservation, but in the last century also because of their health benefits. They are regarded as GRAS (generally regarded as safe) organisms for human consumption, and even benefit human health as in the case of probiotics. Probiotics are defined as "Life microorganisms, which, when administered in adequate amounts, confer a health benefit to the host".

Despite several decades of research on probiotics, the mechanisms of action of the bacteria are still far from being elucidated or understood. Topics covered in this book include such simple things as 'what is an adequate amount?', and 'are multi-strain products better than single-strain products'? Probably, because health effects are strain dependent, and diseases and disorders are multifactorial, there are no straightforward answers to even these fundamental questions (Chapter 2). Lactic acid bacteria can also be used to enrich the vitamin B status of foods, through *in situ* production (Chapter 7), or they can interact with vitamin D and its receptor, through influencing the gut microbiota and thereby induce immunomodulatory effects (Chapter 6).

They can also impact gut diseases and disorders such as inflammatory bowel disease and irritable bowel syndrome (Chapter 10). But also diseases and disorders distant from the gut, such as cardiovascular diseases (Chapter 12) and associated with that, obesity (Chapter 9), have been shown to be modulated by probiotics. Moreover, it has recently been discovered that they influence the brain and neurodegenerative pathologies too (Chapter 11).

Several possible modes of action have been entertained, which range from the production of antimicrobial peptides called bacteriocins (Chapter 5); cell-cell communication through the production of quorum sensing molecules (Chapter 1); through (modulation of) production of short-chain fatty acids which play a role in gut homeostasis (Chapter 8) or metabolites of polyphenols with anti-inflammatory or even anticancer activity (Chapter 13). Furthermore, lactic acid bacteria may impact the bioavailability and digestibility of food components, such as minerals or proteins (Chapter 3), and in one particular application may be used to decrease the allergenicity of food by fermenting the immunostimulatory epitopes (Chapter 4).

Apart from human health, they also impact animal health (and if these are used for human food consumption, then indirectly also again human health; Chapter 15). And because of their versatility, they can even be used as biological sensors for improving food safety (Chapter 14) and preventing spoilage, amongst others through their production of bacteriocins and acids.

In this timely book, expert international authors have reviewed these selected hot-topics to provide an up-to-date overview. This book is essential reading for everyone aiming to functionally apply lactic acid bacteria/probiotics for human health, from the PhD student to the experienced scientist.

Prof. Dr. Koen Venema

Founder and CEO of Beneficial Microbes® Consultancy
Chair of Gut Microbiology at the Centre for Healthy Eating & Food Innovation
of Maastricht University – campus Venlo

Contents

1

Cell-Cell Communication in Lactic Acid Bacteria

Potential Mechanisms

Emília Maria França Lima,[1] *Beatriz Ximena Valencia Quecán,*[1] *Luciana Rodrigues da Cunha,*[2] *Bernadette Dora Gombossy de Melo Franco*[1] and *Uelinton Manoel Pinto*[1,*]

Lactic Acid Bacteria

Lactic acid bacteria (LAB) are a diverse group of bacteria, yet with similar properties and all produce lactic acid as an end product of the fermentation process (Ferreira 2012). Taxonomically, the species are found in the phylum *Firmicutes,* Class *Bacilli* and order *Lactobacillales*, and include the genera *Lactobacillus, Lactococcus, Leuconostoc, Oenococcus, Pediococcus, Streptococcus, Enterococcus, Tetragenococcus, Aerococcus, Carnobacterium, Vagococcus* and *Weissella* (De Angelis et al. 2007, Reddy et al. 2008) which are all low guanine-cytosine (GC) content organisms (< 50%). However, some authors also consider *Atopobium* and *Bifidobacterium* genera, from the *Actinobacteria* phylum, as belonging to the LAB group for sharing some similar characteristics (Ferreira 2012, Wedajo 2015), despite the higher GC content.

Phenotypically, LAB are Gram-positive rods or cocci, catalase negative (some species can produce pseudocatalase), chemoorganotrophs and non-spore forming bacteria (Dellagio et al. 1994, Holt et al. 1994). They are aerotolerant, microaerophilic or facultative anaerobic microorganisms with optimum temperature of growth between 30°C and 40°C, but some strains are able to grow at temperatures lower than 5°C or higher than 45°C (Gorner and Valik 2004).

[1] University of São Paulo, Food Research Center, Faculty of Pharmaceutical Sciences, São Paulo, SP, Brazil.
Emails: emiliamflima@usp.br; ximenavalencia@usp.br; bfranco@usp.br
[2] Federal University of Ouro Preto, Nutrition School, Ouro Preto, MG, Brazil.
Email: lrcunha@ufop.edu.br
* Corresponding author: uelintonpinto@usp.br

The essential feature of LAB metabolism is efficient carbohydrate fermentation coupled to substrate-level phosphorylation. This group of bacteria exhibits an enormous capacity to degrade different carbohydrates and related compounds (Mozzi et al. 2010). Based on the end products of glucose metabolism they are classified as homofermentative and heterofermentative microorganisms (Ribeiro et al. 2014). Those that produce lactic acid as the major or sole products of glucose fermentation are designated homofermenters and include some *Lactobacillus* and most *Enterococcus* species, *Lactococcus, Pediococcus, Streptococcus, Tetragenococcus* and *Vagococcus*. Meanwhile, those that produce equal molar amounts of lactate, carbon dioxide, and ethanol from hexoses are designated heterofermentative, including *Leuconostoc,* some species of *Lactobacillus, Oenococcus* and *Weissella* (Jay et al. 2005). The apparent difference on the enzyme level between these two categories is the presence or absence of the key cleavage enzymes of the Embden-Meyerhof pathway (fructose 1,6-diphosphate) and the PK pathway (phosphoketolase).

Lactic bacteria represent one of the most important groups of microorganisms for humans, both for their role in the production and preservation of food, as well as for the aspects related to human health (Ferreira 2012). Some members of this group, isolated from the intestinal tract of humans, possess probiotic characteristics, which when ingested in adequate amounts offer innumerous benefits to the host's health, such as prevention of colon cancer (Dallal et al. 2015, Nouri et al. 2016), attenuation of intestinal constipation (Riezzo et al. 2018, Ou et al. 2019), reduction of serum cholesterol (Wang et al. 2018), improvement of lactose digestion (Dhama et al. 2016), stimulation of the immune system (Imani et al. 2013) and protection from allergy (Wu et al. 2016) and intestinal infections (Collado et al. 2006).

Besides the human health benefits, LAB play a significant role in the food industries, both in food production and preservation. These cultures have been used as starter or adjunct cultures for the fermentation of foods and beverages, accelerating and directing its fermentation process (Leroy and Vuyst 2004). In addition, they contribute to the sensorial characteristics of these products by the acid production, degradation of proteins and lipids, and production of alcohols, aldehydes, acids, esters and sulphur compounds (Zotta et al. 2009). LAB also help to increase the safety and the shelf life of products by producing metabolites, such organic acids, bacteriocins and hydrogen peroxide which possess an inhibitory effect on the growth of pathogenic microorganisms (Servin 2004). The organic acid safety effect is related to the non-dissociated form of the molecule (Podolak et al. 1996), which being lipophilic and apolar can passively diffuse through the membrane (Kashket 1987) and promote acidification of the cellular cytoplasm and impairment of the metabolic function of the competing microorganism (Vasseur et al. 1999). Hydrogen peroxide promotes oxidation of sulfhydryl groups and inactivation of various enzymes. Moreover, it may alter the permeability of the cell membrane by peroxidation of lipids and further cause damages to DNA by the formation of free radicals such as (O_2^-) and hydroxyl (OH^-) (Byezkowski and Gessner 1988). On the other hand, bacteriocins act promoting collapse of the membrane potential by means of electrostatic bonds with the phospholipids (negatively charged). After bonding, the hydrophobic portion of the bacteriocin is inserted in the membrane forming pores allowing the out flow of ions, mainly potassium and magnesium. This promotes dissipation of the proton motive force, compromising synthesis of macromolecules and production of energy and resulting in cell death (Montville et al. 1995). Studies have shown that regulation of bacteriocin production is contingent on cell density in a phenomenon known as 'quorum sensing' (Kuipers et al. 1998, Kleerebezem 2004, Johansen and Jespersen 2017).

Quorum Sensing in Bacteria

Bacteria can communicate and regulate the expression of several genes according to cell density. This mechanism is a type of communication known as quorum sensing (QS), which is based on the production, secretion, and detection of small signaling molecules, whose concentration correlates with the cell density of microorganisms secreting these molecules in the surroundings (Choudhary and Schmidt-Dannert 2010). All quorum sensing systems usually involve the synthesis of a biomolecule with low molecular weight, also termed autoinducer (AI), which is recognized by the responder cell (Declerck et al. 2007, LaSarre and Federle 2013). As bacterial population density increases, the concentration of autoinducer molecules in the environment also rises (Johansen and Jespersen 2017).

Quorum sensing plays a role in many complex processes such as secretion of virulence factors, biofilm formation, sporulation, production of bacteriocins and antimicrobial compounds, among others; activities that would not be possible or beneficial in low population density (Rocha-Estrada et al. 2010, Saeidi et al. 2011, Smid and Lacroix 2013, Monnet and Gardan 2015, Johansen and Jespersen 2017, Quecán et al. 2018).

The signaling molecules may be different according to each bacterial group, as shown in Table 1. The phenomenon has been extensively studied in Gram-negative bacteria, in which signaling is commonly mediated by acylated homoserine lactone molecules (AHLs), known as auto-inducer-1 (AI-1) (Kuipers 1998, Williams 2007, Papenfort and Bassler 2016). However, in Gram-positive bacteria, signaling usually occurs by auto-inducing peptides (AIP) (Kleerebezem et al. 1997, Banerjee and Ray 2017). New molecules have been discovered with the advancement of studies in the area, indicating the existence of alternative types of QS signaling mechanisms besides AHL and AIP (LaSarre and Federle 2013, Zhao et al. 2018). The molecule known as autoinducer 2 (AI-2) is associated with the quorum sensing in the two bacterial groups (Miller and Bassler 2001, Fuqua and Greenberg 2002), and there are also other molecules apart from these classes, such as the *Pseudomonas* quinolone signal (PQS) and the autoinducer-3 (AI-3) in Gram-negative bacteria.

Autoinducer molecules clearly differ in structure and composition (Table 1), but have common characteristics such as high receptor specificity and transport across the cell membrane which may be active or passive (Vadakkan et al. 2018). Some signaling molecules can have a dual function, acting mainly as antimicrobial compounds, besides being involved in the QS system as inducer molecules (Sturme et al. 2007, Smid and Lacroix 2013). This is the case of bacteriocins, which will be discussed in more detail later in this chapter. Table 2 shows the most common signaling molecules involved in quorum sensing by LAB, like bacteriocin production.

The mechanism of quorum sensing was first observed in *Aliivibrio fischeri* (formely *Vibrio fischeri*), a Gram-negative marine bacterium that produces bioluminescence and lives in symbiosis with the Hawaiian squid *Euprymna scolopes*

Table 1: Autoinducer molecules in quorum sensing.

QS system	Signaling molecule	Bacterial group	Reference
AI-1	Many types of acyl homoserine lactones (AHLs) which vary in acyl chain length (4 to 18C) and substitution on C3 (H, OH, or O)	Gram-negative	Fuqua and Greenberg 2002
AI-2	Furanosyl borate diester or (2R,4S)-2-methyl-2,3,3,4-tetrahydroxytetrahidrofuran	Gram-negative and Gram-positive	Park et al. 2016
AI-3	Aromatic aminated, but final structure has not yet been elucidated	*Salmonella, E. coli*	Kendall and Sperandio 2014
AIPs	Autoinducer peptides (ex. Nisin A)	Gram-positive	Kleerebezem 2004
PQS	2-heptyl-3-hydroxy-4-quinolone	*Pseudomonas aeruginosa*	Heeb et al. 2011, Allegretta et al. 2017
HAQs	Hydroxyl-2-alkyl-quinolines or 2-alkyl-4(1H)-quinolones	*Pseudomonas aeruginosa*	Heeb et al. 2011, Maura and Rahme 2017
HHQs	2-Heptyl-3*H*-4-Quinolone	*Pseudomonas aeruginosa*	Baker et al. 2017
IQS	2-(2-hydroxyphenyl)-thiazole-4-carbaldehyde	*Pseudomonas aeruginosa*	Lee et al. 2013
Pyrones	α-pyrones produced by a ketosynthase (PpyS)	*Photorhabdus luminescens*	Brachmann et al. 2013, Zhao et al. 2018
DARs and CHDs	Dialkylresorcinols (DAR) and Cyclohexanediones (CHD)	*Photorhabdus asymbiotica*	Brameyer et al. 2015
Epinephrine and Noradrenaline	Neurotransmitter produced by the host. Catecholamine with a methyl group (epinephrine) or hydrogen (noradrenaline) binds to nitrogen	Gut pathogenic bacteria (*E. coli, S.* Typhimurium, *V. parahaemolyticus*)	Kendall and Sperandio 2014, Bäumler and Sperandio 2016
DSF	cis-11-methyl-2-dodecenoic acid (or diffusible signal factor, DSF)	*Xanthomonas campetris* pv. *campestris*	He and Zhang 2008

(Lazdunski et al. 2004). In this microorganism, the communication is mediated by the N-3-oxohexanoyl-L-homoserine lactone (OHHL), which is produced by a LuxI-type enzyme. When OHHL molecules are in high concentration due to bacterial high cell density, they bind to LuxR-receptor proteins, activating the transcription of genes responsible for bioluminescence (Fuqua and Greenberg 2002). In absence of AHL, the LuxR transcriptional factor is unstable and is rapidly degraded by intracellular proteases (Nealson and Hastings 1979, Miller and Bassler 2001). In other words, the bioluminescence process is associated with the high cell density of *V. fisheri*.

Nowadays, it is common knowledge that quorum sensing signaling is widespread in the microbial world and the signaling molecule repertoire is quite diverse (Williams et al. 2007). Thus, it is perceived that bacterial communication has direct consequences in medicine, agriculture, and potentially food safety and food quality (Pinto et al. 2007, Bai and Rai 2011, Skandamis and Nychas 2012, Rul and Monnet 2015, Zhao et al. 2018). Due to the importance of the phenotypes regulated by quorum sensing, there is a broad interest in developing mechanisms that can interfere in bacterial communication (Defoirdt et al. 2013, Kalia 2013, Maura et al. 2016).

Table 2: Updates in LAB quorum sensing in different products and mechanisms.

Product/ mechanism	Lactic acid bacteria	Signaling molecule	QS effect	Reference
Kimchi (Korean fermented vegetable)	*Weissella, Leuconostoc, Lactobacillus*	AI-2	AI-2 activity in kimchi samples. AI-2 production varied among strains of LAB	Park et al. 2016
Chinese fermented meat	*Lactobacillus* sp., *L. sakei* and *L. plantarum*	AI-2	Signals increased in stress conditions– Important in the regulation of microbial succession in fermented meat	Lin et al. 2015
Shrimp spoilage by co-cultivation	*L. plantarum* and *L. casei*	AI-2	Higher bacteriocin production and inhibition of shrimp spoilage	Li et al. 2019
Gastrointestinal tract	*Lactobacillus rhamnosus* and *L. plantarum*	AI-2	Adapting to host's ecosystem and interacting within intestinal environment	Yeo et al. 2015
Nisin production	*Lactococcus lactis*	Nisin	Antimicrobial activity, reducing biofilms	Jung et al. 2018
Bacteriocin production by co-cultivation	*L. plantarum* and *L. helveticus*	AI-2	Bacteriocin synthesis	Jia et al. 2017
Intestinal microbiota composition	Bacterial microbiota	AI-2	Intestinal AI-2 could increase colonization of healthy microbiota	Sun et al. 2015, Zhao et al. 2018

QS in Gram-Positive Bacteria

There are several processes modulated by QS in Gram-positive bacteria that have been described as exemplified by the regulation of virulence factors in *Staphylococcus aureus*, genetic competence in *Bacillus subtilis* and *Streptococcus pneumoniae*, and the production of bacteriocins in several bacteria (Dunny and Leonard 1997, Kleerebezem et al. 1997, Cotter et al. 2005). The communication via QS in Gram-positive bacteria, in contrast to Gram-negative microbes, is not mediated by AHL signaling molecules and does not use the LuxI/LuxR-type signaling.

In Gram-positive bacteria, signaling is mediated by small peptides used as auto-inducer molecules with variations in the type and complexity of additional regulatory factors (Kuipers et al. 1998, Bassler 1999, Novick and Geisinger 2008, Monnet et al. 2016). These molecules vary drastically depending on the cell and they allow each strain to communicate intra-specifically. In this process, Gram-positive bacteria secrete peptides processed generally by ABC-type exporting proteins, since the oligopeptides do not diffuse freely through the plasma membrane. Thus, they need a specific transporter which generally modifies the structure of the autoinducer during the secretion process (Dunny and Leonard 1997). These signals are then recognized by two-component sensor kinases located in the membrane of the cell, which monitor one or more signaling molecules, and transfer this information via phosphorylation to intracellular response regulators, which modulate the expression of specific genes as

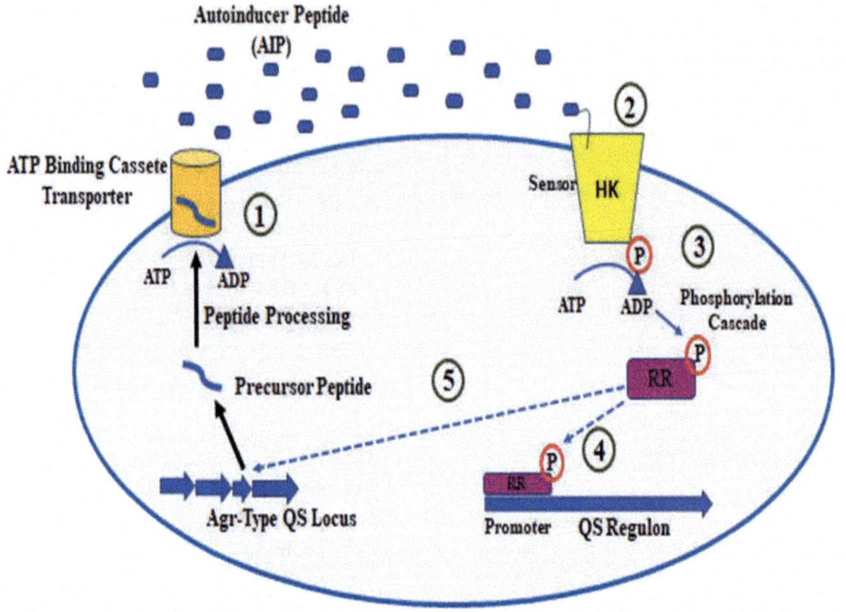

Figure 1: General mechanism of quorum sensing in Gram-positive bacteria: **(1)** The ABC transporter system allows the removal of the N-terminal leader sequence of the peptide and the secretion out of the bacterial cell. **(2)** The signal oligopeptides (AIP) are detected by a sensor protein of the histidine-kinase family (HK) when they reach a significant concentration. **(3)** The protein becomes phosphorylated upon binding to the AIP and the phosphate group is transferred to a response regulator protein (RR). **(4)** The phosphorylated regulator protein binds to promoter regions of genes regulated by QS (the QS regulon) and alter its transcription. **(5)** The autoinduction process is depicted by the positive feedback loop, in which the RR activates the synthesis of more AIPs. Source: Authors. Based on several studies including: Kleerebezem et al. 1997, Novick and Geisinger 2008, Ziemichód and Skotarczak 2017.

a mechanism of signal transduction, as indicated in Figure 1 (Bassler 1999, Declerck et al. 2007, Pang et al. 2016).

One of the best studied models of the QS system in Gram-positive bacteria is the one described for *S. aureus*. When this bacterium is at low cellular concentration, different protein factors are expressed, allowing the adherence and colonization of the microorganism in surfaces. However, at high cellular concentrations, the synthesis of these factors is repressed, and virulence factors secretion is activated (Yarwood et al. 2004, Novick and Geisinger 2008, Paulander et al. 2018).

All these processes are regulated by the Agr QS system, which is constituted by four genes that encode the expression of four proteins: (i) AgrD, the autoinducer peptide (AIP); (ii) AgrC, the histidin-kinase responsible for transmitting the signal from the plasma membrane to the transcriptional regulator located in the cytoplasm; (iii) AgrA, the transcriptional regulator that gets phosphorylated by AgrC in response to high cell density; and (iv) AgrB: an export protein that modifies the thiolactone ring of the precursor peptide autoinducer (see Figure 1) (Peng et al. 1988, Paulander et al. 2018).

When the levels of *S. aureus* cells increase, the peptide autoinducer binds to the membrane histidine kinase AgrC and induces its phosphorylation, which then transfers the phosphate group to the transcriptional regulator AgrA that binds to DNA with several effects: one to induce the transcription of RNA III that regulates the expression of virulence factors; second to repress the expression of colonization and adhesion factors to surfaces, and third to induce the expression of the *agr* operon (D, C, A, B) to autoregulate the system, being a classical example of a positive feedback loop (Kleerebezem et al. 1997, Díaz et al. 2010, Wang and Muir 2016, Paulander et al. 2018).

In the case of QS in other Gram-positive bacteria, there are similarities to the system described for *S. aureus* and even homologous systems in several bacteria (Thoendel et al. 2011). The QS communication in LAB is increasingly calling the interest of many researchers around the world as described in the following section.

Quorum Sensing in Lactic Acid Bacteria—Production of Bacteriocins

Bacteriocins produced by LAB comprise a diverse group of antimicrobial peptides; however, two main categories have been established based on their structural modifications, their size, their thermostability, and mechanisms of action. The first category refers to Class I (lantibiotics) which are small cationic membrane-active peptides from 19 to 38 amino acid residues that contain lanthionine or ß-methyl lanthionine, which undergo post-translational modifications and exert their effect at the level of the membrane and the cell wall. In this group, nisin is the most representative bacteriocin (Cotter et al. 2005, López et al. 2008, Rea et al. 2011, Silva et al. 2018). On the other hand, class II (non-lantibiotics) is defined as heat-stable bacteriocins, constituted by 30 to 70 amino acid residues that do not contain lanthionine. They induce pore formation in the membrane of the target cells. These bacteriocins are divided into subclasses: subclass IIa anti-listeria, pediocin-like bacteriocins is the largest of all, and its members have the amino terminal motif containing one or two disulfide bridges (Nes et al. 2013); subclass IIb is formed by bacteriocins that require the combined action of two peptides to have activity because they cannot act individually (López et al. 2008, Rea et al. 2011); subclass IIc the cyclic bacteriocins due to the covalent attachment of the carboxy and amino terminal ends and; subclass IId which is formed by a variable group of linear non pediocin-like peptides (Sánchez et al. 2003, López et al. 2008, Nes et al. 2013).

Studies indicate that the production of bacteriocins in some bacteria is partly regulated by quorum-sensing (Kuipers et al. 1998, Kleerebezem 2004, Cotter et al. 2015, Johansen and Jespersen 2017). Bacteriocins are produced by various Gram-positive and Gram-negative bacteria, and one of the best cases of QS in lactic acid bacteria is provided by the knowledge on bacteriocin biosynthesis (Kleerebezem 2004, Sturme et al. 2007, Jia et al. 2017). Different strains of LAB produce a variety of bacteriocins and sometimes the active molecular species is a non-modified peptide or a combination of two peptides (Dunny and Leonard 1997). However, the nisin production by *Lactococcus lactis* stands out as the best studied mechanism (Hilmi et al. 2006, Arauz et al. 2009, Rul and Monnet 2015, Jung et al. 2018).

Nisin is a polypeptide member of the antibiotic family called lantibiotics. Thus, it possesses the amino-acid lanthionine in its composition, as the other bacteriocins of this group, besides having methyl-lanthionine, dehydroalanine (Dha) and dehydrobutyrine (Dhb) residues (Dunny and Leonard 1997, Kleerebezem and Quadri 2001, Willians and Delves-Broughton 2003, Jung et al. 2018). The precursor for nisin consists of two parts: a leader peptide, and a modifiable core peptide. The leader peptide contributes to the interaction between nisin and dedicated enzymes (NisB and NisC), to perform post-translational modification in the molecule. NisB dehydrates serines and threonines, while NisC covalently couples thiol groups of Cys to Dha or Dhb. These modified enzymes form a complex with the ABC transport protein (NisT) which is likely involved in the secretion of the modified pre-nisin molecule. Thus, NisB, NisC and NisT consist in a complex of nisin modification enzymes (Kleerebezem and Quadri 2001, Khusainov et al. 2013). The autoregulation process in *Lactococcus lactis* is mediated by two component regulatory system: the histidine kinase NisK that acts as a sensor for nisin, and the response regulator NisR (Table 2). Once the signal is transduced by autophosphorylation of NisK and subsequent phospho-transfer to NisR, the transcription of target genes is activated (Kuipers et al. 1998, Hilmi et al. 2006, Jung et al. 2018). A similar mechanism is depicted in Figure 1.

Nisin was the first bacteriocin produced by a LAB to be approved for use in foods, and has been considered as an excellent preservative, because it has the capacity to act as an inhibitor of several types of Gram-positive bacteria such as strains of *Streptococcus, Enterococcus, Staphylococcus, Micrococcus, Pediococcus, Lactobacillus, Listeria* and *Mycobacterium*, but not in Gram-negative bacteria due to their outer membrane barrier (Williams and Delves-Broughton 2003, Arauz et al. 2009, Punyauppa-path et al. 2015, Jung et al. 2018). This is an antimicrobial agent considered as GRAS (Generally Recognized as Safe) and it was evaluated to be safe for use in foods by the Joint Food and Agriculture Organization/World Health Organization (FAO/WHO) (Arauz et al. 2009, Rul and Monnet 2015, Jung et al. 2018).

Researchers have found that the antimicrobial action of nisin is mediated by the passage through the cell wall and the interaction with lipid II (Punyauppa-path et al. 2015). In the first case, nisin induces the lysis of vegetative cells due to permeabilization of the cytoplasmic membrane, helping in the formation of pores that promote the extravasation of low molecular weight intracellular molecules such as ATP, with consequent collapse of cells, mainly by deprivation of the energy source. In the second, nisin binds to lipid II, preventing the incorporation of new monomers and consequently the growing peptidoglycan chain (Williams and Delves-Broughton 2003, Arauz et al. 2009, Jung et al. 2018).

In other bacteria, bacteriocin production has also been shown to be induced by QS. For instance, plantaricin production in *Lactobacillus* (*L.*) *plantarum* happens at high population density in a coordinated way, using specific autoinducing signaling peptides (AIPs) that are often post-translationally modified and exported by dedicated transport systems, similar to the mechanism described for nisin. The AIP often involves a two-component regulatory system, comprising the histidine protein kinase (HPK) that monitors environmental factors, and the cytoplasmic response regulator (RR) that regulates gene expression, as previously described (Sturme et al. 2007). The secreted pheromone plantaricin A (PlnA) is used as a measure of bacterial cell density: PlnA triggers phosphorylation reactions of HPK and RR. Thus, phosphorylated RR binds to the promoters of the bacteriocin regulon and activates the genes involved in its biosynthesis. The antimicrobial activity of PlnA can be explained by its interaction with membrane lipids, changing the PlnA conformation. The interaction with the cell membrane also allows PlnA to bind the receptor which mediates the autoinducing effect (Kristiansen et al. 2005, Sturme et al. 2007, Di Cagno et al. 2010).

Authors have suggested the need for more experimental studies on the regulatory mechanisms of *Lactobacillus* QS systems and their effects on gene regulation. It is also important to study other factors such as stressful conditions and co-cultivation with competing bacteria which could provide new insights into the mechanisms of bacterial adaptation (Sturme et al. 2007, Di Cagno et al. 2010).

Co-Cultivation in LAB and Quorum Sensing

Many processes including food production are affected when bacteria are in co-cultivation in a manner that depends upon QS communication. Table 2 shows a large number of studies on this topic. Co-cultivation may favor the production of bacteriocin in *L. plantarum*. This fact can be explored to increase the viability of this species in fermented food processing and feeds, as well as to contribute in gastrointestinal tract colonization, conferring a probiotic function (Maldonado-Barragán et al. 2013). In *L. plantarum* NC8, plantaricin production was increased in the presence of other bacteria. According to the Maldonado and collaborators (2004), this regulatory mechanism may be part of a defense strategy that includes the detection of other bacteria in the same environment and a subsequent response leading to bacteriocin production. In *L. plantarum* DC400, the pheromone PlnA is produced either under mono or co-culture conditions, with different responses depending on the microbial partner. Except for *L. pentosus* 12H5, co-cultivation with several LAB species induces de plantaricin synthesis, being highest with *L. sanfranciscensis* DPPMA174 partner, which leads to lethal conditions for the later (Di Cagno et al. 2010).

The enhancement of QS effect in co-culture was also observed with *L. plantarum* and *L. casei*, resulting in higher bacteriocin production and inhibiting shrimp spoilage (Li et al. 2019). LAB co-cultivation has also been studied in production of AIP signaling molecules in fermented foods, and this process generally influences the effect and the production of AIPs and AI-2 (Johansen and Jespersen 2017).

LAB Quorum Sensing in Fermented Foods

Currently, there is an interest in the study of cell-cell communication in fermented foods for the control of microbial communities (Johansen and Jespersen 2017).

The typical microbial populations presented in fermented food are directly and/or indirectly associated with the beneficial effects that these products confer to human gastrointestinal and metabolic health (Park et al. 2016, Johansen and Jespersen 2017). Although the understanding of AI-2 signaling in these products is deficient, researchers expect a dynamic AI-2 signaling from microbial food cultures in non-heat treated fermented foods, and that the microbial interactions should affect the AI-2 signaling accumulation (Park et al. 2016). Quorum sensing in LAB is involved in the production of different fermented foods and beverages such as vegetables, dairy products and wine, as demonstrated in Table 2.

As the studies continue to progress, an increased body of knowledge on quorum sensing regulation in fermented products will help the food industry to actively pursue QS to improve the quality and safety of foods. This could be accomplished by favoring, for example, the growth of LAB and probiotic strains in order to promote the beneficial effects that these bacteria have on human health and well-being (Park et al. 2016, Johansen and Jespersen 2017).

Conclusion

LAB have been playing a significant role in the food industries since they have been used as starter cultures, promoting fermentation and contributing to the sensorial characteristics of the products. Additionally, they are also implicated in increasing the safety and the shelf life of these products by producing metabolites, such as hydrogen peroxide, organic acids and bacteriocins that have an inhibitory effect on the growth of pathogenic microorganisms. Many of these processes are modulated by QS, in a process mediated by signals recognized by a two-component regulatory system that modulates the expression of specific genes as a mechanism of signal transduction. Nisin, a bacteriocin produced by *Lactococcus lactis* is one of the most studied bacteriocins, and its regulation, like others, depends upon QS. In fact, studies indicate that fermentation of different products is influenced by bacterial interactions and co-cultivation, generally mediated by signaling molecules including AI-2 mediated communication. Even though there are studies demonstrating the importance of QS in LAB, more research is necessary to optimize their potential in food production.

References

Allegretta, G., C.K. Maurer, J. Eberhard, D. Maura, R.W. Hartmann, L. Rahme and M. Empting. 2017. In-depth profiling of MvfR-regulated small molecules in *Pseudomonas aeruginosa* after quorum sensing inhibitor treatment. Front. Microbiol. 8: 924.

Arauz, L.J., A.F. Jozala, P.G. Mazzola and T.C.V. Penna. 2009. Nisin biotechnological production and application: a review. Trend. Food Sci. Technol. 20: 146–154.

Bai, A.J. and V.R. Rai. 2011. Bacterial quorum sensing and food industry. Compr. Rev. Food Sci. Food Saf. 10: 183–193.

Baker, Y.R., J.T. Hodgkinson, B.I. Florea, E. Alza, W.R.J.D. Galloway, L. Grimm, S.M. Geddis, H.S. Overkleeft and M. Welch. 2017. Identification of new quorum sensing autoinducer binding partners in *Pseudomonas aeruginosa* using photoaffinity probes. Chem. Sci. 11: 7403–7411.

Banerjee, G. and A.K. Ray. 2017. Quorum-sensing network-associated gene regulation in Gram-positive bacteria. Acta Microbiol. Imm. Hung. 64: 439–453.

Bassler, B.L. 1999. How bacteria talk to each other: regulation of gene expression by quorum sensing. Curr. Opin. Microbiol. 2: 582–587.

Bäumler, A.J. and V. Sperandio. 2016. Interactions between the microbiota and pathogenic bacteria in gut. Nature. 7610: 85–93.

Brachman, A.O., S. Brameyer, D. Kresovic, I. Hitkova, Y. Kopp, C. Manske, K. Schubert, H.B. Bode and R. Heermann. 2013. Pyrones as bacterial signaling molecules. Nat. Chem. Biol. 9: 573.

Brameyer, S., D. Kresovic, H.B. Bode and R. Heermann. 2015. Dialkylresorcinols as signaling molecules. Proc. Natl. Acad. Sci. 112: 572–577.

Byezkowski, J. and T. Gessner. 1988. Biological role of superoxide íon-radical. Int. J. Biochem. 20: 569–580.

Choudhary, S. and C. Schmidt-Dannert. 2010. Applications of quorum sensing in biotechnology. Appl. Microbiol. Biotechnol. 86: 1267–1279.

Collado, M.C., L. Jalonen, J. Meriluoto and S. Salminen. 2006. Protection mechanism of probiotic combination against human pathogens: *in vitro* adhesion to human intestinal mucus. Asia Pac. J. Clin. Nutr. 15: 570–575.

Cotter, P.D., C. Hill and R.P. Ross. 2005. Bacterial lantibiotics: strategies to improve therapeutic potential. Curr. Protein Pept. Sci. 6: 61–75.

Dallal, M.M.S., M. Mojarrad, F. Baghbani, R. Raoofian, J. Mardaneh and Z. Salehipour. 2015. Effects of probiotic *Lactobacillus acidophilus* and *Lactobacillus casei* on colorectal tumor cells activity (CaCo-2). Arch. Iran. Med. 18: 167–72.

De Angelis, M., R. Di Cagno, G. Gallo, M. Curci, S. Siragusa, C. Crecchio, E. Parente and M. Gobbetti. 2007. Molecular and functional characterization of *Lactobacillus sanfranciscensis* strains isolated from sourdoughs. Int. J. Food Microbiol. 114: 69–82.

Declerck, N., L. Bouillaut, D. Chaix, N. Rugani, L. Slamti, F. Hoh, D. Lereclus and S.T. Arold. 2007. Structure of PlcR: Insights into virulence regulation and evolution of quorum sensing in Gram-positive bacteria. Proc. Natl. Acad. Sci. 104: 18490–18495.

Defoirdt, T., G. Brackman and T. Coenye. 2013. Quorum sensing inhibitors: How strong is the evidence? Trends Microbiol. 21: 619–624.

Dellagio, F., H. Roissart, S. Torriani, M.C. Curk and D. Janssens. 1994. Caracteristiquesgeneralesdês bacteries lactiques. pp. 25–116. *In*: H. Roissart and F.M. Liquet (eds.). Bacteries Lactiques. Uriage, France.

Dhama, K., S.K. Latheef and A.K. Munjal. 2016. Probiotics in curing allergic and inflammatory conditions—research progress and futuristic vision. Recent Pat. Inflamm. Allergy Drug Discov. 10: 21–33.

Di Cagno, F., M.D. Angelis, M. Calasso, O. Vincentini, P. Vernocchi, M. Ndagijimana, M.D. Vincenzi, M.R. Dessi, M.E. Guerzoni and M. Gobbetti. 2010. Quorum sensing in sourdough *Lactobacillus plantarum* DC400: Induction of plantaricin A (PlnA) under co-cultivation with other lactic acid bacteria and effect of PlnA on bacterial and Caco-2 cells. Proteomics. 10: 2175–2190.

Díaz, D.M. and A.S. de la Sem. 2010. Sistemas de quorum sensing en bacterias. Reduca (Biología) 3: 39–55.

Dunny, G.M. and B.A. Leonard. 1997. Cell-cell communication in Gram-positive bacteria. Annu. Rev. Microbiol. 51: 527–564.

Ferreira, C.L.L.F. 2012. Grupo de bactérias lácticas e aplicação tecnológica de bactérias probióticas. pp. 1–27. *In*: Ferreira, C.L.L.F. (ed.). Prebióticos e Probióticos – Atualização e Prospecção. Rubio, Rio de Janeiro, Brasil.

Fuqua, C. and E.P. Greenberg. 2002. Listening in on bacteria: Acyl-homoserine lactone signalling. Nat. Rev. Mol. Cell. Biol. 3: 685–695.

Gorner, F. and L. Valik. 2004. Aplikovaná mikrobiológia požívatín. Malé Centrum, Bratislava.

He, Y.W. and L.H. Zhang. 2008. Quorum sensing and virulence regulation in *Xanthomonas campestris*. FEMS Microbiol. Rev. 32: 842–857.

Heeb, S., M.P. Fletcher, S.R. Chhabra, S.P. Diggle, P. Williams and M. Cámara. 2011. Quinolones: from antibiotics to autoinducers. FEMS Microbiol. Rev. 35: 247–274.

Hilmi, H.T.A., K. Kylä-Nikkilä, R. Ra and P.E. Saris. 2006. Nisin induction without nisin secretion. Microbiology 152: 1489–1496.

Holt, J.G., N.R. Krieg, P.H.A. Sneath, J.T. Staley and S.T. Williams. 1994. Gram positive cocci. pp. 527–588. *In*: Holt, J.C., N.R. Krieg, P.H.A. Sneath, J.T. Staley and S.T. Williams (eds.). Bergey's Manual of Determinative Bacteriology. Ed. Baltimore: Williams & Wilkins.

Imani, A.A., H. Mahmoodzadeh, M.R. Nourani, S. Khani and S.M. Alavian. 2013. Probiotic as a novel treatment strategy against liver disease. Hepat. Mon. 13: e7521.

Jay, J.M., M.J. Loessner and D.A. Golden. 2005. Fermentation and dairy fermented products. pp. 131–149. *In:* Jay, J.M., M.J. Loessner and D.A. Golden (eds.). Modern Food Microbiology. Springer Science + Business Media, Inc., USA.

Jia, F.F., X.H. Pang, D.Q. Zhu, Z.T. Zhu, S.R. Sun and X.C. Meng. 2017. Role of the luxS gene in bacteriocin biosynthesis by *Lactobacillus plantarum* KLDS1.0391: A proteomic analysis. Sci. Rep. 7: 13871.

Johansen, P. and L. Jespersen. 2017. Impact of quorum sensing on the quality of fermented foods. Curr. Opin. Food Sci. 13: 16–25.

Jung, Y., A.B. Alayande, S. Chae and I.S. Kim. 2018. Applications of nisin for biofouling mitigation of reverse osmosis membranes. Desalination. 429: 52–59.

Kashket, E.R. 1987. Bioenergetics of lactic acid bacteria: Cytoplasmic pH and osmotolerance. FEMS Microbiol. Rev. 46: 233–244.
Kendall, M.M. and V. Sperandio. 2014. Cell-to-cell signaling in *E. coli* and *Salmonella*. EcoSal Plus. 6(1). doi:10.1128/ecosalplus.ESP-0002-2013.
Khusainov, R., G.N. Moll and O.P. Kuipers. 2013. Identification of distinct nisin leader peptide regions that determine interactions with the modification enzymes NisB and NisC. FEBS Open Bio. 3: 237–242.
Kleerebezem, M., L.E. Quadri, O.P. Kuipers and W.M. De Vos. 1997. Quorum sensing by peptide pheromones and two-component signal-transduction systems in Gram-positive bacteria. Mol. Microbiol. 24: 895–904.
Kleerebezem, M. and L.E. Quadri. 2001. Peptide pheromone-dependent regulation of antimicrobial peptide production in Gram-positive bacteria: a case of multicellular behavior. Peptides 22: 1579–1596.
Kleerebezem, M. 2004. Quorum sensing control of lantibiotic production; nisin and subtilin autoregulate their own biosynthesis. Peptides 25: 1405–1414.
Kristiansen, P.E., G. Fimland, D. Mantzilas and J. Nissen-Meyer. 2005. Structure and mode of action of the membrane-permeabilizing antimicrobial peptide pheromone plantaricin A. J. Biol. Chem. 280: 22945–22950.
Kuipers, O.P., P.G.G.A. de Ruyter and M.K.W.M. de Vos. 1998. Quorum sensing-controlled gene expression in lactic acid bacteria. J. Biotechnol. 64: 15–21.
LaSarre, B. and M.J. Federle. 2013. Exploiting quorum sensing to confuse bacterial pathogens. Microbiol. Mol. Biol. Rev. 77: 73–111.
Lazdunski, A.M., I. Ventre and J.N. Sturgis. 2004. Regulatory circuits and communication in Gram-negative bacteria. Nat. Rev. Microbiol. 2: 581–592.
Lee, J., J. Wu, Y. Deng, J. Wang, C. Wang, J. Wang, C. Chang, Y. Dong, P. Williams and L.H. Zhang. 2013. A cell-cell communication signal integrates quorum sensing and stress response. Nat. Chem. Biol. 9: 339–343.
Leroy, F. and L. De Vuyst. 2004. Lactic acid bacteria as functional starter cultures for the food fermentation industry. Trends Food Sci. Technol. 15: 67–78.
Li, J., X. Yang, G. Shi, J. Chang, Z. Liu and M. Zeng. 2019. Cooperation of lactic acid bacteria regulated by the AI-2/LuxS system involve in the biopreservation of refrigerated shrimp. Food Res. Int. 120: 679–687.
Lin, M., G. Zhou, Z. Wang and B. Yun. 2015. Functional analysis of AI2/LuxS from bacteria in Chinese fermented meat after high nitrate concentration shock. Eur. Food Res. Technol. 240: 119–127.
López, J.E., A. Ochoa, G. Santoyo, J.L. Anaya, E. Medina, M. Martínez and P.D. Loeza. 2008. Bacteriocinas de bacterias Gram positivas: una fuentepotencial de nuevostratamientosbiomédicos. Ver. Mex. Cienc. Farm. 39: 49–57.
Maldonado, A., J.L. Ruiz-Barba and R. Jiménez-Díaz. 2004. Production of plantaricin NC8 by *Lactobacillus plantarum* NC8 is induced in the presence of different types of gram-positive bacteria. Arch. Microbiol. 181: 8–16.
Maldonado-Barragán, A., B. Caballero-Guerrero, H. Lucena-Padrós and J.L. Ruiz-Barba. 2013. Induction of bacteriocin production by coculture is widespread among plantaricin-producing *Lactobacillus plantarum* strains with different regulatory operons. Food Microbiol. 33: 40–47.
Maura, D., A.E. Ballok and L. Rahme. 2016. Considerations and caveats in anti-virulence drug development. Curr. Opin. Microbiol. 33: 41–46.
Maura, D. and L. Rahme. 2017. Pharmacological inhibition of the *Pseudomonas aeruginosa* MvfR quorum-sensing system interferes with biofilm formation and potentiates antibiotic-mediated biofilm disruption. Antimicrob. Agents Chemother. 61: e01362–17.
Miller, M.B. and B.L. Bassler. 2001. Quorum sensing in bacteria. Annu. Rev. Microbiol. 55: 165–199.
Monnet, V. and R. Gardan. 2015. Quorum-sensing regulators in Gram-positive bacteria: 'cherchez le peptide'. Mol. Microbiol. 97: 181–184.
Monnet, V., V. Juillard and R. Gardan. 2016. Peptide conversations in Gram-positive bacteria. Crit. Rev. Microbiol. 42: 339–351.
Montville, T.J., K. Winkowski and R.D. Ludescher. 1995. Models and mechanisms for bacteriocin action application. Int. Dairy J. 5: 797–814.

Mozzi, F.F., R.R. Raya and G.M. Vignolo. 2010. Biotechnology of Lactic Acid Bacteria Novel Applications. Ames: Wiley Blackwell.
Nealson, K.H. and J.W. Hastings. 1979. Bacterial bioluminescence: its control and ecological significance. Microbiol. Rev. 43: 496–518.
Nes, I.F., D.A. Brede and D.B. Diep. 2013. Class II non-lantibiotic bacteriocins. pp. 85–92. *In*: Kastin, A.J. (ed.). Handbook of Biologically Active Peptides. Academic Press.
Nouri, Z., F. Karami, N. Neyazi, M.H. Modarressi, R. Karimi, M.R. Khorramizadeh, B. Taheri and E. Motevaseli. 2016. Dual anti-metastatic and antiproliferative activity assessment of two probiotics on HeLa and HT-29 cell lines. Cell J. 18: 127–134.
Novick, R.P. and E. Geisinger. 2008. Quorum sensing in staphylococci. Ann. Rev. Genet. 42: 541–564.
Ou, Y., S. Chen, F. Ren, M. Zhang, S. Ge, H. Guo, H. Zhang and L. Zhao. 2019. *Lactobacillus casei* strain Shirota alleviates constipation in adults by increasing the pipecolinic acid level in the gut. Front. Microbiol. 10: 324.
Pang, X., C. Liu, P. Lyu, S. Zhang, L. Liu, J. Lu, C. Ma and J. Lv. 2016. Identification of quorum sensing signal molecule of *Lactobacillus delbrueckii* subsp. *bulgaricus*. J. Agric. Food Chem. 64: 9421–9427.
Papenfort, K. and B.L. Bassler. 2016. Quorum sensing signal-response systems in Gram-negative bacteria. Nat. Rev. Microbiol. 14: 576–588.
Park, H., H. Shin, K. Lee and W. Holzapfel. 2016. Autoinducer-2 properties of kimchi are associated with lactic acid bacteria involved in its fermentation. Int. J. Food Microbiol. 225: 38–42.
Paulander, W., A.N. Varming, M.S. Bojer, C. Friberg, K. Bæk and H. Ingmer. 2018. The *agr* quorum sensing system in *Staphylococcus aureus* cells mediates death of sub-population. BMC Research Notes 11: 503.
Peng, H.L., R.P. Novick, B. Kreiswirth, J. Kornblum and P.M. Schlievert. 1988. Cloning, characterization, and sequencing of an accessory gene regulator (*agr*) in *Staphylococcus aureus*. J. Bacteriol. 170: 4365–4372.
Pinto, U.M., E.S. Viana, M.L. Martins and M.C.D. Vanetti. 2007. Detection of acylated homoserine lactones in gram-negative proteolytic psychrotrophic bacteria isolated from cooled raw milk. Food Control. 18: 1322–1327.
Podolak, P.K., J.F. Zayas, C.L. Kastner and D.Y.C. Fung. 1996. Inhibition of *Listeria monocytogenes* and *Escherichia coli* O157:H7 on beef by application of organic acids. J. Food Prot. 59: 370–373.
Punyauppa-path, S., P. Phumkhachorn and P. Rattanachaikunsopon. 2015. Nisin: production and mechanism of antimicrobial action. Int. J. Cur. Res. Rev. 7: 47–53.
Quecán, B.X.V., M.L.C. Rivera and U.M. Pinto. 2018. Bioactive phytochemicals targeting microbial activities mediated by quorum sensing. pp. 397–416. *In*: Kalia, V.C. (ed.). Biotechnological Applications of Quorum Sensing Inhibitors. Springer Nature Singapore.
Rea, M.C., R.P. Ross, P.D. Cotter and C. Hill. 2011. Classification of bacteriocins from Gram-positive bacteria. pp. 29–53. *In*: Drider, D. and S. Rebuffat (eds.). Prokaryotic Antimicrobial Peptides. Springer, New York, USA.
Reddy, G., M.D. Altaf, B.J. Naveena, M. Venkateshwar and E.V. Kumar. 2008. Amylolytic bacterial lactic acid fermentation—a review. Biotechnol. Adv. 26: 22–34.
Ribeiro, S.C., M.C. Coelho, S.D. Teodorov, B.D.G.M. Franco, M.L.E. Dapkevicius and C.C.G. Silva. 2014. Technological properties of bacteriocin-producing lactic acid bacteria isolated from Pico cheese an artisanal cow's milk cheese. J. Appl. Microbiol. 116: 573–585.
Riezzo, G., A. Orlando, B. D'Attoma, M. Linsalata, M. Martulli and F. Russo. 2018. Randomised double blind placebo controlled trial on *Lactobacillus reuteri* DSM 17938: improvement in symptoms and bowel habit in functional constipation. Benef. Microbes 9: 51–59.
Rocha-Estrada, J., A.E. Aceves-Diez, G. Guarneros and M. De La Torre. 2010. The RNPP family of quorum-sensing proteins in Gram-positive bacteria. Appl. Microbiol. Biotechnol. 87: 913–923.
Rul, Françoise and V. Monnet. 2015. How microbes communicate in food: a review of signaling molecules and their impact on food quality. Curr. Opin. Food Sci. 2: 100–105.
Saeidi, N., C.K. Wong, T. Lo, H.X. Nguyen, H. Ling, S.S.J. Leong, C.L. Poh and M.W. Chang. 2011. Engineering microbes to sense and eradicate *Pseudomonas aeruginosa*, a human pathogen. Mol. Syst. Biol. 7: 1–11.

Sánchez, B.M.J., R.M. Martínez, A. Gálvez, E. Valdivia, M. Maqueda, V. Cruz and A. Albert. 2003. Structure of bacteriocin AS-48: From soluble state to membrane bound state. J. Mol. Biol. 334: 541–549.

Servin, A.L. 2004. Antagonistic activities of lactobacilli and bifidobacteria against microbial pathogens. Microbiol. Rev. 28: 405–440.

Silva, C., S.P.M. Silva and S.S. Ribeiro. 2018. Application of bacteriocins and protective cultures in dairy food preservation. Front. Microbiol. 9: 1–15.

Skandamis, P.N. and G.J.E. Nychas. 2012. Quorum sensing in the context of food microbiology. Appl. Environ. Microbiol. 78: 5473–5482.

Smid, E.J. and C. Lacroix. 2013. Microbe-microbe interactions in mixed culture food fermentations. Curr. Opin. Biotechnol. 24: 148–154.

Sturme, M.H.J., C. Francke, R.J. Siezen, W.M. de Vos and M. Kleerebezem. 2007. Making sense of quorum sensing in lactobacilli: a special focus on *Lactobacillus plantarum* WCFS1. Microbiol. 153: 3939–3947.

Sun, Z.K., V. Grimm and C.U. Riedel. 2015. Al-2 to the rescue against antibiotic-induced intestinal dysbiosis? Trends Microbiol. 23: 327–328.

Thoendel, M., J.S. Kavanaugh, C.E. Flack and A.R. Horswill. 2011. Peptide signaling in Staphylococci. Chem. Rev. 111: 117–151.

Vadakkan, K., A.A. Choudhury, R. Gunasekaran, J. Hemapriya and S. Vijayanand. 2018. Quorum sensing intervened bacterial signaling: Pursuit of its cognizance and repression. J. Genet. Eng. Biotechnol. 16(2): 239–252.

Vasseur, C., L. Bavarel, M. Hebraud and J. Labadie. 1999. Effect of osmotic alkaline, acid or thermal stresses on the growth and inihibition of *Listeria monocytogenes*. J. Appl. Microbiol. 86: 469–476.

Wang, B. and T.W. Muir. 2016. Regulation of virulence in *Staphylococcus aureus*: molecular mechanisms and remaining puzzles. Cell. Chem. Biol. 23: 214–224.

Wang, L., M.J. Guo, Q. Gao, J.F. Yang, L. Yang, X.L. Pang and X.J. Jiang. 2018. The effects of probiotics on total cholesterol: A meta-analysis of randomized controlled trials. Medicine. 97(5).

Wedajo, B. 2015. Lactic acid bacteria: benefits, selection criteria and probiotic potential in fermented food. J. Prob. Health. 3(2).

Williams, G.C. and J.D. Broughton. 2003. *Nisin*. Danisco Innovation, Beaminster, Dorset, UK. 4128–4135.

Williams, P. 2007. Quorum sensing, communication and cross-kingdom signalling in the bacterial world. Microbiology 153: 3923–3938.

Wu, C.T., P.J. Chen, Y.T. Lee, J.L. Ko and K.H. Lue. 2016. Effects of immunomodulatory supplementation with *Lactobacillus rhamnosus* on airway inflammation in a mouse asthma model. J. Microbiol. Immunol. Infect. 49: 625–635.

Yarwood, J.M., D.J. Bartels, E.M. Volper and E.P. Greenberg. 2004. Quorum sensing in *Staphylococcus aureus* biofilms. J. Bacteriol. 186: 1838–1850.

Yeo, S., H. Park, Y. Ji, S. Park, J. Yang, J. Lee, J.M. Mathara, H. Shin and W. Holzapfel. 2015. Influence of gastrointestinal stress on autoinducer-2 activity of two *Lactobacillus* species. FEMS Microbiol. Ecol. 91: 1–9.

Zhao, J., C. Quan, L. Jin and M. Chen. 2018. Production, detection and application perspectives of quorum sensing autoinducer-2 in bacteria. J. Biotechnol. 268: 53–60.

Ziemichód, A. and B. Skotarczak. 2017. QS-systems communication of Gram-positive bacterial cells. Acta Biologica. 24: 51–56.

Zotta, T., E. Parente and A. Ricciardi. 2009. Viability staining and detection of metabolic activity of sourdough lactic acid bacteria under stress conditions. World J. Microbiol Biotechnol. 25: 1119–1124.

2

Probiotic Dose-Response and Strain Number

Arthur C Ouwehand

Introduction

Lactic acid bacteria (LAB) perform many roles; they are involved in the fermentation of foods and drinks where they contribute to improved flavour, nutritional value and preservation. However, LAB and bifidobacteria may also provide health benefits to the consumer in addition to their nutritional contribution. In this case we are referring to them as probiotics. The most widely accepted definition of probiotics is 'Live microbes that, when administered in adequate amounts, confer a health benefit on the host' (FAO/WHO 2002, Hill et al. 2014). This definition thus stipulates the administration of an 'adequate amount' in order to obtain a health benefit. However, from the definition it is unclear what this 'adequate' dose should be; this is unfortunate, but also understandable. A common assumption is that there is a dose-response effect on the health benefit. Besides efficacy, there may also be other benefits to the consumption of higher doses; such as an earlier health benefit and/or a longer health benefit when consumption seizes. The effective dose of probiotics is likely influenced by a multitude of variables; it is possibly influenced by the targeted health benefit and the strain (or combination of strains). There is also a substantial subject-to-subject variation in the response to probiotics. These and many other factors make it difficult to determine a general optimal dose for probiotic effects; but considering the large number of human intervention studies on probiotic efficacy, the available data may give some hints on this topic (Ouwehand 2017). Probiotic dose is not only relevant for efficacy; it is potentially also relevant from a safety perspective. Is there any risk for the consumer with high doses?

In addition to the question of dose, there is the topic of the number of strains included in a probiotic product. Products can be single-strain or multi-strain probiotics, where the latter may consist of two or more probiotics strains from the same, but

Global Health and Nutrition Sciences, DuPont Nutrition & Biosciences, 02460 Kantvik, Finland.
Email: arthur.ouwehand@dupont.com

usually different species and/or genera (Ouwehand et al. 2018a). The probiotic definition does not treat these two categories different. However, the challenges in determining whether the difference between the categories suffers from similar challenges to the dose-response question.

This chapter will address first the evidence on dose-response and second the discussion of the data on single versus multi-strain probiotics.

Dose-Response

There are different ways by which dose-response relations can be studied. Here the focus will be placed on human intervention studies as these are most relevant. For the present chapter, human trials that enable addressing dose-response effects have been divided into two categories.

1) Meta-analyses enable the comparison of a large number of probiotic studies for a given health benefit. The included studies will often have used a wide range of doses enabling a sub-group analysis focussing on dose-response. The weakness is that meta-analyses usually combine a variety of strains and strain-combinations. The studies will also have been performed in different populations, with different protocols, etc. These variations may impact the ability to draw conclusions on dose-response.
2) Preferred of course are studies where a given probiotic (combination) was compared head-to-head at different doses for a given health benefit.

To this end, PubMed was searched using 'meta-analysis' and 'probiotic' as search terms. Meta-analyses were checked for the presence of a dose-response sub-group analysis. For strain-specific dose-response studies, PubMed was searched using the terms 'probiotic', 'dose-response' and filtering for 'human'. The results from this latter search were complemented, with articles on different doses reported in the meta-analyses.

Meta-analyses with dose-response sub-group analyses

Several meta-analyses have done sub-group analysis on dose as shown in Table 1. The most convincing evidence for a dose-response effect was observed for Antibiotic Associated Diarrhoea (AAD). Interestingly, no such effect was observed for *Clostridium* (*C.*) *difficile* Associated Diarrhoea (CDAD) or authors concluded there was insufficient data; this despite the fact that 26 (Lau and Chamberlain 2016) or 31 (Lau and Chamberlain 2016) studies were included. Similarly, for Necrotising Enterocolitis (NEC) with 26 included studies, the authors conclude there was insufficient data (Aceti et al. 2015). Notwithstanding this, Deshpande and co-workers (Deshpande et al. 2011) recommend a dose of at least 3×10^9 CFU; this, based on the median dose used in studies. Meta-analyses on reducing blood pressure and weight management indicate doses of at least 10^{11} CFU and 3×10^{10} CFU, respectively, for benefits. For other health benefits, the authors conclude there is either no dose-response effect and/or there is insufficient data. Another complicating factor may be that the range of tested doses is relatively narrow; making it unlikely to actually observe a difference even if it would exist.

Table 1: Meta-analyses with dose-response sub-group analyses.

Age group	Number of studies analysed	Outcome	Reference
Antibiotic Associated Diarrhoea (AAD)			
Children	8	Most effective at 10^{10} CFU or more	(Van Niel et al. 2002)
Children and adults	25	Doses less than 10^{10} CFU tend to be unsuccessful	(McFarland 2006)
Children	6	Doses larger than 5×10^9 CFU are more effective	(Johnston et al. 2006)
Children	16	Doses larger than 5×10^9 CFU are more effective	(Johnston et al. 2011)
***Clostridium difficile* Associated Diarrhoea (CDAD)**			
Children and adults	6	No dose-response effect	(McFarland 2006)
Children and adults	31	No dose-response effect	(Goldenberg et al. 2013)
Children and adults	21	No dose-response effect	(McFarland 2015)
Children and adults	26	Insufficient data	(Lau and Chamberlain 2016)
Children and adults	31	Insufficient data	(Goldenberg et al. 2017)
Adults	19	No dose-response effect	(Shen et al. 2017)
Necrotising Enterocolitis (NEC)			
Premature infants	26	Insufficient data	(Aceti et al. 2015)
Gastrointestinal diseases			
Children and adults	64	No dose-response effect	(Ritchie and Romanuk 2012)
Irritable bowel syndrome (IBS)			
Children and adults	14	Insufficient data	(Hoveyda et al. 2009)
Not given	17	No dose-response effect	(Zhang et al. 2016b)
Colonic transit			
Adults	13	No dose-response effect	(Miller and Ouwehand 2013)
Adults	15	No dose-response effect	(Miller et al. 2016a)
Colorectal cancer			
Adults	6	Insufficient data	(He et al. 2013)
Adjunct to *H. pylori* eradication therapy			
All age groups	19	No dose-response effect	(McFarland et al. 2016)
Blood Pressure			
Adults	9	No effect below 10^{11} CFU	(Khalesi et al. 2014)
Primary prevention of atopic dermatitis			
New born infants	13	No dose-response effect	(Pelucchi et al. 2012)
New born infants	12	No dose-response effect	(Zhang et al. 2016a)

Table 1 contd. ...

... *Table 1 contd.*

Age group	Number of studies analysed	Outcome	Reference
Weight management			
Not given	21	Doses larger than 3×10^{10} CFU are more effective	(John et al. 2018)
Rheumatoid arthritis			
Adults	5	Insufficient data	(Aqaeinezhad Rudbane et al. 2018)
Gestational diabetes			
Adults	4	Insufficient data	(Taylor et al. 2017)
Post-operative sepsis			
Adults	15	Insufficient data	(Arumugam et al. 2016)

Thus, the fact that there is no observed dose-response in meta-analyses does not mean it may not exist; rather currently available data is not sufficient or appropriate to draw any conclusion.

Dose-response studies on specific probiotic (combinations)

Some human intervention studies have been performed comparing different doses of probiotics in the same study; in most cases two different doses, but in some cases up to four different doses, plus placebo. Studies like this are challenging; not only are more volunteers required to fill the extra dose group. Increasing the number of groups in a study has statistical implications and may require a further increase in the number of volunteers.

Six studies, published in seven articles traced faecal recovery of the fed probiotic (combination); four reported a dose-dependent effect (Table 2). Two studies reported changes in microbiota composition, i.e., levels of *Lactobacillus* and *Bifidobacterium*; this likely represents the fed probiotics. One of them (Gianotti et al. 2010) using among others *Lactobacillus* (*L.*) *johnsonii* indicated that the highest dose was associated with increased faecal lactobacilli; this likely reflects the observed increase in the administered *L. johnsonii*. Faecal recovery therefore seems dose dependent. To avoid any doubt, faecal recovery is not considered a health benefit but is considered here as a physiological outcome for the purpose of understanding potential dose-response relations.

Four studies reported on immune markers (Table 2), two of these did not show an immune modulating effect. This is likely related to the study population, healthy subjects where a change in immune markers is less likely to occur. Only one study found for one parameter, influenza specific IgG, an improved effect for the higher dose. This makes it difficult to draw any conclusion on dose-response for immune marker modulation.

For general health and well-being, one study did not observe an effect; this is likely related to the study set-up, healthy adults and a relatively short (4 weeks) intervention. Two of the remaining studies reported a dose-dependent reduction in infection

Table 2: Human intervention studies comparing two or more doses of the same probiotic product in parallel in the same study.#

Strain (combination)	Doses tested (CFU/day)	Study population	Population size	Physiological effect/health target	Outcome	Reference
Faecal recovery						
L. rhamnosus GG	1.6×10^8, 1.2×10^{10}	Healthy adults	20	Faecal recovery	Higher dose resulted in more people with detectable *L. rhamnosus* GG.	(Saxelin et al. 1995)
B. lactis BB-12 + *L. paracasei* CRL-341	Placebo, 10^8, 10^9, 10^{10}, 10^{11}	Healthy adults	75	Faecal recovery	Increasing dose resulted in more people with detectable *B. lactis* BB-12 and higher mean levels. *L. paracasei* CRL-341 was not detected.	(Larsen et al. 2006, Christensen et al. 2006)
B. lactis BB-12	Placebo, 10^9, 10^{10}	Healthy adults	58	faecal recovery	Similar increase of *B. lactis* levels for both doses.	(Savard et al. 2011)
B. longum 536 + *L. johnsonii* La1	Placebo, 10^7, 10^9	Patients elective colorectal resection for cancer	31	Faecal recovery	High dose had more subjects with detectable *L. johnsonii* La1; not *B. longum* 536.	(Gianotti et al. 2010)
B. subtilis R0179	Placebo, 10^8, 10^9, 10^{10}	Healthy adults	81	Faecal recovery	Dose dependent faecal recovery.	(Hanifi et al. 2015)
L. acidophilus La-14, *L. pararcasei* LPC-37, *L. plantarum* Lp-115, *B. lactis* Bl-04	7×10^9, 7×10^{10}	Healthy adults	40	Faecal recovery	Dose dependent faecal recovery of all strains.	(Taverniti et al. 2019)
Microbiota modulation						
B. lactis HN019	Placebo, 6.5×10^7, 10^9, 5×10^9	Healthy elderly	80	Faecal *Bifidobacterium* and *Lactobacillus* levels	Similar significant increase in bifidobacteria, Trend for dose-dependent increase in lactobacilli.	(Ahmed et al. 2007)

Table 2 contd. ...

... *Table 2 contd.*

Strain (combination)	Doses tested (CFU/day)	Study population	Population size	Physiological effect/health target	Outcome	Reference
B. longum 536 + *L. johnsonii* La1	Placebo, 10^7, 10^9	Patients elective colorectal resection for cancer	31	Microbiota composition	High dose had higher faecal *Lactobacillus* levels.	(Gianotti et al. 2010)
Immune markers						
B. longum 536 + *L. johnsonii* La1	Placebo, 10^7, 10^9	Patients elective colorectal resection for cancer	31	Immune markers	No effect on measured immune markers.	(Gianotti et al. 2010)
B. lactis BB-12 + *L. paracasei* CRL-341	Placebo, 10^8, 10^9, 10^{10}, 10^{11}	Healthy adults	75	Immune markers	No effect on measured immune markers.	(Christensen et al. 2006)
L. plantarum CECT7315 + *L. plantarum* CECT7316	Placebo, 5×10^8, 5×10^9	Healthy elderly	50	Immune markers	No difference between doses for measured markers.	(Mane et al. 2011)
L. plantarum CECT7315 + *L. plantarum* CECT7316	Placebo, 5×10^8, 5×10^9	Healthy elderly	60	Influenza vaccination	5×10^9 increased influenza specific serum IgG, similar increase in influenza specific serum IgA for both doses.	(Bosch et al. 2012)
General health/immune function						
L. plantarum CECT7315 + *L. plantarum* CECT7316	Placebo, 5×10^8, 5×10^9	Healthy elderly	50	General health	Similar mortality risk for both doses, greater reduction in infection risk in 5×10^9 group.	(Mane et al. 2011)
Heat-killed *L. pentosus* b240	Placebo, 2×10^9, 2×10^{10}	Healthy elderly	300	Respiratory tract infection risk	Dose dependent reduction in common cold risk; though no influence on duration or symptom scores.	(Shinkai et al. 2013)
B. subtilis R0179	Placebo, 10^8, 10^9, 10^{10}	Healthy adults	81	General wellbeing	No effect on general wellbeing.	(Hanifi et al. 2015)

E. coli Nissle 1917	Placebo, 10^9, 2×10^9, 4×10^9	Patients ulcerative colitis	90	Time to remission	Dose dependent shorter time to remission.	(Matthes et al. 2010)
Irritable Bowel Syndrome (IBS)						
B. infantis 35624	Placebo, 10^6, 10^8, 10^{10}	Adult women	362	IBS	10^8 effective, 10^6 and 10^{10} not different from placebo.	(Whorwell et al. 2006)
L. acidophilus NCFM	Placebo, 10^9, 10^{10}	Adult IBS patients	391	IBS	No difference between doses and placebo. Both doses improved similarly intestinal pain.	(Lyra et al. 2016)
Bowel function						
L. plantarum CECT7484 + *L. plantarum* CECT7485 + *P. acidilactici* CECT7483	Placebo, 3–6 × 10^9, 1–3 × 10^{10}	Adult IBS-D patients	84	IBS-D	Both doses improved quality of life similar but better than placebo.	(Lorenzo-Zuniga et al. 2014)
B. lactis BB-12 + *L. paracasei* CRL-341	Placebo, 10^8, 10^9, 10^{10}, 10^{11}	Healthy adults	75	Stool consistency	Increasing dose correlated with increase in stool consistency.	(Larsen et al. 2006)
B. lactis BB-12	Placebo, 10^9, 10^{10}	Healthy adults with low defecation frequency	1248	Defecation frequency	Both doses improved bowel function similarly.	(Eskesen et al. 2015)
B. lactis HN019	Placebo, 1.8 × 10^9, 1.72 × 10^{10}	Adults with self-reported constipation	100	Colonic transit time, gastrointestinal symptoms	Shortening of colonic transit time in a dose dependent manner, Similar relief of symptoms for both doses.	(Waller et al. 2011)
B. lactis HN019	Placebo, 10^9, 10^{10}	Adults with functional constipation	228	Colonic transit time; bowel movement frequency	No difference in colonic transit time between doses and placebo. Both doses improved similarly bowel movement frequency.	(Ibarra et al. 2018)
Heat-killed *L. kunkeei* YB38	Placebo, 2, 10, 50 mg	Adult females with low defaecation frequency	30	Defecation frequency, microbiota, stool characteristics	10 and 50 mg reduced *B. fragilis* group and reduced faecal pH, increased faecal acetate with 50 mg, all doses tended to increase defecation.	(Asama et al. 2016)

Table 2 contd. ...

... Table 2 contd.

Strain (combination)	**Doses tested (CFU/day)**	**Study population**	**Population size**	**Physiological effect/health target**	**Outcome**	**Reference**
Diarrhoea						
L. rhamnosus 35	Placebo, 2×10^8, 6×10^8	Children with rotavirus diarrhoea	23	Faecal rotavirus shedding	Only high dose reduced faecal rotavirus shedding compared to placebo.	(Fang et al. 2009)
L. acidophilus CL1285 + *L. casei* LBC80R	Placebo, 5×10^{10}, 10^{11}	Adult patients	225	Antibiotic associated diarrhoea, *C. difficile* associated diarrhoea	Dose-dependent reduced antibiotic associated diarrhoea and *C. difficile* associated diarrhoea incidence.	(Gao et al. 2010)
L. acidophilus NCFM + *L. paracasei* Lpc-37 + *B. lactis* Bi-07 + *B. lactis* Bl-04	Placebo, 4.17×10^9, 1.7×10^{10}	Adult patients	503	Antibiotic associated diarrhoea, *C. difficile* associated diarrhoea	Dose dependent reduction in antibiotic associated diarrhoea risk, number of loose stools and duration of diarrhoea. similar reduction in *C. difficile* associated diarrhoea for both doses.	(Ouwehand et al. 2014)
C. butyricum CBM588	Placebo, 10^7, 2×10^7	*H. pylori* positive peptic ulcer patients	19	Antibiotic associated diarrhoea	Dose dependent reduction in diarrhoea.	(Imase et al. 2008)
S. cerevisiae Boulardii	Placebo, 5×10^9, 10^{10}	Tourists	1231	Travelers diarrhoea	Similar reduction in diarrhoea risk for both doses.	(Kollaritsch et al. 1989)
S. cerevisiae Boulardii	Placebo, 5×10^9, 2×10^{10}	Tourists	1016	Travelers diarrhoea	Dose dependent reduction in diarrhoea risk.	(Kollaritsch et al. 1993)
Women's health						
L. rhamnosus GR1 + *L. reuteri* RC-14	8×10^8, 1.6×10^9, 6×10^9	Healthy women	33	Vaginal microbiota restoration (as Nugent score)	All doses improved Nugent score, intermediate dose seemed most effective.	(Reid et al. 2001)

L. crispatus CTV-05	5×10^8, 10^9, 2×10^9	Healthy women	12	Safety of product upon vaginal application	No differences in adverse events incidence (all mild to moderate), no difference in tolerance and acceptance.	(Hemmerling et al. 2009)
L. fermentum CECT5716	Placebo, 3×10^9, 6×10^9, 9×10^9	Lactating women with painful breastfeeding	148	*Staphylococcus* counts in milk, milk microbiota and breast feeding pain	Similar reduction in milk *Staphylococcus* load and relief of pain.	(Maldonado-Lobon et al. 2015)
Weight management/metabolic syndrome						
B. bifidum W23, *B. lactis* W51, *B. lactis* W52, *L. acidophilus* W37, *L. brevis* W63, *L. casei* W56, *L. salivarius* W24, *Lc. lactis* W19, *Lc. lactis* W58	Placebo, 2.5×10^9, 10^{10}	Obese postmenopausal women	81	Lipopolysaccharide level and cardiometabolic profile	Beneficial effect in both groups, but more so in the high dose group; lipopolysaccharide, uric acid, triglycerides, and glucose.	(Szulinska et al. 2018)
L. gasseri BNR17	Placebo, 10^9, 10^{10}	Overweight-obese adults	90	Body weight, waist circumference	Reduced visceral adipose tissue in high dose group.	(Kim et al. 2018)
L. reuteri DSM 17938	Placebo, 10^8, 10^{10}	Adults with Type 2 diabetes	46	Glycated haemoglobin (HbA1c), insulin sensitivity, liver fat content, body composition, body fat distribution, faecal microbiota composition and serum bile acids	No difference between groups; no effect.	(Mobini et al. 2017)

[#]Modified and expanded after (Ouwehand et al. 2018a).

risk and one further study reported a dose-dependent shorter time to remission (see Table 2). Thus, the number of studies is limited but suggests a potential dose-dependence.

Two studies investigated irritable bowel syndrome (IBS) as shown in Table 2. Not all doses were found to be efficacious as compared to the placebo. So, no conclusions can be drawn on any dose dependence.

Of the six dose-response studies on bowel function, two reports observed dose-dependence (Table 2). Four did not observe dose-dependence, but did report improved bowel function. Interestingly, two strains *Bifidobacterium* (*B.*) *lactis* BB-12 and *B. lactis* HN019, both reported dose-dependence and non-dose-dependence in separate studies. Hence inconclusive results do not allow for conclusions on potential dose-dependent effects for probiotics and bowel function.

Six intervention studies reported on different dose in management of various forms of diarrhoea (Table 2); five of them suggest a dose-dependent effect; in particular all three of the studies dealing with AAD. There is only one rotavirus diarrhoea study suggesting a dose-response effect. The two traveller's diarrhoea studies are contradictory. Three studies in women's health do not suggest a clear dose-response effect (Table 2). This may partially be due to the chosen population in two of the studies; healthy women. Two of the three dose-response studies in weight management/metabolic health, suggest a better effect in the higher dose (Table 2). The third study was not effective.

The most convincing results for a dose-response can thus be observed for faecal recovery and AAD. Especially the latter is in agreement with the results from the meta-analyses as well. Here too, the fact that there is no observed dose-response for other health benefits does not mean it may not exist; currently available data is not sufficient or appropriate to draw any conclusion.

Safety consideration related to probiotic dosing

Safety of probiotics is a prime issue for regulators and manufacturers. The aim of this section is not to give an in-depth review of the overall safety of probiotics; for this reader is advised to consult other sources, e.g. (Sanders et al. 2016, Didari et al. 2014). Here the question is related to dose and hence if there is a potential risk for overconsumption.

As can be seen from Table 3, acute toxicity studies in mice and rats have failed to observe an LD50 or no observed adverse effect level (NOAEL). While for some organisms the limit is at least 1 g/kg body weight and for others up to 50 g/kg body weight, this should not be interpreted as the latter being safer. The doses rather indicate the highest dose tested.

If we assume the lowest dose tested in Table 3; 1 g/kg body weight; this amount corresponds to approximately 5×10^{11} CFU. When translating toxicity data from rodents to humans, a safety margin of 10 to 100 is used. If we assume the latter, the NOAEL for humans is 5×10^{9} CFU/kg body weight; which for a 70 kg human translates into 3.5×10^{11} CFU. One of the highest doses studied in humans is 3.6×10^{12} CFU/day of the probiotic combination VSL#3 in Ulcerative Colitis patients. No adverse events were observed (Tursi et al. 2010).

Table 3: Non-exhaustive list of acute toxicity of probiotics in mice.[#]

Species	**Strain**	**LD50 or NOAEL (g/kg body weight)**	**Reference**
B. coagulans	BC30	> 1.0‡ ‡‡	(Endres et al. 2009)
B. lactis	Bi-07	> 5	(Morovic et al. 2017)
B. lactis	Bl-04	> 5	(Morovic et al. 2017)
B. lactis	HN019	> 50	(Zhou et al. 2000)
B. longum	BB536	> 50	(Ouwehand and Salminen 2003)
B. longum	*	25	(Ouwehand and Salminen 2003)
L. acidophilus	HN017	> 50	(Zhou et al. 2000)
L. acidophilus	NCFM	> 5	(Morovic et al. 2017)
L. casei	Shirota	> 2.0	(Ouwehand and Salminen 2003)
L. delbrueckii subsp. *bulgaricus*	*	> 6.0	(Ouwehand and Salminen 2003)
L. helveticus	*	> 6.0	(Ouwehand and Salminen 2003)
L. paracasei	Lpc-37	> 5	(Morovic et al. 2017)
L. plantarum	CECT 7527	> 1‡	(Mukerji et al. 2016)
L. plantarum	CECT 7528	> 1‡	(Mukerji et al. 2016)
L. plantarum	CECT 7529	> 1‡	(Mukerji et al. 2016)
L. rhamnosus	GG	> 6.0	(Ouwehand and Salminen 2003)
L. rhamnosus	HN001	> 50	(Zhou et al. 2000)
S. cerevisiae var. *boulardii*	Unique 28	> 6.5‡‡	(Sudha 2011)

[#] Modified and expanded after (Ouwehand and Salminen 2003)
* Strain not indicated
‡ Chronic toxicity (90 days)
‡‡ Rats

Overall it therefore seems unlikely that an overconsumption of probiotics by otherwise healthy adults is feasible.

Number of Strains

Fermented foods typically contain only a limited number of probiotic strains; even so-called ABC yogurts contain only three probiotic species; typically, *L. acidophilus*, *B. lactis* and *L. (para)casei*; though often with starter cultures *L. delbrueckii* subsp. *bulgaricus* and *Streptococcus* (*S.*) *thermophilus*. However, dietary supplements, may contain anywhere from one to 10 or more strains of different probiotics. In the case of these multi-strain formulations it is important to note the different strains in a formulation may be present at different counts. From a quality control perspective, it may be challenging to follow the viable counts of these different strains throughout the shelf life of the product; especially when dealing with strains from the same species. With complex formulations of probiotics it is therefore common to follow the overall viable count of the product throughout shelf life (Ouwehand et al. 2018a).

There are a number of perceived benefits, but also a potential risk associated with the use of probiotic combinations as compared to single strain products (Ouwehand et al. 2018a).

1. More strains give more chances of success for a beneficial effect of the product and thus a larger success rate for a given health benefit
2. More strains provide a greater diversity and thus more potential niches and a broader efficacy spectrum
3. Additive or synergistic effects
   a. Enhanced adhesion
   b. Creation of favourable environment
   c. Reduced antagonism of endogenous microbiota
4. Antagonistic activity between the combined probiotics

For the above scenarios it would be beneficial to understand the mechanism of action of probiotics so one could predict if a positive interaction can be expected; there are synergies, antagonism or can additive effects be expected. Unfortunately there is currently insufficient understanding of probiotic mechanisms to make such predictions (Kleerebezem et al. 2018).

In this section the data available concerning the potential benefits and drawbacks of combining probiotic strains will be discussed, with a focus on human studies.

To understand what is known about strain number and probiotic efficacy the best answer is provided by studies that have compared single and multi-strain products. Alternatively, meta-analyses may have done sub-group analyses comparing studies with single and multiple strains.

Studies comparing single and multi-strain probiotics

Similar as with the comparison of different doses, studying the effect of single and multi-strain probiotics requires additional study resources. There is, however, an additional challenge. In the simplest case, a two-strain probiotic product will require two study groups for each separate strain, a group for the two-strain combination and a placebo group. More complex probiotic combinations will require even larger study designs. It is therefore not surprising that such studies are limited in number. A further challenge is the dose. A combination of X strains would have X-times the dose of the single strains introducing a dose-effect, which as we have seen above is not completely clear. Of course, the fractions of probiotics in a multi-strain product do not need to be equal; further complicating this matter.

Of the six articles indicated in Table 4, four reports on different aspects of the same study; furthermore, the single strain in this study is not a component of the multi-strain product. Three of the six articles report no difference between single and multi-strain probiotics; two articles report a stronger effect for the multi-strain and one article a smaller effect for the tested multi-strain product. From this data it is not possible to draw conclusions on a general benefit of either single or multi-strain probiotics.

Table 4: Non-exhaustive list of studies comparing single (S) and multi-strain (M) probiotics in human studies.

Objective	**Strains**	**Dose (10^9 CFU)**	**Subjects Age (years) N**	**Comparison**	**Reference**
Respiratory tract infections	*L. acidophilus* NCFM	10	3.7 years N = 110	M ≥ S	(Leyer et al. 2009)
	L. acidophilus NCFM *B. lactis* Bi-07	10	3.8 years N = 112		
Respiratory tract infections	*B. lactis* Bl-04	2	36 years N = 161	M = S	(West et al. 2014b)
	L. acidophilus NCFM *B. lactis* Bi-07	10	36 years N = 155		
Immune markers	*B. lactis* Bl-04	2	50 years N = 46	M = S	(West et al. 2014a)
	L. acidophilus NCFM *B. lactis* Bi-07	10	51 years N = 47		
Routine haematology and clinical chemistry markers	*B. lactis* Bl-04	2	42.2 years N = 39	M ≥ S	(Cox et al. 2014)
	L. acidophilus NCFM *B. lactis* Bi-07	10	37.3 years N = 45		
Peripheral blood regulatory T-cells	*B. lactis* Bl-04	2	39 years N = 30	M = S	(West et al. 2016)
	L. acidophilus NCFM *B. lactis* Bi-07	10	37 years N = 32		
Analgesic receptor expression	*L. acidophilus* NCFM	20	18–70 years N = 13–14	M ≤ S	(Ringel-Kulka et al. 2014)
	L. acidophilus NCFM *B. lactis* Bi-07	20	18–70 years N = 13–14		

Meta-analyses comparing studies with single and multi-strain probiotics

Some meta-analyses have performed sub-group analyses to determine if and what the influence of strain number is for a particular health end-point. A disadvantage with this approach is, as usual with meta-analyses, that different strains and strain combinations are compared. PubMed was searched for: meta-analysis AND probiotic AND (multi- or combination or blend). The results of the search are presented in Table 5. Most of the retrieved meta-analyses report similar activity for single and multi-strain probiotics. However, the four (out of 11) articles that report a difference between single and multi-strain probiotics indicate that the latter are more efficacious. This is a fairly weak suggestion that multi-strain probiotics might sometimes be more efficacious.

Potential negative interactions with multi-strain probiotics

Many probiotics strains produce antimicrobial components and all probiotics from the genera *Lactobacillus* and *Bifidobacterium* produce lactic and acetic acid, which

Table 5: A non-exhaustive list of meta-analyses comparing single- and multi-strain probiotics for selected health benefits in humans.

Age group	Number of studies analysed	Comparison	Reference
Antibiotic Associated Diarrhoea (AAD)			
All age groups	51	S = M	(Cai et al. 2018)
Children	23	S = M	(Goldenberg et al. 2015)
Triple therapy *H. pylori* eradication			
All age groups	29	S = M	(Feng et al. 2017)
All age groups	13	S = M	(Lu et al. 2016)
Post-operative infections			
All age groups	9	S < M	(Liu et al. 2017)
Adults	20	S = M	(Lytvyn et al. 2016)
Ulcerative colitis (UC)			
All age groups	12	S < M	(Ganji-Arjenaki and Rafieian-Kopaei 2018)
Necrotising Enterocolitis (NEC)			
Infants	24	*Lactobacillus* = M	(AlFaleh and Anabrees 2014)
Infants	11	S < M	(Guthmann et al. 2010)
Intestinal transit time			
Adults	15	S = M	(Miller et al. 2016a)
Primary prevention of atopic dermatitis			
Infants	14	S < M	(Dang et al. 2013)

are also powerful antimicrobial components. It is therefore a fair question if there is a potential negative interaction between the probiotic strains in a multi-strain product. In fermented multi-strain products, interactions may occur already during the fermentation process. In dietary supplements, however, the strains are freeze dried and because they are dormant no interaction occurs. After consumption, the strains become active, and then interactions are possible. From Tables 4 and 5 it is clear that most comparisons do not observe a difference between single and multi-strain probiotics, and a fair number observe better effects with the multi-strain probiotics. Only in one case is the single-strain product superior to the multi-strain product (Ringel-Kulka et al. 2014). This indicates that antagonistic activities are rare. A recent *in vitro* simulation of digestive passage has shown that a single strain and a multi-strain product exhibit similar probiotics survival (Forssten and Ouwehand 2017). Further studies demonstrating that antagonistic activities are unlikely to happen in multi-strain probiotics is needed.

Conclusion

The studies discussed here do not give a clear answer to the questions of regarding dose and the number of strains used. What is clear, however, is that there is no concern regarding overdosing of probiotics or that there is a risk for antagonistic interactions.

Hence, there is no argument against higher doses or more different strains. That is of course not enough; the question is if there is a benefit. In the case of dosing, it appears that at least for AAD there seems to be ample evidence from specific dose-response studies as well as from meta-analyses that higher doses provide a greater benefit. This could relate to the, also convincingly documented increased faecal colonisation with higher doses. For other health benefits, the picture is less clear. This does not necessarily mean a dose-response does not exist; rather the current data is insufficient to draw conclusions. More studies are needed that test different doses, with emphasis placed on particular lower doses. This to document potential dose-response effects on one hand and on the other hand to determine if there is a potential saturation effect and higher doses do not further increase efficacy (Eskesen et al. 2015). In the short term, it would be advisable if more meta-analyses performed sub-group analysis on the topic of dose. This could be done by meta-regression or by comparing population above and below the median dose (this is preferred over an arbitrarily set dose as this may lead to an uneven distribution of the high and low-dose groups).

Likewise, for the number of strains, it appears difficult to draw a clear conclusion. Here too, too few studies have compared single versus multi-strain products and the single strain may not have been part of the multi-strain product. Hence, further studies would be necessary before conclusions can be drawn. Also here, a short term solution would be to include the issue of the number of strains as a sub-group analysis in meta-analyses.

Disclosure

The author is an employee of DuPont Nutrition & Biosciences. DuPont Nutrition & Biosciences manufactures and markets probiotics.

References

Aceti, A., D. Gori, G. Barone, M.L. Callegari, A. Di Mauro, M.P. Fantini, F. Indrio, L. Maggio, F. Meneghin, L. Morelli, G. Zuccotti, L. Corvaglia and N. Italian Society of Neonatology. 2015. Probiotics for prevention of necrotizing enterocolitis in preterm infants: systematic review and meta-analysis. Ital. J. Pediatr. 41: 89.

Ahmed, M., J. Prasad, H. Gill, L. Stevenson and P. Gopal. 2007. Impact of consumption of different levels of *Bifidobacterium lactis* HN019 on the intestinal microflora of elderly human subjects. J. Nutr. Health Aging 11(1): 26–31.

AlFaleh, K. and J. Anabrees. 2014. Probiotics for prevention of necrotizing enterocolitis in preterm infants. Cochrane Database Syst. Rev. 4: CD005496.

Aqaeinezhad Rudbane, S.M., S. Rahmdel, S.M. Abdollahzadeh, M. Zare, A. Bazrafshan and S.M. Mazloomi. 2018. The efficacy of probiotic supplementation in rheumatoid arthritis: a meta-analysis of randomized, controlled trials. Inflammopharmacology 26(1): 67–76.

Arumugam, S., C.S. Lau and R.S. Chamberlain. 2016. Probiotics and synbiotics decrease postoperative sepsis in elective gastrointestinal surgical patients: a meta-analysis. J. Gastrointest. Surg. 20(6): 1123–1131.

Asama, T., Y. Kimura, T. Kono, T. Tatefuji, K. Hashimoto and Y. Benno. 2016. Effects of heat-killed *Lactobacillus kunkeei* YB38 on human intestinal environment and bowel movement: a pilot study. Benef. Microbes 7(3): 337–344.

Bosch, M., M. Mendez, M. Perez, A. Farran, M.C. Fuentes and J. Cune. 2012. *Lactobacillus plantarum* CECT7315 and CECT7316 stimulate immunoglobulin production after influenza vaccination in elderly. Nutr. Hosp. 27(2): 504–509.

Cai, J., C. Zhao, Y. Du, Y. Zhang, M. Zhao and Q. Zhao. 2018. Comparative efficacy and tolerability of probiotics for antibiotic-associated diarrhea: Systematic review with network meta-analysis. United European Gastroenterol. J. 6(2): 169–180.

Christensen, H.R., C.N. Larsen, P. Kaestel, L.B. Rosholm, C. Sternberg, K.F. Michaelsen and H. Frokiaer. 2006. Immunomodulating potential of supplementation with probiotics: a dose-response study in healthy young adults. FEMS Immunol. Med. Microbiol. 47(3): 380–390.

Cox, A.J., N.P. West, P.L. Horn, M.J. Lehtinen, G. Koerbin, D.B. Pyne, S.J. Lahtinen, P.A. Fricker and A.W. Cripps. 2014. Effects of probiotic supplementation over 5 months on routine haematology and clinical chemistry measures in healthy active adults. Eur. J. Clin. Nutr. 68(11): 1255–1257.

Dang, D., W. Zhou, Z.J. Lun, X. Mu, D.X. Wang and H. Wu. 2013. Meta-analysis of probiotics and/or prebiotics for the prevention of eczema. J. Int. Med. Res. 41(5): 1426–1436.

Deshpande, G.C., S.C. Rao, A.D. Keil and S.K. Patole. 2011. Evidence-based guidelines for use of probiotics in preterm neonates. BMC Med. 9: 92.

Didari, T., S. Solki, S. Mozaffari, S. Nikfar and M. Abdollahi. 2014. A systematic review of the safety of probiotics. Expert Opin. Drug Saf. 13(2): 227–239.

Endres, J.R., A. Clewell, K.A. Jade, T. Farber, J. Hauswirth and A.G. Schauss. 2009. Safety assessment of a proprietary preparation of a novel Probiotic, *Bacillus coagulans*, as a food ingredient. Food Chem. Toxicol. 47(6): 1231–1238.

Eskesen, D., L. Jespersen, B. Michelsen, P.J. Whorwell, S. Muller-Lissner and C.M. Morberg. 2015. Effect of the probiotic strain *Bifidobacterium animalis* subsp. *lactis*, BB-12(R), on defecation frequency in healthy subjects with low defecation frequency and abdominal discomfort: a randomised, double-blind, placebo-controlled, parallel-group trial. Br. J. Nutr. 114(10): 1638–1646.

Fang, S.B., H.C. Lee, J.J. Hu, S.Y. Hou, H.L. Liu and H.W. Fang. 2009. Dose-dependent effect of *Lactobacillus rhamnosus* on quantitative reduction of faecal rotavirus shedding in children. J. Trop. Pediatr. 55(5): 297–301.

FAO/WHO. 2002. Guidelines for the evaluation of probiotics in food, 2002, at http://www.who.int/foodsafety/publications/fs_management/probiotics2/en/.

Feng, J.R., F. Wang, X. Qiu, L.V. McFarland, P.F. Chen, R. Zhou, J. Liu, Q. Zhao and J. Li. 2017. Efficacy and safety of probiotic-supplemented triple therapy for eradication of Helicobacter pylori in children: a systematic review and network meta-analysis. Eur. J. Clin. Pharmacol. 73(10): 1199–1208.

Forssten, S.D. and A.C. Ouwehand. 2017. Simulating colonic survival of probiotics in single-strain products compared to multi-strain products. Microb. Ecol. Health Dis. 28(1): 1378061.

Ganji-Arjenaki, M. and M. Rafieian-Kopaei. 2018. Probiotics are a good choice in remission of inflammatory bowel diseases: A meta analysis and systematic review. J. Cell Physiol. 233(3): 2091–2103.

Gao, X.W., M. Mubasher, C.Y. Fang, C. Reifer and L.E. Miller. 2010. Dose-response efficacy of a proprietary probiotic formula of *Lactobacillus acidophilus* CL1285 and *Lactobacillus casei* LBC80R for antibiotic-associated diarrhea and *Clostridium difficile*-associated diarrhea prophylaxis in adult patients. The Am. J. Gastroenterol. 105(7): 1636–1641.

Gianotti, L., L. Morelli, F. Galbiati, S. Rocchetti, S. Coppola, A. Beneduce, C. Gilardini, D. Zonenschain, A. Nespoli and M. Braga. 2010. A randomized double-blind trial on perioperative administration of probiotics in colorectal cancer patients. World J. Gastroenterol. 16(2): 167–175.

Goldenberg, J.Z., S.S. Ma, J.D. Saxton, M.R. Martzen, P.O. Vandvik, K. Thorlund, G.H. Guyatt and B.C. Johnston. 2013. Probiotics for the prevention of *Clostridium difficile*-associated diarrhea in adults and children. Cochrane Database Syst. Rev. 5: CD006095.

Goldenberg, J.Z., L. Lytvyn, J. Steurich, P. Parkin, S. Mahant and B.C. Johnston. 2015. Probiotics for the prevention of pediatric antibiotic-associated diarrhea. Cochrane Database Syst. Rev. (12): CD004827.

Goldenberg, J.Z., C. Yap, L. Lytvyn, C.K. Lo, J. Beardsley, D. Mertz and B.C. Johnston. 2017. Probiotics for the prevention of *Clostridium difficile*-associated diarrhea in adults and children. Cochrane Database Syst. Rev. 12: CD006095.

Guthmann, F., C. Kluthe and C. Buhrer. 2010. Probiotics for prevention of necrotising enterocolitis: an updated meta-analysis. Klin Padiatr. 222(5): 284–290.

Hanifi, A., T. Culpepper, V. Mai, A. Anand, A.L. Ford, M. Ukhanova, M. Christman, T.A. Tompkins and W.J. Dahl. 2015. Evaluation of *Bacillus subtilis* R0179 on gastrointestinal viability and general wellness: a randomised, double-blind, placebo-controlled trial in healthy adults. Benef. Microbes 6(1): 19–27.

He, D., H.Y. Wang, J.Y. Feng, M.M. Zhang, Y. Zhou and X.T. Wu. 2013. Use of pro-/synbiotics as prophylaxis in patients undergoing colorectal resection for cancer: a meta-analysis of randomized controlled trials. Clin. Res. Hepatol. Gastroenterol. 37(4): 406–415.

Hemmerling, A., W. Harrison, A. Schroeder, J. Park, A. Korn, S. Shiboski and C.R. Cohen. 2009. Phase 1 dose-ranging safety trial of *Lactobacillus crispatus* CTV-05 for the prevention of bacterial vaginosis. Sex Transm. Dis. 36(9): 564–569.

Hill, C., F. Guarner, G. Reid, G.R. Gibson, D.J. Merenstein, B. Pot, L. Morelli, R.B. Canani, H.J. Flint, S. Salminen, P.C. Calder and M.E. Sanders. 2014. Expert consensus document: The International Scientific Association for Probiotics and Prebiotics consensus statement on the scope and appropriate use of the term probiotic. Nat. Rev. Gastroenterol. Hepatol. 11(8): 506–514.

Hoveyda, N., C. Heneghan, K.R. Mahtani, R. Perera, N. Roberts and P. Glasziou. 2009. A systematic review and meta-analysis: probiotics in the treatment of irritable bowel syndrome. BMC Gastroenterol. 9: 15.

Ibarra, A., M. Latreille-Barbier, Y. Donazzolo, X. Pelletier and A.C. Ouwehand. 2018. Effects of 28-day *Bifidobacterium animalis* subsp. *lactis* HN019 supplementation on colonic transit time and gastrointestinal symptoms in adults with functional constipation: A double-blind, randomized, placebo-controlled, and dose-ranging trial. Gut Microbes 9(3): 236–251.

Imase, K., M. Takahashi, A. Tanaka, K. Tokunaga, H. Sugano, M. Tanaka, H. Ishida, S. Kamiya and S. Takahashi. 2008. Efficacy of *Clostridium butyricum* preparation concomitantly with *Helicobacter pylori* eradication therapy in relation to changes in the intestinal microbiota. Microbiol. Immunol. 52(3): 156–161.

John, G.K., L. Wang, J. Nanavati, C. Twose, R. Singh and G. Mullin. 2018. Dietary alteration of the gut microbiome and its impact on weight and fat mass: A systematic review and meta-analysis. Genes (Basel) 9(3).

Johnston, B.C., A.L. Supina and S. Vohra. 2006. Probiotics for pediatric antibiotic-associated diarrhea: a meta-analysis of randomized placebo-controlled trials. CMAJ 175(4): 377–383.

Johnston, B.C., J.Z. Goldenberg, P.O. Vandvik, X. Sun and G.H. Guyatt. 2011. Probiotics for the prevention of pediatric antibiotic-associated diarrhea. Cochrane Database Syst. Rev. (11): CD004827.

Khalesi, S., J. Sun, N. Buys and R. Jayasinghe. 2014. Effect of probiotics on blood pressure: a systematic review and meta-analysis of randomized, controlled trials. Hypertension 64(4): 897–903.

Kim, J., J.M. Yun, M.K. Kim, O. Kwon and B. Cho. 2018. *Lactobacillus gasseri* BNR17 supplementation reduces the visceral fat accumulation and waist circumference in obese adults: A randomized, double-blind, placebo-controlled trial. J. Med. Food 21(5): 454–461.

Kleerebezem, M., S. Binda, P.A. Bron, G. Gross, C. Hill, J.E. van Hylckama Vlieg, S. Lebeer, R. Satokari and A.C. Ouwehand. 2018. Understanding mode of action can drive the translational pipeline towards more reliable health benefits for probiotics. Curr. Opin. Biotechnol. 56: 55–60.

Kollaritsch, H., P. Kremsner, G. Wiedermann and O. Scheiner. 1989. Prevention of traveller's diarrhea: comparison of different non-antibiotic preparations. Travel Med. Int. 7: 9–18.

Kollaritsch, H., H. Holst, P. Grobara and G. Wiedermann. 1993. Prevention of traveler's diarrhea with *Saccharomyces boulardii*. Results of a placebo controlled double-blind study. Fortschr. Med. 111(9): 152–156.

Larsen, C.N., S. Nielsen, P. Kaestel, E. Brockmann, M. Bennedsen, H.R. Christensen, D.C. Eskesen, B.L. Jacobsen and K.F. Michaelsen. 2006. Dose-response study of probiotic bacteria *Bifidobacterium animalis* subsp. *lactis* BB-12 and *Lactobacillus paracasei* subsp. *paracasei* CRL-341 in healthy young adults. Eur. J. Clin. Nutr. 60(11): 1284–1293.

Lau, C.S. and R.S. Chamberlain. 2016. Probiotics are effective at preventing *Clostridium difficile*-associated diarrhea: a systematic review and meta-analysis. Int. J. Gen Med. 9: 27–37.

Leyer, G.J., S. Li, M.E. Mubasher, C. Reifer and A.C. Ouwehand. 2009. Probiotic effects on cold and influenza-like symptom incidence and duration in children. Pediatrics 124(2): e172–e179.

Liu, P.C., Y.K. Yan, Y.J. Ma, X.W. Wang, J. Geng, M.C. Wang, F.X. Wei, Y.W. Zhang, X.D. Xu and Y.C. Zhang. 2017. Probiotics reduce postoperative infections in patients undergoing colorectal surgery: A systematic review and meta-analysis. Gastroenterol. Res. Pract. 2017: 6029075.

Lorenzo-Zuniga, V., E. Llop, C. Suarez, B. Alvarez, L. Abreu, J. Espadaler and J. Serra. 2014. I.31, a new combination of probiotics, improves irritable bowel syndrome-related quality of life. World J. Gastroenterol. 20(26): 8709–8716.

Lu, M., S. Yu, J. Deng, Q. Yan, C. Yang, G. Xia and X. Zhou. 2016. Efficacy of probiotic supplementation therapy for Helicobacter pylori eradication: A meta-analysis of randomized controlled trials. PLoS One 11(10): e0163743.

Lyra, A., M. Hillilä, T. Huttunen, M.S.M. Taalikka, J. Tennilä, A. Tarpila, S. Lahtinen, A.C. Ouwehand and L. Veijola. 2016. Irritable bowel syndrome symptom severity improves equally with probiotic and placebo. World J. Gastroenterol. 22(48): 10631–10642.

Lytvyn, L., K. Quach, L. Banfield, B.C. Johnston and D. Mertz. 2016. Probiotics and synbiotics for the prevention of postoperative infections following abdominal surgery: a systematic review and meta-analysis of randomized controlled trials. J. Hosp. Infect. 92(2): 130–139.

Maldonado-Lobon, J.A., M.A. Diaz-Lopez, R. Carputo, P. Duarte, M.P. Diaz-Ropero, A.D. Valero, A. Sanudo, L. Sempere, M.D. Ruiz-Lopez, O. Banuelos, J. Fonolla and M. Olivares Martin. 2015. Lactobacillus fermentum CECT 5716 reduces Staphylococcus load in the breastmilk of lactating mothers suffering breast pain: A randomized controlled trial. Breastfeed. Med. 10(9): 425–432.

Mane, J., E. Pedrosa, V. Loren, M.A. Gassull, J. Espadaler, J. Cune, S. Audivert, M.A. Bonachera and E. Cabre. 2011. A mixture of *Lactobacillus plantarum* CECT 7315 and CECT 7316 enhances systemic immunity in elderly subjects. A dose-response, double-blind, placebo-controlled, randomized pilot trial. Nutr. Hosp. 26(1): 228–235.

Matthes, H., T. Krummenerl, M. Giensch, C. Wolff and J. Schulze. 2010. Clinical trial: probiotic treatment of acute distal ulcerative colitis with rectally administered *Escherichia coli* Nissle 1917 (EcN). BMC Complement. Altern. Med. 10: 13.

McFarland, L.V. 2006. Meta-analysis of probiotics for the prevention of antibiotic associated diarrhea and the treatment of *Clostridium difficile* disease. Am. J. Gastroenterol. 101(4): 812–822.

McFarland, L.V. 2015. Probiotics for the primary and secondary prevention of *C. difficile* infections: A meta-analysis and systematic review. Antibiotics (Basel) 4(2): 160–178.

McFarland, L.V., Y. Huang, L. Wang and P. Malfertheiner. 2016. Systematic review and meta-analysis: Multi-strain probiotics as adjunct therapy for *Helicobacter pylori* eradication and prevention of adverse events. United European Gastroenterol. J. 4(4): 546–561.

Miller, L.E. and A.C. Ouwehand. 2013. Short-term probiotic supplementation decreases intestinal transit time in adults: a systematic review and meta-analysis of randomized controlled trials. World J. Gastroent. 19(29): 4718–4725.

Miller, L.E., A.K. Zimmermann and A.C. Ouwehand. 2016a. Contemporary meta-analysis of short-term probiotic consumption on gastrointestinal transit. World J. Gastroenterol. 22(21): 5122–5131.

Mobini, R., V. Tremaroli, M. Stahlman, F. Karlsson, M. Levin, M. Ljungberg, M. Sohlin, H. Berteus Forslund, R. Perkins, F. Backhed and P.A. Jansson. 2017. Metabolic effects of *Lactobacillus reuteri* DSM 17938 in people with type 2 diabetes: A randomized controlled trial. Diabetes Obes. Metab. 19(4): 579–589.

Morovic, W., J.M. Roper, A.B. Smith, P. Mukerji, B. Stahl, J.C. Rae and A.C. Ouwehand. 2017. Safety evaluation of HOWARU((R)) Restore (*Lactobacillus acidophilus* NCFM, *Lactobacillus paracasei* Lpc-37, *Bifidobacterium animalis* subsp. *lactis* Bl-04 and *B. lactis* Bi-07) for antibiotic resistance, genomic risk factors, and acute toxicity. Food Chem. Toxicol. 110: 316–324.

Mukerji, P., J.M. Roper, B. Stahl, A.B. Smith, F. Burns, J.C. Rae, N. Yeung, A. Lyra, L. Svard, M.T. Saarinen, E. Alhoniemi, A. Ibarra and A.C. Ouwehand. 2016. Safety evaluation of AB-LIFE (*Lactobacillus plantarum* CECT 7527, 7528 and 7529): Antibiotic resistance and 90-day repeated-dose study in rats. Food Chem. Toxicol. 92: 117–128.

Ouwehand, A.C. and S. Salminen. 2003. Safety evaluation of probiotics. pp. 316–336. *In*: M. Mattila-Sandholm and M. Saarela (eds.). Functional Dairy Products. Woodhead Publishing. Cambridge (UK).

Ouwehand, A.C., C. DongLian, X. Weijian, M. Stewart, J. Ni, T. Stewart and L.E. Miller. 2014. Probiotics reduce symptoms of antibiotic use in a hospital setting: a randomized dose response study. Vaccine 32(4): 458–463.

Ouwehand, A.C. 2017. A review of dose-responses of probiotics in human studies. Benef. Microbes 8(2): 143–151.

Ouwehand, A.C., M.M. Invernici, F.A.C. Furlaneto and M.R. Messora. 2018a. Effectiveness of Multistrain Versus Single-strain Probiotics: Current Status and Recommendations for the Future. J. Clin. Gastroenterol. 52 Suppl 1, Proceedings from the 9th Probiotics, Prebiotics and New Foods, Nutraceuticals and Botanicals for Nutrition & Human and Microbiota Health Meeting, held in Rome, Italy from September 10–12, 2017: S35–S40.

Pelucchi, C., L. Chatenoud, F. Turati, C. Galeone, L. Moja, J.F. Bach and C. La Vecchia. 2012. Probiotics supplementation during pregnancy or infancy for the prevention of atopic dermatitis: a meta-analysis. Epidemiology 23(3): 402–414.

Reid, G., D. Beuerman, C. Heinemann and A.W. Bruce. 2001. Probiotic *Lactobacillus* dose required to restore and maintain a normal vaginal flora. FEMS Immunol. Med. Microbiol. 32(1): 37–41.

Ringel-Kulka, T., J.R. Goldsmith, I.M. Carroll, S.P. Barros, O. Palsson, C. Jobin and Y. Ringel. 2014. *Lactobacillus acidophilus* NCFM affects colonic mucosal opioid receptor expression in patients with functional abdominal pain—a randomised clinical study. Aliment. Pharmacol. Ther. 40(2): 200–207.

Ritchie, M.L. and T.N. Romanuk. 2012. A meta-analysis of probiotic efficacy for gastrointestinal diseases. PLoS One 7(4): e34938.

Sanders, M.E., D.J. Merenstein, A.C. Ouwehand, G. Reid, S. Salminen, M.D. Cabana, G. Paraskevakos and G. Leyer. 2016. Probiotic use in at-risk populations. J. Am. Pharm. Assoc. 56(6): 680–686.

Savard, P., B. Lamarche, M.E. Paradis, H. Thiboutot, E. Laurin and D. Roy. 2011. Impact of *Bifidobacterium animalis* subsp. *lactis* BB-12 and, *Lactobacillus acidophilus* LA-5-containing yoghurt, on fecal bacterial counts of healthy adults. Int. J. Food Microbiol. 149(1): 50–57.

Saxelin, M., T. Pessi and S. Salminen. 1995. Fecal recovery following oral administration of *Lactobacillus* strain GG (ATCC 53103) in gelatine capsules to healthy volunteers. Int. J. Food Microbiol. 25(2): 199–203.

Shen, N.T., A. Maw, L.L. Tmanova, A. Pino, K. Ancy, C.V. Crawford, M.S. Simon and A.T. Evans. 2017. Timely use of probiotics in hospitalized adults prevents *Clostridium difficile* infection: A systematic review with meta-regression analysis. Gastroenterology 152(8): 1889–1900.

Shinkai, S., M. Toba, T. Saito, I. Sato, M. Tsubouchi, K. Taira, K. Kakumoto, T. Inamatsu, H. Yoshida, Y. Fujiwara, T. Fukaya, T. Matsumoto, K. Tateda, K. Yamaguchi, N. Kohda and S. Kohno. 2013. Immunoprotective effects of oral intake of heat-killed *Lactobacillus pentosus* strain b240 in elderly adults: a randomised, double-blind, placebo-controlled trial. Br. J. Nutr. 109(10): 1856–1865.

Sudha, M.R. 2011. Safety assessment studies of probiotic *Saccharomyces boulardii* strain Unique 28 in Sprague-Dawley rats. Benef. Microbes 2(3): 221–227.

Szulinska, M., I. Loniewski, S. van Hemert, M. Sobieska and P. Bogdanski. 2018. Dose-dependent effects of multispecies probiotic supplementation on the lipopolysaccharide (LPS) level and cardiometabolic profile in obese postmenopausal women: A 12-week randomized clinical trial. Nutrients 10(6).

Taverniti, V., R. Koirala, A. Dalla Via, G. Gargari, E. Leonardis, S. Arioli and S. Guglielmetti. 2019. Effect of cell concentration on the persistence in the human intestine of four probiotic strains administered through a multispecies formulation. Nutrients 11(2).

Taylor, B.L., G.E. Woodfall, K.E. Sheedy, M.L. O'Riley, K.A. Rainbow, E.L. Bramwell and N.J. Kellow. 2017. Effect of probiotics on metabolic outcomes in pregnant women with gestational diabetes: A systematic review and meta-analysis of randomized controlled trials. Nutrients 9(5).

Tursi, A., G. Brandimarte, A. Papa, A. Giglio, W. Elisei, G.M. Giorgetti, G. Forti, S. Morini, C. Hassan, M.A. Pistoia, M.E. Modeo, S. Rodino, T. D'Amico, L. Sebkova, N. Sacca, E. Di Giulio, F. Luzza,

M. Imeneo, T. Larussa, S. Di Rosa, V. Annese, S. Danese and A. Gasbarrini. 2010. Treatment of relapsing mild-to-moderate ulcerative colitis with the probiotic VSL#3 as adjunctive to a standard pharmaceutical treatment: a double-blind, randomized, placebo-controlled study. Am. J. Gastroenterol. 105(10): 2218–2227.

Van Niel, C.W., C. Feudtner, M.M. Garrison and D.A. Christakis. 2002. *Lactobacillus* therapy for acute infectious diarrhea in children: a meta-analysis. Pediatrics 109(4): 678–684.

Waller, P.A., P.K. Gopal, G.J. Leyer, A.C. Ouwehand, C. Reifer, M.E. Stewart and L.E. Miller. 2011. Dose-response effect of *Bifidobacterium lactis* HN019 on whole gut transit time and functional gastrointestinal symptoms in adults. Scand. J. Gastroenterol. 46(9): 1057–1064.

West, N.P., P.L. Horn, S. Barrett, H.S. Warren, M.J. Lehtinen, P.A. Fricker and A.W. Cripps. 2014a. Supplementation with a single and double strain probiotic on the innate immune system for respiratory illness. e-SPEN J. 9: e178–e184.

West, N.P., P.L. Horn, D.B. Pyne, V.J. Gebski, S.J. Lahtinen, P.A. Fricker and A.W. Cripps. 2014b. Probiotic supplementation for respiratory and gastrointestinal illness symptoms in healthy physically active individuals. Clin. Nutr. 33(4): 581–587.

West, N.P., P.L. Horn, D.B. Pyne, H.S. Warren, S. Asad, A.J. Cox, S.J. Lahtinen, M.J. Lehtinen, P.A. Fricker, A.W. Cripps and B. Fazekas de St Groth. 2016. Probiotic supplementation has little effect on peripheral blood regulatory T-cells. J. Allergy Clin. Immunol. 138(6): 1749–1752.e7.

Whorwell, P.J., L. Altringer, J. Morel, Y. Bond, D. Charbonneau, L. O'Mahony, B. Kiely, F. Shanahan and E.M. Quigley. 2006. Efficacy of an encapsulated probiotic *Bifidobacterium infantis* 35624 in women with irritable bowel syndrome. Am. J. Gastroenterol. 101(7): 1581–1590.

Zhang, G.Q., H.J. Hu, C.Y. Liu, Q. Zhang, S. Shakya and Z.Y. Li. 2016a. Probiotics for prevention of atopy and food hypersensitivity in early childhood: A PRISMA-Compliant systematic review and meta-analysis of randomized controlled trials. Medicine (Baltimore) 95(8): e2562.

Zhang, Y., L. Li, C. Guo, D. Mu, B. Feng, X. Zuo and Y. Li. 2016b. Effects of probiotic type, dose and treatment duration on irritable bowel syndrome diagnosed by Rome III criteria: a meta-analysis. BMC Gastroenterol. 16(1): 62.

Zhou, J.S., Q. Shu, K.J. Rutherfurd, J. Prasad, M.J. Britles, P.K. Gopal and H.S. Gill. 2000. Safety assessment of potential probiotic lactic acid bacteria strains *Lactobacillus rhamnosus* HN001, *Lb. acidophilus* HN017, and *Bifidobacterium lactis* HN019 in BALB/c mice. Int. J. Food Microbiol. 56: 87–96.

3

Role of Lactic Acid Bacteria in Impacting Nutrient Bioavailability

Julie M Hess

Introduction

Fermentation involves using bacteria and microorganisms to convert sugars into acids or alcohol. Naturally-occurring lactic acid bacteria (LAB) have been used to ferment and preserve food for thousands of years. Fermentation may have been used over 15,000 years ago to preserve food after the domestication of grains with the first agricultural revolution (Gänzle and Ripari 2016). Written records of human history indicate that sourdough fermentation has been used for at least 5,000 years (Gobbetti et al. 2018). While ancient peoples may not have been aware of it, fermentation enhances the nutritional properties of foods in addition to preserving them.

One way in which LAB fermentation affects the nutritional properties of food is through its impact on the bioaccessibility, bioavailability, and digestibility of different nutrients. Bioaccessibility refers to the amount of a nutrient released from its food source and available for absorption (Rousseau et al. 2019). Bioavailability and digestibility reflect the extent to which a nutrient is absorbed and used or stored, relative to the amount ingested. For instance, minerals like iron, zinc, calcium, phosphorus, and copper are present in some foods but may not be available for use by the body because they are bound to antinutrient complexes that prevent them from being accessible and absorbed. Fermentation can help break down these complexes and free these minerals for absorption, thereby enhancing their bioaccessibility. Unlike bioaccessibility, which can be determined with *in vivo* studies, bioavailability

Department of Food Science and Nutrition, University of Minnesota, 1334 Eckles Ave., St. Paul, MN 55108, USA.
Dairy Management, Inc. Rosemont, IL.
Email: jmhess@umn.edu

and digestibility can also be impacted by an individual's nutrient status. Someone who is iron-replete will, for example, absorb less dietary iron than someone who is iron-deficient (Gropper and Smith 2013).

This chapter focuses primarily on the ability of LAB fermentation to improve mineral bioaccessibility and bioavailability of plant foods, including cereal grains and legumes. It also describes the results of research on the impact of LAB fermentation on the bioaccessibility and bioavailability of nutrients in dairy foods. In addition to its discussion on the impact of LAB fermentation on mineral bioavailability, this chapter also briefly covers how fermentation can affect protein digestibility and the content of bioactive components in certain foods. Organisms in addition to LAB, such as certain fungi, are also capable of fermenting foods. In this chapter, however, the terms LAB fermentation and fermentation will be used interchangeably and, unless otherwise specified, refer specifically to food fermented with LAB.

Cereal Grains

Cereal grains, especially wheat, rice, and maize, are major sources of energy in the human diet, comprising an average of 50% of all calories consumed globally (FAO 1998, Awika 2011). While nutrient profiles vary by cereal type, due to their significant energy contributions to the human diet, cereal grains are important sources of nutrients, including carbohydrates, protein, and micronutrients, that are vital for health maintenance (Novotni et al. 2011). Cereal grains make important nutrient contribution to the diet in high-income countries as well as in low- and middle-income countries. For instance, in the United States, a high-income country, data from the National Health and Nutrition Examination Survey (NHANES) conducted from 2003–2006 indicates that "yeast bread/rolls", especially if they contain whole grains, were one of the most nutrient-dense sources of energy, along with beef and milk, for American adults over the age of 19 (O'Neil et al. 2012). They were also a top source of fiber for all Americans over the age of two (O'Neil et al. 2012, Keast et al. 2013). Data from 2011–2014 NHANES indicate that bread, rolls, and tortillas are still one of the top three sources of fiber for U.S. children aged two to 18 (O'Neil et al. 2018). These foods include both "whole meal" and "refined" breads, rolls, and tortillas.

The term "cereal grains" can refer to both whole grains and refined grains. In 2006, the U.S. Food and Drug Administration (FDA) defined whole grains as "cereal grains that consist of the intact, ground, cracked or flaked caryopsis, whose principal anatomical components—the starchy endosperm, germ and bran—are present in the same relative proportions as they exist in the intact caryopsis" (Center for Food Safety and Applied Nutrition 2006), a definition previously developed in 1999 by the American Association of Cereal Chemists International (American Association of Cereal Chemists International 1999). Caryopsis refers to the fruit of the grain, sometimes also called a kernel, grain, or berry. In contrast, refined grains are milled to remove the bran, which also contains many of the nutrients (Awika 2011), leaving only the starchy endosperm (Figure 1). Because they contain the bran as well as the endosperm, foods containing whole grains are naturally more nutrient-dense than foods that contain primarily refined grains.

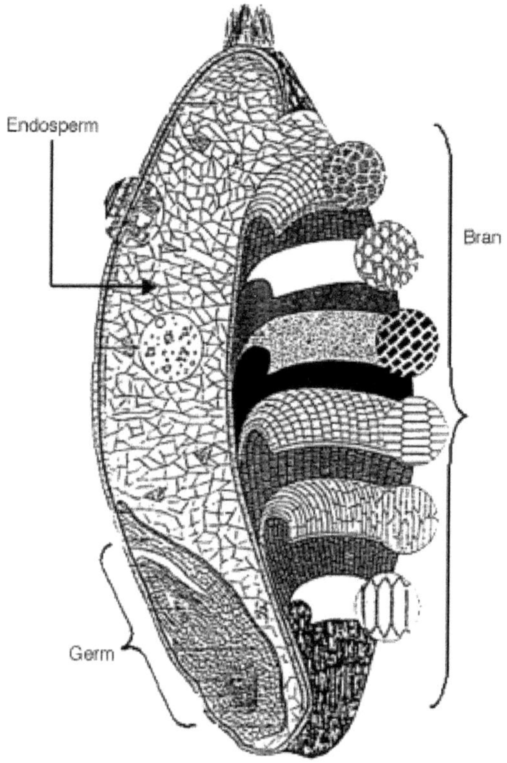

Figure 1: Structure of a whole grain [from (Slavin 2004), reproduced with permission].

Some evidence indicates that fermenting cereal grains, especially whole grains, with LAB can improve their nutrient profile and sensory characteristics. A common method of fermenting grains is with sourdough. Sourdough is a traditional leavening method used for baking bread, a major part of the human diet (Menezes et al. 2018). Making sourdough bread involves fermenting flour made with whole or refined grains, and water with LAB and sometimes yeast (Menezes et al. 2018). There are three primary types of sourdough, which differ by the technology and cultures used to make them (Chavan and Chavan 2011). Type I sourdough bread is made with the back slopping method of fermentation, which involves using a small amount of fermented dough from a previous batch to restart fermentation. Type I sourdough does not always include the addition of yeast and tends to be more common in noncommercial preparations (Chavan and Chavan 2011). Types II and III sourdough are more commonly used commercially and rely on the use of specific, adapted bacterial strains, either liquid or dry, in addition to yeast, for breadmaking (Chavan and Chavan 2011).

This section of the chapter will provide an overview of the impact of LAB fermentation on the mineral bioaccessibility, gluten content, FODMAP content, and the starch digestibility/glycemic response to foods made with cereal grains.

Impact of LAB fermentation

Mineral bioaccessibility and bioavailability

One of the nutritional benefits of consuming whole grains is that they tend to have a higher mineral and fiber content than refined grains (Ferruzzi et al. 2014). Many cereal grains, including wheat, contain minerals like iron, zinc, calcium, manganese, and copper. However, these minerals are often not accessible to uptake by the human gastrointestinal (GI) tract, because they are bound to certain components, antinutrients, that block them from being absorbed (Gobbetti et al. 2014). Fermenting cereal grains using LAB helps to break down the links between antinutrients and minerals, making the mineral more likely to be absorbed. The most well-known antinutrients in grain are phytates.

Phytate

Phytates (*myo*-inositolhexophosphate) or (myo-inositol(1,2,3,4,5,6)-hexa*kis*phosphate) are found in many foods that contain fiber, including grains, vegetables, nuts, and fruits (Schlemmer et al. 2009, Nielsen et al. 2013). Phytate is the primary form in which plants store phosphorus. It accounts for 60–90% of the phosphorus stored in seeds (Nielsen et al. 2013), and over 80% of the phosphorus in cereal grains (Schlemmer et al. 2009, Nielsen et al. 2013). In wheat and rye, phytate is located in the inner most layer of the bran, the aleurone layer (Katina et al. 2005, Schlemmer et al. 2009). The starchy endosperm is nearly free of phytate (Schlemmer et al. 2009), meaning that refined grains are less likely to have phytate but are also less likely to offer many nutrients, which are concentrated in the bran.

Data on the phytate content of plants is often given in amounts of phytic acid (Schlemmer et al. 2009). Phytic acid is not actually present in plants. However, analyzing the phytate content of food requires acidic extraction, meaning that the phytate values must be measured in terms of phytic acid (Schlemmer et al. 2009).

Phytate is the salt of phytic acid (Schlemmer et al. 2009) and exists in plants as mixed salts of several different mineral cations (Nielsen et al. 2013). In wheat and rye, phytate can be found as magnesium-potassium salt (Katina et al. 2005, Schlemmer et al. 2009). Phytate uses oxygen to form insoluble complexes with several different divalent and trivalent metal cations, including iron, zinc, copper, chromium, potassium, magnesium, and calcium (Reale et al. 2007).

The affinity of phytate for different metal cations depends on the pH (Nielsen et al. 2013). At physiological pH (between 6 and 7), phytate is highly negatively charged, and this charge is often counterbalanced by sodium ions or other cations like iron, zinc, and calcium (Schlemmer et al. 2009). In a pH range of 3 to 7 (approximately gastric range), phytate will carry six negative charges (Nielsen et al. 2013). The more phosphate groups the inositol rings have, the stronger the interaction with cations and the lower the mineral solubility (Nielsen et al. 2013). The minerals in the resulting phytate-mineral complexes are poorly absorbed by the human body.

Mechanism for phytate breakdown

Phytate degradation in fermented products depends on a range of factors, including phytase activity, particle size, temperature, water content, and fermentation time

(Poutanen et al. 2009). Phytases, phosphate esterases capable of degrading phytate, act by releasing phosphate from phytate (Schlemmer et al. 2009) and, therefore, releasing the chelated minerals (iron, calcium, and others; Figure 2) in a process called phytase-catalyzed dephosphorylation (Nielsen et al. 2013). Phytases partially hydrolyze phytate to "penta-, tentra-, or lower phosphorylated analogues" of phytate (World Health Organization et al. 1996, Schlemmer et al. 2009). These lower analogues have a reduced capacity to bind minerals and are less likely to inhibit mineral utilization (Reale et al. 2007). The extent of dephosphorylation that needs to occur varies by mineral cation. Phytate needs to be "dephosphorylated to a higher extent for iron and calcium release to occur than for zinc, magnesium, and phosphorus" (Nielsen et al. 2013).

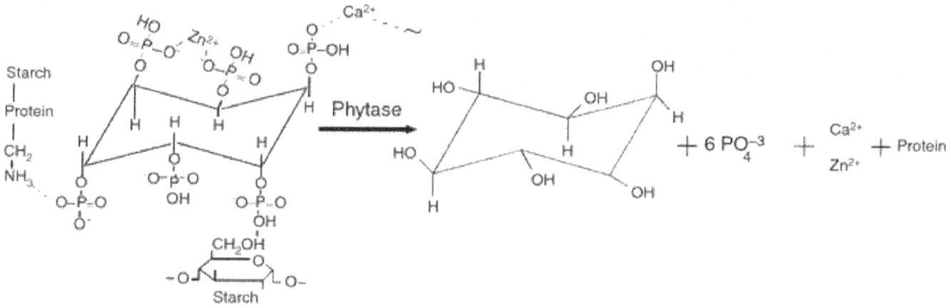

Figure 2: Phytate hydrolysis by phytase to inositol, phosphate, minerals, and protein [adapted from (Yao et al. 2012)]. https://onlinelibrary.wiley.com/terms-and-conditions.

Phytate has to be dephosphorylated to allow mineral cations to be released, and fecal studies indicate that phytate dephosphorylation is limited in the human GI tract (Nielsen et al. 2013). Most mineral absorption occurs in the small intestine, which has a slightly alkaline pH, also the ideal pH for the formation of insoluble phytate-mineral complexes. Because of low phytase activity in the human small intestine, phytate dephosphorylation likely occurs through intake of dietary phytase from the microbial phytases present in fermented foods (Nielsen et al. 2013).

Sourdough fermentation degrades phytate by creating a favorable environment for the activity of both endogenous phytases in cereal grains as well as microbial phytases (D'Alessandro and De Pergola 2014, Schlemmer et al. 2009). This is how fermentation with LAB improves mineral bioavailability of grains. Bacterial activity during fermentation generates organic acids like lactic acid, that lowers the milieu pH, and this lower pH activates phytase (Katina et al. 2005, Schlemmer et al. 2009). A pH of 5.5 has been identified as both acceptable by consumers who dislike acidic tastes and the optimum pH for phytate degradation, as this is the pH at which endogenous plant phytase is most active (Reale et al. 2007). Unfermented plant material, in contrast, has a pH of around 7. LAB fermentation is, therefore, not directly responsible for phytate degradation but rather establishes conditions favorable for endogenous phytase activity, primarily through lowering the pH (Reale et al. 2007).

Leavening bread with yeast also decreases the phytate content. Yeast used to make bread, like *Saccharomyces cerevisiae*, also contains endogenous phytases that can hydrolyze some of the phytate and can liberate some phosphorus for absorption. Sourdough fermentation is more effective at phytate degradation. Sourdough reduces the content of phytate to less than half of the phytate content of whole wheat bread leavened with yeast (Gobbetti et al. 2018).

Sourdough fermentation can improve the bioavailability of the minerals below. Each section also contains an explanation for why improving the bioavailability of these nutrients is important from a global nutrition perspective.

Calcium. Improving calcium absorption is important because, globally, calcium is considered an "important deficiency" (High Level Panel of Experts on Food Security 2017). Even in the U.S., calcium is considered one of four nutrients of public health concern (along with vitamin D, potassium, and fiber) due to low intake (USDA and HHS 2016). Phytate binds calcium and decreases its availability. While cereal grain foods are not naturally important sources of calcium, certain grain foods like breakfast cereals may be calcium-fortified (*A Guide to Calcium-Rich Foods* - National Osteoporosis Foundation). To our knowledge, the ability of phytate breakdown to impact calcium absorption from fortified foods has not been assessed in a clinical study. However, it has been assessed with iron-fortified foods and, in some cases, phytate degradation of fortified cereals did increase absorption of iron added through fortification (Hurrell et al. 2003). Phytate degradation may be important for calcium absorption as well, whether through improving absorption from fortified foods or allowing absorption from marginal sources of calcium in instances where natural calcium-rich foods may not be available.

Iron. Non-heme iron, the iron found in plant foods, has a lower uptake by the body than heme iron from animal foods (Nielsen et al. 2013). Heme iron is 50–60% of the iron in animal foods and its average bioavailability is 15–35% (Nielsen et al. 2013). Non-heme iron, on the other hand, has a bioavailability ranging from 1–22% (Nielsen et al. 2013). Some cereal-based foods have absorption levels as low as 2–3% (Nielsen et al. 2013). Non-heme iron is found in a variety of chemical forms and, in some foods like wheat, the complex it forms with phytate (ferric phytate) is responsible for its lower bioavailability (Figure 3) (Nielsen et al. 2013).

Iron absorption, including from plant foods, is important, because iron deficiency is another one of the most common micronutrient deficiencies in the world. It is also one of greatest global health concerns, along with vitamin A and iodine deficiencies (High Level Panel of Experts on Food Security 2017). Anemia, often caused by iron deficiency, affects one third of women of reproductive ages (15–49 years) as well as 300 million children globally (WHO 2018). Iron deficiency among women and

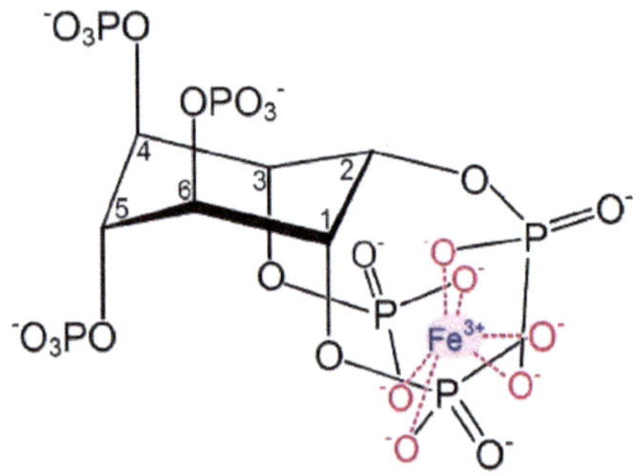

Figure 3: Iron phytate complex [adapted from (Nielsen et al. 2013, Schlemmer et al. 2009)]. https://www.mdpi.com/openaccess

children is common in low-, middle-, and high-income countries alike. For women around the globe, iron deficiency is often due to both menstrual loses and insufficient dietary intake (WHO 2018). In children, iron deficiency can "result from inadequate intake or absorption of dietary iron, increased need during periods of growth, and blood loss from parasitic infection or menstruation in adolescent girls" (WHO 2018). Further research to understand nonheme iron absorption is important to the health of populations around the world.

There is little evidence as to how iron bound to phytate is broken down and absorbed. Non-heme iron is often found in foods in its oxidized form (Fe^{3+}) but is absorbed in its reduced form (Fe^{2+}) (Nielsen et al. 2013). The reduced form of iron is also less likely to complex with phytate (Nielsen et al. 2013). Given the available evidence, it seems iron must be both released from the phytate and reduced to be absorbed (Nielsen et al. 2013). Solubility of the iron and phytate complexes varies with degree of phosphorylation, which means that the bioavailability of iron from the different iron phytate complexes also varies (Nielsen et al. 2013). A 2005 study assessed the impact of meeting daily recommendations for fiber-rich wheat bread (with or without phytase) on the iron status of 41 iron-replete young women in Denmark. Meeting the Nordic Nutrition Recommendations by consuming 300 g of whole grain bread daily, whether it was prepared with phytase or not, led to a significant decrease in the women's hemoglobin concentrations after four months of the intervention (Bach Kristensen et al. 2005). Consuming fiber-rich foods is recommended for health by several authoritative public health organizations (WHO 2003, USDA and HHS 2016) but may also affect the iron status of certain populations. As noted by Nielsen et al. (2013), however, a major global nutrition company applied for a patent on an iron fortification blend that includes phytase eight years after this study of fiber-rich bread. There may still be potential for phytase added to whole grain foods to improve iron bioavailability (Nielsen et al. 2013).

Phosphorus. Bioavailability of phosphorus in phytate is limited to about 50%, because humans lack phytase. However, the lower pH generated by fermentation enhances phosphorus solubility (Lopez et al. 2001, Poutanen et al. 2009). While not a nutrient of concern from a deficiency perspective, phosphorus intake can enhance calcium absorption and, like calcium, comprises a major portion of hydroxyapatite, the mineral component of bone (Guéguen and Pointillart 2000, Ilich and Kerstetter 2000).

Zinc. The importance of fermenting grains to improve zinc absorption was discovered in parallel with the determination of zinc as a vital nutrient. In the mid-20th century (1960s), there were several cases in Egypt, Iran, and Turkey of young boys displaying growth retardation and immature sexual development (Prasad 2013). Unleavened bread was a primary component of the boys' diet. Because the bread was not leavened with either yeast or bacteria, its zinc content was bound to phytate and not absorbable. These boys suffered from chronic zinc deficiency, which can retard growth and sexual development in childhood and adolescence (Prasad 2013). Like calcium, zinc is still considered a nutrient of important global deficiency (High Level Panel of Experts on Food Security 2017). In places where diets are based on unrefined or unleavened cereal grains or pulses, zinc absorption from grains remains a concern (World Health Organization et al. 1996).

Other Minerals: Copper, Magnesium, Manganese, Chromium. Phytates can also complex with copper, magnesium, manganese, and chromium. The increase in acidity with sourdough fermentation does increase magnesium solubility, according to results of one *in vitro* study (Lopez et al. 2001). It has not been assessed, to our knowledge, in a human study (Lopez et al. 2001, Poutanen et al. 2009). Results of an animal study indicate that phytate may actually improve copper intake (Lee et al. 1988), and results of one human intervention study suggest that lowering the phytate content of a soy beverage may enhance manganese uptake (Davidsson et al. 1995). To our knowledge, no human studies have been conducted on the links between phytate and chromium bioavailability. Given the rarity of deficiencies in any of these minerals, it seems unlikely that the impacts of phytate on the bioavailability of any of these minerals would have clinical consequences.

Gluten content

In addition to nutrient deficiencies and insufficiencies, LAB fermentation of grains may also have implications for individuals with celiac disease (CD). CD is a serious chronic autoimmune disorder in which the intestinal villi become damaged with ingestion of the gliadin fraction of gluten, or wheat proteins (Poutanen et al. 2009). CD affects 1% of the population worldwide (Kelly et al. 2015, See et al. 2015). The immune response to gluten in an individual with CD causes inflammation and damage to the small intestine, where most nutrient absorption occurs (Leonard et al. 2017). Damage from CD inhibits overall nutrient absorption (Wierdsma et al. 2013). Currently, the only treatment for CD is a gluten-free diet (Wierdsma et al. 2013, Leonard et al. 2017). Some individuals who do not test positive for CD but have severe abdominal issues from a systemic immune reaction to wheat exposure may also choose to adopt a gluten-free diet (Rizzello et al. 2016). These individuals may have non-celiac gluten sensitivity (NCGS) (Uhde et al. 2016). The increasing prevalence of both CD and NCGS has led to a proliferation of research into and production of gluten-free bakery products, including breads (Rizzello et al. 2016). Gluten-free per European Commission (The European Commission 2014) and FDA (Center for Food Safety and Applied Nutrition 2018) guidance means no more than 20 mg/kg or <20 ppm gluten, respectively. Following a gluten-free diet typically requires removing gluten-containing grains (wheat, barley, rye and related grains like spelt, kamut, emmer, and farro) and foods that contain them. For reference, a typical Western diet contains 15–20 g/d of gluten (See et al. 2015).

LAB fermentation could be a useful technology for developing palatable gluten-free bakery products. Some reviews postulate that sourdough fermentation may degrade gluten and therefore naturally reduce the gluten content of bread (Katina et al. 2005, Gobbetti et al. 2014, Gobbetti et al. 2018, Muir et al. 2019). Some LAB exhibit proteolytic activity, which can degrade gluten and may help make sourdough breads acceptable to individuals with CD and NCGS. This effect, however, appears to be strain-specific and does not occur with all sourdough fermentations (Muir et al. 2019). A study by Rizzello et al. found that long duration fermentation (48 hr) of semolina durum flour with *Lactobacillus* (*L.*) *sanfranciscensis* 7A, LS47, LS38, LS23, LS19, LS10, and LS3, *L. hilgardii* 51B, *L. brevis* 14G, and *L. alimentarius* decreased gluten content to less than 10 ppm (from 100,157 ppm). In this study, the fermented bread had concentrations of phosphorus, calcium, magnesium, potassium, iron, and zinc in similar or significantly higher amounts than the values observed in conventional gluten-free breads (Rizzello et al. 2016). Panelists who evaluated the sensory characteristics of the breads in this study described the rendered gluten-free sourdough as having a salty flavor, "acidic flavor and taste", and low sweetness, but its scores were comparable to those of other wheat sourdough breads containing gluten (Rizzello et al. 2016). Compared to conventional gluten-free breads, though, the rendered gluten-free bread had significantly better scores for elasticity, acid taste and flavor, and salty taste (Rizzello et al. 2016). Like all gluten-free breads, breads rendered gluten-free with LAB fermentation lose gluten's viscoelastic qualities that contribute to bread "volume and quality". Severely decreasing gluten content with fermentation still requires adding back in structuring agents (Rizzello et al. 2016, Muir et al. 2019). Gluten-free bread made by hydrolyzing gluten is commercially available in Italy and may signal another important area for further research with LAB fermentation (Rizzello et al. 2016).

FODMAPs

Some clinicians posit that a group of poorly absorbed carbohydrates in wheat- and gluten-containing foods may be responsible for the symptoms of NCGS rather than the actual gluten (Muir et al. 2019). These carbohydrates are often referred to as FODMAPs, or fermentable oligo- di- monosaccharides and polyols, and are present in many gluten-containing products.

FODMAPs are poorly absorbed in both healthy individuals and those with GI conditions. However, in healthy individuals, FODMAP consumption generally causes few adverse effects and may be beneficial for digestive and gut health. In individuals with certain GI conditions like irritable bowel syndrome (IBS), FODMAP consumption may exacerbate symptoms (Gibson and Shepherd 2010). Due to their small size, FODMAPs exert an osmotic effect in the small intestine and cause luminal distension. Once they reach the large intestine, FODMAPs begin to ferment, which generates gas and leads to symptoms including flatulence, bloating, and stretching of the intestinal walls, also common symptoms of NCGS (Biesiekierski et al. 2013, Halmos et al. 2014). In many foods, gluten co-occurs with oligosaccharides, and the symptoms of IBS and gluten sensitivity can overlap and complicate the diagnostic process. Some practitioners and researchers believe that FODMAPs, rather than wheat or gluten, may be responsible for the symptoms of NCGS (Biesiekierski et al. 2013, Giorgio et al. 2016), and low-FODMAP diets are an emerging short-term nutrition therapy treatment for individuals suffering from IBS (Giorgio et al. 2016).

Low-FODMAP diets restrict the intake of foods known to contain excess amounts of fermentable oligo- di- monosaccharides and polyols like lactose, fructose, fructans, galactans, and sugar alcohols like sorbitol and mannitol, among others. Foods restricted on the low-FODMAP diet come from several different food groups, including several that contain nutrient-rich foods, like fruits (apples, pears, peaches), dairy foods (cow's milk, yogurt, fresh cheeses), vegetables (broccoli, garlic, onions), legumes (lentils, baked beans), and cereals (wheat and rye-based foods). The cut-off levels for FODMAPs in foods acceptable on low-FODMAP diets remain poorly defined. Amounts of FODMAPs in individual foods are less important, in terms of symptom occurrence, compared to total FODMAP intake at any single eating occasion. Due to the restrictive nature of the low-FODMAP diet and its unknown long-term effects on intestinal health, a low-FODMAP diet should only be undertaken under the guidance of a registered dietitian and followed for a limited amount of time. On this eating pattern, FODMAPs should only be restricted to a level that is necessary for "symptomatic comfort" (Giorgio et al. 2016).

A prominent criticism of the low-FODMAP eating pattern is that it tends to be low in fiber, since many commonly consumed foods that contain FODMAPs, like wheat-based products, are significant sources of fiber in the American diet and worldwide (Gibson and Shepherd 2010, Menezes et al. 2018, Loponen and Gänzle 2018, O'Neil et al. 2018). Whole grain bakery products that are also low in FODMAPs provide options for fiber-containing foods on a low-FODMAP diet.

Since whole grains like wheat also contain FODMAPS, they have to be processed to lower their FODMAP content. Wheat is considered a food source of FODMAPs because of its fructan content (Gibson and Shepherd 2010, Loponen and Gänzle 2018). An early study on FODMAPs suggested that wheat may also provide polyols, especially mannitol, in the diet (Gibson and Shepherd 2010), but a recent analytical study failed to find mannitol in a serving size of several cereal and grain foods, including wheat bran and spelt bread (Yao et al. 2014). While not found in unprepared grain foods, mannitol can be generated during fermentation of wheat with certain cultures that can degrade fructans to fructose, which is then partially converted to mannitol (Loponen and Gänzle 2018). However, fermenting grains for a long period of time (> 3 hr) can also reduce the FODMAP content (both fructans and mannitol) and therefore make them more digestible to people with NCGS and IBS (Menezes et al. 2018). Sourdough breads provide "slowly fermented and well-tolerated dietary fiber" to individuals following a low-FODMAP diet that otherwise tends to provide little fiber (Loponen and Gänzle 2018). Ensuring that individuals with IBS consume fiber with low-FODMAP content may prevent "the depletion of intestinal bifidobacterial that has been observed on other low-FODMAP diets" (Loponen and Gänzle 2018).

Producing low-FODMAP sourdough breads necessitates using LAB cultures capable of both degrading fructans to fructose and consuming any subsequently generated mannitol (Laatikainen et al. 2016, Loponen and Gänzle 2018). Therefore, producing low-FODMAP bread requires choosing cultures with extracellular fructanase activity from either lactobacilli expressing fructanases (*L. amylovorus*, *L. crispatus*, *L. paracasei*) or fructanase-positive yeasts (Loponen and Gänzle 2018). The extent of FODMAP reduction in sourdough bread depends on the "fermentation organisms, the fermentation process, the grain raw material, and the sourdough dosage to the final bread dough" (Loponen and Gänzle 2018).

Results of one intervention study suggest that eating sourdough bread, regardless of culture, may not exacerbate IBS symptoms (Laatikainen et al. 2016). A crossover study assessed the differences in symptom severity in participants with IBS consuming 7–8 slices per day of either regular sourdough rye bread or a low-FODMAP sourdough rye bread for four weeks and found fewer participant-rated adverse GI symptoms on the low-FODMAP intervention than the control (Laatikainen et al. 2016). However, because there were no differences in IBS symptom scores or quality of life ratings between the two treatments, a similar study is needed to assess efficacy of low-

FODMAP bread as part of an overall low-FODMAP dietary pattern (Laatikainen et al. 2016).

Processing method may be more important than wheat variety in creating wheat-based low-FODMAP bakery products. A key factor to ensuring low-FODMAP content in breads appears to be the proofing time. The longer the proofing time, the lower the fructan and mannitol content (Ziegler et al. 2016). The fermentation process depletes carbohydrates in order of polymerization. Monosaccharides and disaccharides are used first, followed by carbohydrates like fructans, with a higher degree of polymerization (Menezes et al. 2018). Prolonging dough proofing times (> 4 hr) may lower FODMAP levels in bread by nearly 90% (Ziegler et al. 2016). Yeast leavening can also decrease the FODMAP content of breads but is not long enough (0.5 to 3 hr) for extensive hydrolysis of FODMAPs (Menezes et al. 2018).

The grains used to make sourdough matter as well. Different grains contain different levels of FODMAPs. For instance, einkorn contains significantly more FODMAPs than emmer (Ziegler et al. 2016). Modifying the type of grain and the microorganisms can help develop breads optimized in both flavor and other sensory characteristics as well as being low in FODMAPs. More research is needed in both nutrition science and food science on FODMAP-lowering processes with different grain varieties, impacts of FODMAP lowering on bread quality, and evaluation of FODMAP content in foods including sourdough breads.

Starch digestibility and the glycemic response

Starch digestibility refers to the rate at which starches are broken down by the body into their constituent glucose units. Starch digestibility can affect the amount of glucose present in the blood after eating, also known as the glycemic response (GR) (Anderson et al. 2010). Both starch digestibility and the glycemic response depend on a wide range of factors, including starch structure, processing, and acidity level (Poutanen et al. 2009). Whole grain breads tend to produce a lower glycemic response than white breads, because the structure of the grain kernel protects the starch from being hydrolyzed by GI amylolytic enzymes (Katina et al. 2005).

Fermenting grains with LAB, as in sourdough bread, can slow starch digestion and decrease starch availability, or the amount of starch available for amylolysis, leading to a lower glycemic impact (Lopez et al. 2001). Sourdough breads made with whole grain wheat flour provoke a lower glycemic response than breads made with unfermented whole grain wheat flour (Novotni et al. 2011). Sourdough breads also have a lower rate of starch hydrolysis than chemically acidified breads (Novotni et al. 2011). Current literature suggests that this effect may be due primarily to the organic acids generated during the fermentation process decreasing starch digestion (lactic acid) and/or slowing gastric emptying (propionic and acetic acids) (Liljeberg and Björck 1996, Liljeberg and Björck 1998, Poutanen et al. 2009, Novotni et al. 2011).

These acids (lactic acid, propionic acid, and acetic acid) may inhibit the amylolytic enzymes responsible for starch breakdown (Lopez et al. 2001, D'Alessandro and De Pergola 2014, Poutanen et al. 2009). One study served subjects either a white bread meal or a white bread meal supplemented with vinegar (acetic acid) to determine if the addition of acid affected glucose and insulin responses to the meal (Liljeberg and Björck 1998). Adding acetic acid to the white bread meal significantly lowered postprandial glucose and insulin responses (Liljeberg and Björck 1998). Because this effect occurs with foods that have added acids (i.e., pickled foods and vinegar) as well as with sourdough breads, lower starch digestibility and glycemic response from LAB fermented grains appears to be due to acid formation rather than to an effect unique to the action of LAB alone.

The presence of these organic acids in fermented breads may also lead to a lower rise in blood glucose than yeast-leavened breads (Liljeberg and Björck 1996, Liljeberg and Björck 1998, Scazzina et al. 2009). A 2009 study assessed differences in starch digestibility from breads made with white (refined grain) and wheat (whole grain) flours leavened either with sourdough or with *S. cerevisiae* yeast. *In vitro* assessments of starch hydrolysis rates indicated no differences among the four breads (Scazzina et al. 2009). However, the sourdough breads (white and wheat) led to significantly lower blood glucose responses in an *in vivo* assessment in healthy adults (Scazzina et al. 2009). A similar study conducted in 2010 measured the differences in postprandial glucose and insulin response to sourdough bread compared to yeast-leavened wheat and white wheat breads (Lappi et al. 2010). The sourdough wheat bread led to postprandial glucose and insulin responses even lower than that of the reference standard white wheat bread in 11 insulin-resistant Finnish subjects (Lappi et al. 2010). The mechanism to explain these findings remains unknown but could be due to the organic acids delaying gastric emptying without affecting starch digestibility (Scazzina et al. 2009).

The impact of sourdough bread on glycemic response may also depend on the sourdough culture used. A recent study found that most, but not all, sourdough cultures led to a lower glycemic index scores than yeast-fermented breads (Novotni et al. 2011). Yet, while sourdough bread may result in a lower glycemic response than yeast-fermented breads, the clinical implications of consuming foods with a lower glycemic index remain controversial (Vega-López et al. 2018). While the premise of the study of low-glycemic index diets is that different glycemic responses to different diets will affect other metabolic responses, a recent systematic review indicates that current evidence from both observational and intervention studies does not support this hypothesis (Vega-López et al. 2018). Fermenting grains with LAB cultures may lead to lower postprandial glucose and insulin responses compared to yeast-leavened breads, but the clinical implications of this finding lack scientific consensus.

Legumes

Many impacts of LAB fermentation on the bioavailability and digestibility of legumes mirror those of cereal grains. Like grains, legumes contain phytate, which prevents mineral absorption but can be broken down with fermentation. Fermentation can also affect the vitamin content of legumes. In addition, legume fermentation may generate bioactive compounds including antihypertensive peptides and antioxidants. Due to the similarities between legume and grain fermentation as well as the comparatively small body of literature published specifically on legume fermentation, the impact of LAB fermentation on legumes will be reviewed only briefly here.

Impact of LAB fermentation

Mineral bioavailability and vitamin content

Legumes refers to both pulse grains as well as legumes used for oil extraction (soybeans), harvested green for food (peas, green beans, sprouts), and groundnuts. Pulses are the "dry, edible variety of beans, peas, and lentils" (Havemeier et al. 2017). According to the Food and Agriculture Organization of the United Nations (FAO), pulses include dry beans, broad beans, peas, chickpeas, cow peas, pigeon peas, lentils, Bambara beans, vetches (mainly used for animal feed), lupins, and minor pulses (FAO 1994). The legumes category is broader than pulses, because it also includes soybeans and peanuts as well as other crops grown for a purpose other than dry grains (FAO 2016). This chapter describes the impacts of LAB fermentation, broadly, on beans, peas, and soybeans.

While both the 2015–2020 Dietary Guidelines for Americans (DGA) (USDA and HHS 2016) and the World Health Organization (WHO 2003, WHO 2004) list legumes as a nutrient-dense food, legumes, like many other plant-based protein sources, contain phytate. In addition to cereal grains and legumes, phytate is also found in small amounts in nuts, including hazelnuts, walnuts, almonds, and cashew nuts, and some vegetables and tubers (Sandberg 2002, Schlemmer et al. 2009). Phytate is the main storage form of phosphorus in legumes, as it is in grains. In legumes, phytate can be found in the cotyledon, where protein is also found (Sandberg 2002, Schlemmer et al. 2009). As with grains, phytate prevents the uptake of certain minerals from legumes as well. The lower pH generated by legume fermentation enables the activity of endogenous and bacterial phytases, which lowers the phytate content and makes their mineral content more bioavailable (Sandberg 2002, Starzyńska-Janiszewska and Stodolak 2011). Improving the mineral bioavailability of legumes may not be important for mixed diets that contain sufficient servings of animal-source foods that contain bioavailable zinc and iron, for instance (Hotz and Gibson 2007). However, in vegetarian and vegan diets as well as in other diets low in animal-source foods, enhancing the bioavailability of minerals from plant-based foods becomes important (Sandberg 2002, Dror and Allen 2011).

While some studies have been conducted to assess the differences in phytate content between unfermented and fermented legumes as well as the amount of bioactive components in fermented legumes, little research has been published on either *in vivo* mineral absorption following fermentation or the impact of fermentation-generated bioactive components on health. A 2011 study evaluated the impact of grass pea flour fermentation with *L. plantarum* (Starzyńska-Janiszewska and Stodolak 2011). Grass peas, commonly consumed in Africa and Asia, were assessed in part for adoption in the Eastern European diet, which tends to contain insufficient servings of legumes like the American diet (Starzyńska-Janiszewska and Stodolak 2011, USDA and HHS 2016). When grass pea flour was fermented until the pH dropped below 4.0, the amount of inositol phosphates was significantly lower than in the raw flour, especially in the samples inoculated with 5% (v/v) inoculum compared to 10% (v/v) inoculum (Starzyńska-Janiszewska and Stodolak 2011). In this study, the decrease in phytate appeared to be due to the activation of endogenous phytases. However, like a similar study on kidney beans (Limón et al. 2015), this study did not evaluate the *in vivo* effects of decreased phytate content to see if it was the primary factor responsible for low mineral absorption.

Fermentation can change the vitamin content of legumes and legume-based food products such as the amounts of B vitamins, vitamin C, and tocopherol. These differences in vitamin content with fermentation may be due to the microbes used—some synthesize certain nutrients, some degrade them—as well as the substrate. A 2017 review found that fermentation with fungi as well as with *L. acidophilus* increased "thiamin, riboflavin, niacin or pyridoxine in soymilk" but also decreased the vitamin C content in soymilk as well as lupin seeds. Fermentation increased vitamin C content of red beans. Fermenting soy with *L. plantarum* or with autochthonous bacteria (natural fermentation) decreased the α-, β-, and γ-tocopherol content, but increased the δ-tocopherol content (Gan et al. 2017). Lupins fermented with *L. plantarum*, however, had lower tocopherol levels for all tocopherol homologs (α-, γ-, and δ-tocopherol) (Gan et al. 2017). Fermentation with *L. delbrueckii* sp. *bulgaricus* increased the vitamin E content of red beans (Gan et al. 2017). These results are based on a limited pool of studies, however, more directed research is needed to determine mechanisms for the impact of fermentation with different LAB on different legumes as well (Gan et al. 2017).

Combination of grains and legumes

Making sourdough bread with legume flours may have bioavailability benefits beyond those of wheat-based sourdough alone. Fermenting wheat and legume flours (chickpea, lentil, and bean mixture) together led to greater phytase activity, higher antioxidant content, and improved *in vitro* protein digestibility compared to both yeasted and fermented breads made with wheat flour alone (Rizzello et al. 2014). Combining wheat and legume flours also improves the amino acid profile of the bread. Legumes and cereal grains tend to be deficient in some amino acids (sulfur-containing amino acids and lysine, respectively) but together can contain the balanced amount of essential amino acids needed for a complete protein (Rizzello et al. 2014). Consuming sourdough breads made with a combination of wheat and legume flours may also result in a lower rise in postprandial blood glucose than wheat breads due to their levels of carbohydrates, dietary fiber, and resistant starch (Gobbetti et al. 2018).

GABA and bioactive peptides

Edible seeds are rich in proteins, which can be hydrolyzed into small peptides during the fermentation process. Fermenting legumes with LAB may increase their amounts of gamma-aminobutyric acid (GABA) and angiotensin-converting enzyme (ACE) inhibitory peptides, both of which have been linked with lower blood pressure.

GABA is an inhibitory neurotransmitter linked with hypotensive properties, formed by the removal of the α-carboxyl group from glutamate by glutamate decarboxylase (GAD). LAB are among the most well-known GABA producers (Diana et al. 2014). GABA has been linked to a wide range of health benefits. However, the discussion in this chapter will focus on its ability to help regulate blood pressure. Production of GABA in fermented foods is due to GAD activity in LAB cells. If the glutamic acid content of a food is high enough, "GABA-producing LAB can be used to develop fermented, health-oriented food" (Diana et al. 2014). GABA is generated in a range of LAB-fermented foods, including dairy, grains, and legumes.

ACE plays a crucial role in the regulation of blood pressure. ACE-inhibitors act as vasodilators, and synthetic ACE inhibitors have long been used as antihypertensive agents. However, some ACE-inhibitors occur naturally with proteolysis, which occurs when dairy foods (Madureira et al. 2010) as well as some legumes (Gan et al. 2017) are fermented. Legumes and legume products including lentils, mung beans, pea seeds, kidney beans, black soybeans, navy beans, and soy beverage and pudding, when fermented with LAB, exhibit varying levels of ACE-inhibitory activity, primarily from the generation of ACE-inhibitory peptides or GABA (Limón et al. 2015, Gan et al. 2017). Additional research is needed to determine if LAB fermentation can be an effective method of creating functional foods with antihypertensive benefits.

Dairy Foods

As with other foods, use of LAB to ferment milk dates back to the Early Neolithic Period (Price et al. 2012). LAB are autochthonous in milk, and raw milk will sour if left at room temperatures because of LAB activity (Walstra et al. 2005). Several commercial dairy products, notably yogurt, cheese, and cultured butter, are made using naturally occurring or added LAB cultures (Walstra et al. 2005). Fermentation with LAB changes the nutritional profile and physiological impact of milk (Adolfsson et al. 2004). Fermented dairy foods, especially yogurt, are discussed broadly in this section to reflect the current evidence base; however, the generalizability of this evidence may be limited by the considerable variations in yogurt and other fermented dairy food production. Milk composition varies inherently by cow breed and individual animal, time of year, geographic location, and feed composition, among other factors. These factors as well as the species and strain of bacteria and temperature and duration of the fermentation process may affect fermented dairy foods and their impact on health (Adolfsson et al. 2004, Thorning et al. 2017).

Milk fermentation through LAB growth depends on lactose as a primary energy source for bacteria (Walstra et al. 2005). Of the 12 genera of LAB, 4 have organisms used to ferment milk: *Lactococcus*, *Leuconostoc*, *Streptococcus*, and *Lactobacillus* (Walstra et al. 2005). These LAB primarily grow by metabolizing lactose into lactic acid (Walstra et al. 2005). Certain LAB strains can also metabolize citrate, another component of milk, into diacetyl, an important flavor compound in fermented dairy products (Walstra et al. 2005). In addition to creating the characteristic textures and tastes of different fermented dairy foods, fermenting milk with LAB also leads to the partial breakdown and, subsequently, better digestibility and improved absorption of milk's macro- and micronutrients (Miller et al. 2007). Because of the breakdown of macro- and micronutrients that occurs with fermentation, yogurt and cheese have both been referred to as "easy to digest" (Marette and Picard-Deland 2014, Rizzoli 2014, Chen et al. 2017, Rizzoli and Biver 2018). Unlike milk, fermented dairy products like yogurt, cottage cheese and cheese can be introduced in young children's diets as early as 6 months of age (American Academy of Pediatrics et al. 2013). This section will focus primarily on the impact of fermentation on the bioavailability of milk's micronutrients, especially in yogurt. The breakdown of the carbohydrates and protein in milk are detailed in other chapters and will only be reviewed briefly here.

Impact of LAB fermentation

Mineral bioavailability

In the U.S., yogurt must be made with at least two LAB cultures, *L. delbrueckii* subsp. *bulgaricus* and *Streptococcus thermophilus*, according to its standard of identity as defined by the FDA (FDA 2016). Yogurt is considered a good source of calcium, protein, phosphorus, riboflavin, vitamin B_{12}, pantothenic acid, and zinc, according to FDA labeling guidelines (CFR—Code of Federal Regulations Title 21 101.54, USDA 2018).

While milk is also an important source of most of these nutrients, especially in the American diet (O'Neil et al. 2012, Keast et al. 2013, O'Neil et al. 2018), fermenting milk with LAB to produce yogurt creates an acidic environment that may enhance the bioavailability of some minerals, like calcium, phosphorus, zinc, and magnesium, in a manner similar to what occurs in fermented grains. To be absorbed, calcium must be in its soluble ionized form (Ca^{2+}) or bound to another soluble organic molecule to cross the intestinal wall (Vavrusova and Skibsted 2014, Hess et al. 2016). Fermenting milk solubilizes calcium and phosphorus (Fisberg and Machado 2015). The lactic acid generated by LAB during fermentation gives yogurt a lower pH than milk. The pH of raw milk is approximately 6.7, while the pH of yogurt is around 4.0 (de la Fuente et al. 2003). As the pH decreases, the calcium and phosphorus in the milk serum (skim milk) increase as the calcium and phosphorus in the micellar proteins decreases (de la Fuente et al. 2003). While calcium is present in milk as colloidal calcium phosphate (Guéguen and Pointillart 2000), the lower pH of yogurt maintains calcium and magnesium in their ionic forms, which may allow for greater intestinal absorption (Adolfsson et al. 2004, El-Abbadi et al. 2014).

Similarly, zinc in cow's milk is primarily found bound to casein, which may render it unavailable for absorption (de la Fuente et al. 2003). In environments with a pH of ~ 4.6, 95% of the zinc is soluble and may be more available for absorption (de la Fuente et al. 2003). According to Walstra et al. (2005), yogurt has a pH around 4.2, low enough to solubilize zinc. The lower pH also allows soluble zinc to bind to ligands that can facilitate transportation across the intestinal wall and may prevent zinc from being chelated by other substances in the GI tract (de la Fuente et al. 2003, El-Abbadi et al. 2014). Fermenting milk to make yogurt may facilitate zinc absorption as well. Making milk into yogurt may also increase the percentage of total magnesium in a soluble form by 20% or more (de la Fuente et al. 2003). While magnesium is primarily found in a soluble form in cow's milk, even the magnesium that is found in the casein micelles (35%) is progressively released as pH decreases until 88 to 97% of the magnesium is in a soluble form.

While it is feasible that the lower pH leads to increased mineral bioavailability from fermented milk, to our knowledge, only two human studies have compared mineral absorption from fermented dairy versus milk (Nickel et al. 1996, Rosado et al. 2005). A clinical study conducted in premenopausal women assessed the differences in calcium absorption from milk, cheese, and yogurt consumption and found no differences in nutrient absorption between milk and fermented milk products (Nickel et al. 1996). A 2005 study assessed the impact of adding milk or yogurt to plant-based diets to see if they affected zinc or iron bioavailability. Iron absorption did not differ between groups. However, zinc absorption increased by 70% when milk was added to the diet and by 78% when yogurt was added (Rosado et al. 2005). The results of these studies indicate that there may be a small difference between mineral bioavailability from milk compared to fermented dairy. To our knowledge, similar studies have not been conducted to compare phosphorus or magnesium absorption from yogurt or other fermented dairy foods compared to milk.

Macronutrient bioavailability

The evidence showing a benefit for lactose digestion with fermentation is well established in scientific literature and has more of an evidence base than benefits to protein or fat digestibility with fermentation (NIH Lactose Intolerance Conference–Panel Statement, Walstra et al. 2005, EFSA NDA 2010, Savaiano 2014). Yogurt is often described as containing less lactose than milk. While commercial yogurts can contain as much lactose as a glass of milk (~ 12 g) (Savaiano 2014), the activity of LAB after digestion can improve lactose digestion (EFSA NDA 2010, Savaiano 2014). If yogurt contains "live and active cultures" at retail (Live and Active Culture Yogurt Facts), then it contains cultures capable of producing lactase *in vivo* and helping with lactose digestion after consumption (Adolfsson et al. 2004).

In addition to lactose hydrolysis, LAB in yogurt may also exert some proteolytic activity to help digest milk proteins, increasing the amount of free amino acids and allowing for better protein digestibility (Adolfsson et al. 2004, El-Abbadi et al. 2014). Heat treatment with yogurt fermentation may also contribute to greater protein digestibility of yogurt compared to milk (Adolfsson et al. 2004). Proteolytic digestion by LAB could also release bioactive peptides in dairy proteins (German 2014). While a few studies have explored the potential for some of these bioactive peptides to act as antihypertensive agents (Madureira et al. 2010, Qin et al. 2013, Dugan and Fernandez 2014) similar to what occurs with LAB-fermented legumes, other potential benefits of these peptides need further study (Hess et al. 2016).

Conclusion

Fermentation is important for bioavailability and digestibility of minerals, especially from plant-based foods. In high-income countries, where certain foods are fortified to prevent mineral deficiencies and nutrient deficiencies are rare, the enhanced bioavailability of LAB-fermented grains and legumes may not significantly impact micronutrient intake. However, diets in low- and middle-income countries are more likely to rely on staple foods like grains and legumes with less variety (Schlemmer et al. 2009). In these areas, the amount of phytate in the diet could have serious implications for micronutrient malnutrition. A 2014 review of fermented foods in Africa identifies LAB-rich food as vital for malnourished populations or populations suffering from diarrheal diseases (Franz et al. 2014). Their argument centers on the ability of traditional foods, if prepared with a specific combination of health-promoting bacteria, to serve as probiotics. However, the authors also acknowledge the ability of fermentation to increase the bioavailability of minerals and, therefore, the ability of fermented foods to contribute to a decrease in mineral malnourishment, especially when animal-source foods are not readily available, accessible, or affordable (Dror and Allen 2011).

In addition, while the nutrient-density of foods like legumes and grains are generally well-accepted, including by authoritative international health organizations like FAO and WHO as well as by the national governments of many countries (Ancellin et al. 2011, Nordic Nutrition Recommendations 2012, Ministry of Health of Brazil- Secretariate of Health Care and Primary Health Care Department 2014, USDA and HHS 2016, Health Canada 2019), few studies have investigated the potential for adverse health effects of recommending these foods. While legume and whole grain recommendations are seldom met in high-income countries (Starzyńska-Janiszewska and Stodolak 2011, USDA and HHS 2016), a recent push for sustainable eating patterns as well as consumer trends have renewed an interest in plant-based proteins. For instance, an independent group of academics developed recommendations for sustainable diets (Willett et al. 2019), which include primarily plant-based protein sources like legumes and grains. For some populations, following recommendations like these without regard for food processing, digestibility, and nutrient bioavailability could increase malnutrition and nutrient insufficiencies and deficiencies. How plant-based protein foods are prepared is an important consideration to pair with recommendations to ensure that the nutrients in these foods, especially nutrients of concern, are bioaccessible, bioavailable, and digestible.

References

A Guide to Calcium-Rich Foods - National Osteoporosis Foundation. Available from: https://www.nof.org/patients/treatment/calciumvitamin-d/a-guide-to-calcium-rich-foods/.

Adolfsson, O., S.N. Meydani and R.M. Russell. 2004. Yogurt and gut function. Am. J. Clin. Nutr. 80: 245–256.

American Association of Cereal Chemists International. 1999. Whole Grain Definition. Available from: http://www.aaccnet.org/initiatives/definitions/pages/wholegrain.aspx.

Ancellin, R., D. Baelde, L. Barthelemy, F. Bellisle, J.-L. Berta, D. Boute, K. Castetbon, M. Chauliac, C. Duchene, C. Dumas et al. 2011. La Santé vient en mangeant. Le Guide Alimentaire Pour Tous. Saint-Yrieix-la-Perche. Available from: http://www.inpes.sante.fr/CFESBases/catalogue/pdf/581.pdf.

Anderson, G.H., C.E. Cho, T. Akhavan, R.C. Mollard, B.L. Luhovyy and E.T. Finocchiaro. 2010. Relation between estimates of cornstarch digestibility by the Englyst *in vitro* method and glycemic response, subjective appetite, and short-term food intake in young men. Am. J. Clin. Nutr. 91: 932–939.

Awika, J.M. 2011. Major cereal grains production and use around the world. pp. 1–13. *In*: Awika, J.M., V. Piironen and S. Bean (eds.). Advances in Cereal Science: Implications to Food Processing and Health Promotion. American Chemical Society, Park Ridge, USA.

Bach Kristensen, M., I. Tetens, A.B. Alstrup Jørgensen, A. Dal Thomsen, N. Milman, O. Hels, B. Sandström† and M. Hansen. 2005. A decrease in iron status in young healthy women after long-term daily consumption of the recommended intake of fibre–rich wheat bread. Eur. J. Nutr. 44: 334–340.

Biesiekierski, J.R., S.L. Peters, E.D. Newnham, O. Rosella, J.G. Muir and P.R. Gibson. 2013. No effects of gluten in patients with self-reported non-celiac gluten sensitivity after dietary reduction of fermentable, poorly absorbed, short-chain carbohydrates. Gastroenterology 145: 320–328.

Center for Food Safety and Applied Nutrition. 2006. Regulation - Draft Guidance for Industry and FDA Staff: Whole Grain Label Statements. Available from: https://www.fda.gov.

Center for Food Safety and Applied Nutrition. 2018. Allergens - Questions and Answers: Gluten-Free Food Labeling Final Rule. Available from: https://www.fda.gov.

CFR - Code of Federal Regulations Title 21 101.54. Available from: https://www.accessdata.fda.gov/scripts/cdrh/cfdocs/cfcfr/CFRSearch.cfm?fr=101.54.

Chavan, R.S. and S.R. Chavan. 2011. Sourdough technology—a traditional way for wholesome foods: A review. Compr. Rev. Food. Sci. Food. Saf. 10: 169–182. Available from: http://doi.wiley.com/10.1111/j.1541-4337.2011.00148.x.

Chen, G.-C., Y. Wang, X. Tong, I.M.Y. Szeto, G. Smit, Z.-N. Li and L.Q. Qin. 2017. Cheese consumption and risk of cardiovascular disease: a meta-analysis of prospective studies. Eur. J. Nutr. 56: 2565–2575.

D'Alessandro, A. and G. De Pergola. 2014. Mediterranean diet pyramid: A proposal for Italian people. Nutrients 6: 4302–4316.

Davidsson, L., A. Almgren, M.A. Juillerat and R.F. Hurrell. 1995. Manganese absorption in humans: the effect of phytic acid and ascorbic acid in soy formula. Am. J. Clin. Nutr. 62: 984–987.

de la Fuente, M.A., F. Montes, G. Guerrero and M. Juárez. 2003. Total and soluble contents of calcium, magnesium, phosphorus and zinc in yoghurts. Food Chem. 80: 573–578.

Diana, M., J. Quílez and M. Rafecas. 2014. Gamma-aminobutyric acid as a bioactive compound in foods: a review. J. Funct. Foods 10: 407–420.

Dror, D.K. and L.H. Allen. 2011. The importance of milk and other animal-source foods for children in low-income countries. Food Nutr. Bull. 32: 227–243.

Dugan, C.E. and M.L. Fernandez. 2014. Effects of dairy on metabolic syndrome parameters: a review. Yale J. Biol. Med. 87: 135–147.

EFSA Panel on Dietetic Products, Nutrition and Allergies (NDA). 2010. Scientific Opinion on the substantiation of health claims related to live yoghurt cultures and improved lactose digestion (ID 1143, 2976) pursuant to Article 13(1) of Regulation (EC) No. 1924/2006. EFSA J. 8.

El-Abbadi, N.H., M.C. Dao and S.N. Meydani. 2014. Yogurt: role in healthy and active aging. Am. J. Clin. Nutr. 99: 1263S–1270S.

FAO. 1994. Definition and classification of commodities: pulses and derived products. Rome. Available from: http://www.fao.org/es/faodef/fdef04e.htm.

FAO/WHO. 1998. Carbohydrates in Human Nutrition: Report of aJoint FAO/WHO Expert Consultation, 14–18 April 1997, Rome. FAO Food and Nutrition Paper No. 66. Rome. Available from: http://www.fao.org/tempref/docrep/fao/X2650T/X2650t02.pdf.

FAO. 2016. About | 2016 International Year of Pulses. Available from: http://www.fao.org/pulses-2016/about/en/.

FDA. 2016. CFR—Code of Federal Regulations Title 21. Available from: http://www.accessdata.fda.gov.

Ferruzzi, M.G., S.S. Jonnalagadda, S. Liu, L. Marquart, N. McKeown, M. Reicks, G. Riccardi, C. Seal, J. Slavin, F. Thielecke et al. 2014. Developing a standard definition of whole-grain foods for dietary recommendations: Summary report of a multidisciplinary expert roundtable discussion. Adv. Nutr. 5: 164–176.

Fisberg, M. and R. Machado. 2015. History of yogurt and current patterns of consumption. Nutr. Rev. 73: 4–7.

Franz, C.M., M. Huch, J.M. Mathara, H. Abriouel, N. Benomar, G. Reid, A. Galvez and W.H. Holzapfel. 2014. African fermented foods and probiotics. Int. J. Food Microbiol. 190: 84–96.

Gan, R.-Y., H.-B. Li, A. Gunaratne, Z.Q. Sui and H. Corke. 2017. Effects of fermented edible seeds and their products on human health: Bioactive components and bioactivities. Compr. Rev. Food Sci. Food Saf. 16: 489–531.

Gänzle, M. and V. Ripari. 2016. Composition and function of sourdough microbiota: From ecological theory to bread quality. Int. J. Food Microbiol. 239: 19–25.

German, J.B. 2014. The future of yogurt: scientific and regulatory needs. Am. J. Clin. Nutr. 99: 1271S–8S.

Gibson, P.R. and S.J. Shepherd. 2010. Evidence-based dietary management of functional gastrointestinal symptoms: The FODMAP approach. J. Gastroenterol. Hepatol. 25: 252–258.

Giorgio, R. De, U. Volta and P.R. Gibson. 2016. Sensitivity to wheat, gluten and FODMAPs in IBS: facts or fiction? Gut. 65: 169–178.

Gobbetti, M., C.G. Rizzello, R. Di Cagno and M. De Angelis. 2014. How the sourdough may affect the functional features of leavened baked goods. Food Microbiol. 37: 30–40.

Gobbetti, M., M. De Angelis, R. Di Cagno, M. Calasso, G. Archetti and C.G. Rizzello. 2018. Novel insights on the functional/nutritional features of the sourdough fermentation. Int. J. Food Microbiol.

Gropper, S.A.S. and J.L. Smith. 2013. Advanced nutrition and human metabolism. Wadsworth/ Cengage Learning, Belmont. USA.

Guéguen, L. and A. Pointillart. 2000. The bioavailability of dietary calcium. J. Am. Coll. Nutr. 19: 119S–136S.

Halmos, E.P., V.A. Power, S.J. Shepherd, P.R. Gibson and J.G. Muir. 2014. A diet low in FODMAPs reduces symptoms of irritable bowel syndrome. Gastroenterology 146: 67–75.

Havemeier, S., J. Erickson and J. Slavin. 2017. Dietary guidance for pulses: the challenge and opportunity to be part of both the vegetable and protein food groups. Ann. NY Acad. Sci. 1392: 58–66.

Health Canada. 2019. Canada's Dietary Guidelines.

Hess, J.M., S.S. Jonnalagadda and J.L. Slavin. 2016. Dairyfoods: Current evidence of their effects on bone, cardiometabolic, cognitive, and digestive health. Compr. Rev. Food Sci. Food Saf. 15: 251–268.

High Level Panel of Experts on Food Security. 2017. HLPE High Level Panel of Experts T on Food Security and Nutrition. Available from: www.fao.org/cfs/cfs-hlpe.

Hotz, C. and R.S. Gibson. 2007. Traditional food-processing and preparation practices to enhance the bioavailability of micronutrients in plant-based diets. J. Nutr. 137: 1097–1100.

Hurrell, R.F., M.B. Reddy, M.-A. Juillerat and J.D. Cook. 2003. Degradation of phytic acid in cereal porridges improves iron absorption by human subjects. Am. J. Clin. Nutr. 77: 1213–1219.

Ilich, J.Z. and J.E. Kerstetter. 2000. Nutrition in bone health revisited: A story beyond calcium. J. Am. Coll. Nutr. 19: 715–737.

Katina, K., E. Arendt, K.-H.- Liukkonen, K. Autio, L. Flander and K. Poutanen. 2005. Potential of sourdough for healthier cereal products. Trends Food Sci. Technol. 16: 104–112.

Keast, D., V. Fulgoni, T. Nicklas and C. O'Neil. 2013. Food sources of energy and nutrients among children in the United States: National Health and Nutrition Examination Survey 2003–2006. Nutrients 5: 283–301.

Kelly, C.P., J.C. Bai, E. Liu and D.A. Leffler. 2015. Advances in diagnosis and management of celiac disease. Gastroenterology 148: 1175–1186.

Kleinman, R.E. and F.R. Greer. 2013. Pediatric Nutrition. American Academy of Pediatrics, Itasca. USA.

Laatikainen, R., J. Koskenpato, S.-M. Hongisto, J. Loponen, T. Poussa, M. Hillilä and R. Korpela. 2016. Randomised clinical trial: low-FODMAP rye bread vs. regular rye bread to relieve the symptoms of irritable bowel syndrome. Aliment. Pharmacol. Ther. 44: 460–470.

Lappi, J., E. Selinheimo, U. Schwab, K. Katina, P. Lehtinen, H. Mykkänen, M. Kolehmainen and K. Poutanen. 2010. Sourdough fermentation of wholemeal wheat bread increases solubility of arabinoxylan and protein and decreases postprandial glucose and insulin responses. J. Cereal Sci. 51: 152–158.

Lee, D.-Y., J. Schroeder and D.T. Gordon. 1988. Enhancement of Cu bioavailability in the rat by Phytic. Acid. J. Nutr. 118: 712–717.

Leonard, M.M., A. Sapone, C. Catassi and F. Fasano. 2017. Celiac disease and nonceliac gluten sensitivity. JAMA 318: 647.

Liljeberg, H. and I. Björck. 1998. Delayed gastric emptying rate may explain improved glycaemia in healthy subjects to a starchy meal with added vinegar. Eur. J. Clin. Nutr. 52: 368–371.

Liljeberg, H.G. and I.M. Björck. 1996. Delayed gastric emptying rate as a potential mechanism for lowered glycemia after eating sourdough bread: studies in humans and rats using test products with added organic acids or an organic salt. Am. J. Clin. Nutr. 64: 886–893.

Limón, R.I., E. Peñas, M.I. Torino, C. Martínez-Villaluenga, M. Dueñas and J. Frias. 2015. Fermentation enhances the content of bioactive compounds in kidney bean extracts. Food Chem. 172: 343–352.

Live and Active Culture Yogurt Facts. Available from: http://aboutyogurt.com.

Lopez, H.W., V. Krespine, C. Guy, A. Messager, C. Demigne and C. Remesy. 2001. Prolonged fermentation of whole wheat sourdough reduces phytate level and increases soluble magnesium. J. Agric. Food Chem. 49: 2657–2662.
Loponen, J. and M.G. Gänzle. 2018. Use of sourdough in low FODMAP baking. Foods 7: E96.
Madureira, A.R., T. Tavares, A.M.P. Gomes, M.E. Pintado and F.X. Malcata. 2010. Invited review: physiological properties of bioactive peptides obtained from whey proteins. J. Dairy Sci. 93: 437–455.
Marette, A. and E. Picard-Deland. 2014. Yogurt consumption and impact on health: focus on children and cardiometabolic risk. Am. J. Clin. Nutr. 99: 1243S–1247S.
Menezes, L.A.A., F. Minervini, P. Filannino, M.L.S. Sardaro, M. Gatti and J.D.D. Lindner. 2018. Effects of sourdough on FODMAPs in bread and potential outcomes on irritable bowel syndrome patients and healthy subjects. Front. Microbiol. 9: 1972.
Miller, G.D., J.K. Jarvis, L.D. McBean and National Dairy Council. 2007. Handbook of Dairy Foods and Nutrition. CRC Press, Boca Raton. USA.
Ministry of Health of Brazil-Secretariate of Health Care and Primary Health Care Department. 2014. Dietary Guidelines for the Brazilian Population.
Muir, J.G., J.E. Varney, M. Ajamian and P.R. Gibson. 2019. Gluten-free and low-FODMAP sourdoughs for patients with coeliac disease and irritable bowel syndrome: A clinical perspective. Int. J. Food Microbiol. 290: 237–246.
Nickel, K.P., B.R. Martin, D.L. Smith, J.B. Smith, G.D. Miller and C.M. Weaver. 1996. Calcium bioavailability from bovine milk and dairy products in premenopausal women using intrinsic and extrinsic labeling techniques. J. Nutr. 126: 1406–1411.
Nielsen, A.V.F., I. Tetens and A.S. Meyer. 2013. Potential of phytase-mediated iron release from cereal-based foods: a quantitative view. Nutrients 5: 3074–3098.
NIH Lactose Intolerance Conference–Panel Statement. Available from: https://consensus.nih.gov.
Nordic Nutrition Recommendations. 2012. Available from: http://norden.diva-portal.org.
Novotni, D., D. Ćurić, M. Bituh, I. Colić Barić, D. Škevin and N. Čukelj. 2011. Glycemic index and phenolics of partially-baked frozen bread with sourdough. Int. J. Food Sci. Nutr. 62: 26–33.
O'Neil, C.E., D.R. Keast, V.L. Fulgoni and T.A. Nicklas. 2012. Food sources of energy and nutrients among adults in the US: NHANES 2003–2006. Nutrients 4: 2097–2120.
O'Neil, C.E., T.A. Nicklas and V.L. Fulgoni. 2018. Food Sources of energy and nutrients of public health concern and nutrients to limit with a focus on milk and other dairy foods in children 2 to 18 years of age: National Health and Nutrition Examination Survey, 2011–2014. Nutrients 10: 8.
Poutanen, K., L. Flander and K. Katina. 2009. Sourdough and cereal fermentation in a nutritional perspective. Food Microbiol. 26: 693–699.
Prasad, A.S. 2013. Discovery of human zinc deficiency: its impact on human health and disease. Adv. Nutr. 4: 176–190.
Price, C.E., A. Zeyniyev, O.P. Kuipers and J. Kok. 2012. From meadows to milk to mucosa—adaptation of *Streptococcus* and *Lactococcus* species to their nutritional environments. FEMS Microbiol. Rev. 36: 949–971.
Qin, L.-Q., J.-Y. Xu, J.-Y. Dong, Y. Zhao, P. van Bladeren and W. Zhang. 2013. Lactotripeptides intake and blood pressure management: a meta-analysis of randomised controlled clinical trials. Nutr. Metab. Cardiovasc. Dis. 23: 395–402.
Reale, A., U. Konietzny, R. Coppola, E. Sorrentino and R. Greiner. 2007. The importance of lactic acid bacteria for phytate degradation during cereal dough fermentation. J. Agric. Food Chem. 55: 2993–2997.
Rizzello, C.G., M. Calasso, D. Campanella, M. De Angelis and M. Gobbetti. 2014. Use of sourdough fermentation and mixture of wheat, chickpea, lentil and bean flours for enhancing the nutritional, texture and sensory characteristics of white bread. Int. J. Food Microbiol. 180: 78–87.
Rizzello, C.G., M. Montemurro and M. Gobbetti. 2016. Characterization of the bread made with durum wheat semolina rendered gluten free by sourdough biotechnology in comparison with commercial gluten-free products. J. Food Sci. 81: H2263–H2272.
Rizzoli, R. 2014. Dairy products, yogurts, and bone health. Am J. Clin. Nutr. 99: 1256S–62S.
Rizzoli, R. and E. Biver. 2018. Effects of fermented milk products on bone. Calcif. Tissue Int. 102: 489–500.

Rosado, J.L., M. Díaz, K. González, I. Griffin, S.A. Abrams and R. Preciado. 2005. The addition of milk or yogurt to a plant-based diet increases zinc bioavailability but does not affect iron bioavailability in women. J. Nutr. 135: 465–468.

Rousseau, S., C. Kyomugasho, M. Celus, M.E.G. Hendrickx and T. Grauwet. 2019. Barriers impairing mineral bioaccessibility and bioavailability in plant-based foods and the perspectives for food processing. Crit. Rev. Food Sci. Nutr. 1–18.

Sandberg, A.-S. 2002. Bioavailability of minerals in legumes. Br. J. Nutr. 88: 281.

Savaiano, D.A. 2014. Lactose digestion from yogurt: mechanism and relevance. Am. J. Clin. Nutr. 99: 1251S–1255S.

Scazzina, F., D. Del Rio, N. Pellegrini and F. Brighenti. 2009. Sourdough bread: Starch digestibility and postprandial glycemic response. J. Cereal Sci. 49: 419–421.

Schlemmer, U., W. Frølich, R.M. Prieto and F. Grases. 2009. Phytate in foods and significance for humans: Food sources, intake, processing, bioavailability, protective role and analysis. Mol. Nutr. Food Res. 53: S330–S375.

See, J.A., K. Kaukinen, G.K. Makharia, P.R. Gibson and J.A. Murray. 2015. Practical insights into gluten-free diets. Nat. Rev. Gastroenterol. Hepatol. 12: 580–591.

Slavin, J. 2004. Whole grains and human health. Nutr. Res. Rev. 17: 99.

Starzyńska-Janiszewska, A. and B. Stodolak. 2011. Effect of inoculated lactic acid fermentation on antinutritional and antiradical properties of grass pea (Lathyrus sativus 'Krab') flour. Polish J. Food Nutr. Sci. 61.

The European Commission. 2014. Commission Implementing Regulation (EU) No. 828/2014 on the requirements for the provision of information to consumers on the absence or reduced presence of gluten in food. Off. J. Eur. Union. 228/5–228/8.

Thorning, T.K., H.C. Bertram, J.-P. Bonjour, L. de Groot, D. Dupont, E. Feeney, R. Ipsen, J.M. Lecerf, A. Mackie, M.C. McKinley et al. 2017. Whole dairy matrix or single nutrients in assessment of health effects: current evidence and knowledge gaps. Am. J. Clin. Nutr. 105: 1033–1045.

Uhde, M., M. Ajamian, G. Caio, R. De Giorgio, A. Indart, P.H. Green, E.C. Verna, U. Volta and A. Alaedini. 2016. Intestinal cell damage and systemic immune activation in individuals reporting sensitivity to wheat in the absence of coeliac disease. Gut. 65: 1930–1937.

USDA. 2018. USDA National Nutreint Database for Standard Reference Legacy Release. Leg Release [Internet].: Nutrient Data Labaratory Home Page. Available from: https://ndb.nal.usda.gov.

USDA, HHS. 2016. 2015–2020 Dietary Guidelines - health.gov. Available from: http://health.gov/dietaryguidelines/2015/guidelines/.

Vavrusova, M. and L.H. Skibsted. 2014. Calcium nutrition. Bioavailability and fortification. LWT - Food Sci. Technol. 59: 1198–1204.

Vega-López, S., B. Venn, J. Slavin, S. Vega-López, B.J. Venn and J.L. Slavin. 2018. Relevance of the glycemic index and glycemic load for body weight, diabetes, and cardiovascular disease. Nutrients 10: 1361.

Walstra, P., P. Walstra, J.T.M. Wouters and T.J. Geurts. 2005. Dairy Science and Technology, Second Edition. CRC Press, Bota Raton. USA.

WHO | Daily iron supplementation in children and adolescents 5–12 years of age. 2018. Available from: http://www.who.int.

WHO | Diet, nutrition and the prevention of chronic diseases. 2003. Report of the joint WHO/FAO expert consultation. Available from: http://www.who.int.

WHO | Vitamin and mineral requirements in human nutrition. 2004. Available from: http://www.who.int.

WHO | Weekly iron and folic acid supplementation as an anaemia-prevention strategy in women and adolescent girls. 2018. Available from: http://www.who.int.

WHO, Food and Agriculture Organization of the United Nations. International Atomic Energy Agency. 1996. Trace elements in human nutrition and health. Available from: http://www.who.int.

Wierdsma, N.J., M.A.E. van Bokhorst-de van der Schueren, M. Berkenpas, C.j.J. Mulder and A.A. van Bodegraven. 2013. Vitamin and mineral deficiencies are highly prevalent in newly diagnosed celiac disease patients. Nutrients 5: 3975–3992.

Willett, W., J. Rockström, B. Loken, M. Springmann, T. Lang, S. Vermeulen, T. Garnett, D. Tilman, F. DeClerck, A. Wood et al. 2019. Food in the Anthropocene: the EAT–Lancet Commission on healthy diets from sustainable food systems. Lancet. 393(10170): 447–492.

Yao, C.K., H.-L. Tan, D.R. van Langenberg, J.S. Barrett, R. Rose, K. Liels, P.R. Gibson and J.G. Muir. 2014. Dietary sorbitol and mannitol: food content and distinct absorption patterns between healthy individuals and patients with irritable bowel syndrome. J. Hum. Nutr. Diet. 27: 263–275.

Yao, M.-Z., Y.-H. Zhang, W.-L.Lu, M.-Q. Hu, W. Wang and A.-H. Liang. 2012. Phytases: crystal structures, protein engineering and potential biotechnological applications. J. Appl. Microbiol. 112: 1–14.

Ziegler, J.U., D. Steiner, C.F.H. Longin, T. Würschum, R.M. Schweiggert and R. Carle. 2016. Wheat and the irritable bowel syndrome—FODMAP levels of modern and ancient species and their retention during bread making. J. Funct. Foods 25: 257–266.

4

Lactic Acid Bacteria Application to Decrease Food Allergies

Vanessa Biscola,[1,*] *Marcela Albuquerque Cavalcanti de Albuquerque,*[2] *Tatiana Pacheco Nunes,*[3] *Antonio Diogo Silva Vieira*[2] and *Bernadette Dora Gombossy de Melo Franco*[2]

Introduction

Food allergies are defined as abnormal immune responses to a normally tolerated food that occurs in susceptible hosts. Proteins are recognized as the main allergens present in foods and are usually responsible to trigger the immunological adverse reactions. Based on the mechanism involved, allergies can be classified into three categories: Immunoglobulin E (IgE)-mediated; non-IgE mediated (or cell-mediated); and mixed IgE mediated and cell-mediated reactions, when both mechanisms are involved.

Concerning the prevalence and the impact of this problem, food allergy is recognized as a public health issue affecting a great number of individuals worldwide. Its exact prevalence in the general population is unknown, but studies carried out in both developed and developing countries revealed that the rates of challenge-diagnosed cases of food allergy are as high as 10% amongst young children and about 2% in adults. This prevalence has increased over the last decades. Although individuals can be allergic to a variety of foods, a small group accounts for 90% of

[1] Clock-T (Cryolog) R&D Department 44, rue d'Allonville, 44000, Nantes, France.
[2] FoRC, Food Research Center, University of São Paulo, Rua do Lago 250, Butantã, São Paulo, SP, 05508-080, Brazil.
Emails: malbuquerque17@gmail.com; antdiogovieira@gmail.com; bfranco@usp.br
[3] Federal University of Sergipe, Food Technology Department, Av. Marechal Rondon s/n-University Campus, São Cristóvão-SE, 49100-000, Brazil.
Email: tpnunes@ufs.br
* Corresponding author: v_biscola@yahoo.com.br

all cases of food allergy. This group is often referred to as the "Big 8" and comprises milk, eggs, fish, crustacean shellfish, tree nuts, peanuts, wheat and soybean (Loh and Tang 2018).

Currently, the only effective method to prevent food allergy is to avoid the responsible allergen. However, this strategy is particularly difficult due to the ubiquitous nature of food allergens and the risks of cross-contamination. Furthermore, it leads to dietary restrictions that may cause relevant nutritional deficiencies, especially for young children, the most affected population. Therefore, the study of new strategies to reduce food allergy is of great scientific interest. Trying to find new alternatives to fulfill this purpose, many studies have addressed the role of food processing in reducing allergic reactions. The principle of these approaches relies on the modification of the structural conformation of the antigenic proteins present in the food, causing alterations on their specific IgE-binding sites (called IgE-binding epitopes) and preventing allergy to set off. Some of the modifications caused by food processing include unfolding, aggregation, cross-linking between ingredients, oxidation, glycosylation, and disruption of IgE-binding epitopes. However, the efficacy of food processing on allergy reduction is affected by its nature, the exposure time and the intensity of the treatment applied. Since food allergens differ in molecular and structural properties, the techniques used can reduce allergy, but can also present no effect or even increase allergenicity, by exposing hidden epitopes or contributing to the formation of neo-allergens (Verhoeckx et al. 2015, Okolie et al. 2018).

Amongst the different processing strategies used to reduce food allergy the most promising are fermentation and hydrolysis, both presenting the potential to cleave sequential and conformational epitopes to such an extent that symptoms will not be triggered. In this context, the use of proteolytic lactic acid bacteria (LAB) has risen as an interesting method to decrease the antigenicity of various food proteins (Verhoeckx et al. 2015).

Another interesting approach in the study of LAB contribution to reducing food allergy refers to the beneficial effects associated with the consumption of probiotic strains. Clinical reports demonstrated that probiotics can promote the differentiation of T regulatory (Treg) cells and rebalance T helper (Th) responses (Th1/Th2), stimulating Th1-dominant state. However, the underlying mechanisms are not fully understood, and the immunomodulatory effects seem to be strain dependent (Żukiewicz-Sobczak et al. 2014, Fu et al. 2019).

This chapter discusses the use of lactic acid bacteria to reduce food allergy. In this matter, the chapter addresses the potential application of proteolytic LAB in the manufacturing of hypoallergenic products as well as the role of probiotics on immune system regulation and allergy suppression. The focus is the main food allergens, belonging to the Big 8 group, and to which the use of LAB as a strategy for allergy decrease has already been reported in the literature (i.e., milk, soybean, wheat, peanuts, and eggs).

Potential Applications of Lactic Acid Bacteria to Reduce Food Allergies

To better discuss the potential application of LAB in the reduction of food allergies it is important to understand the mechanisms involved in the symptoms set off. As mentioned before, there are two mechanisms involved in allergy development, those that are IgE-mediated and those that are non-IgE-mediated. IgE-mediated reactions are the most common and best-characterized. They involve two stages, called sensitization and induction steps. In a first moment, the presence of antigens (food proteins) stimulates the differentiation of CD4+T cells into Th2 cells, which produce interleukins (ILs) IL-4, IL-5, and IL-13. These cytokines induce B cells to produce antigen-specific IgE antibodies, which attach to receptors present in the mast cells and basophils (sensitization step). In a second moment, when re-exposure to the allergen occurs, these antigenic macromolecules bind to the pre-existing IgE, which stimulate mast cell degranulation and trigger the release of pro-inflammatory mediators, such as histamines, prostaglandins, and leukotrienes, resulting in food allergy symptoms (induction phase). Non-IgE-mediated food allergies are less frequent and not as well characterized as the IgE-mediated ones. The exact mechanism of action is still unclear but it involves specific cells from the immune system and causes acute or chronic inflammation in the gastrointestinal tract, where eosinophils and T cells seem to play a major role (Okolie et al. 2018, Fu et al. 2019).

The strategies applied in the attempt to prevent the triggering of allergy symptoms are usually based on either antigen elimination or modulation of the host's immune system. LAB can be used for both purposes. The proteolytic enzymes produced by some LAB strains can provoke hydrolysis of antigenic proteins, leading to modifications on the recognition sites of those antigens and preventing IgE binding. Some probiotic strains can exert immune-stimulating effects, acting on the balance of Th1/Th2 responses, stimulating Th1-dominant ones. Probiotics are also reported to stimulate intestinal barrier functions and to modulate gut microbiota. Consumption of these microorganisms is frequently associated with allergy alleviation, prevention and tolerance acquisition (Fu et al. 2019).

The implications of proteolytic and probiotic LAB on allergy reduction are discussed in more details below.

Proteolytic lactic acid bacteria

The ability to produce proteolytic enzymes is an important characteristic for LAB, allowing their growth in food matrices poor in small peptides and free amino acids. The proteolytic activity of these microorganisms is also important to the development of the unique organoleptic properties of fermented products and contributes to the release of different molecules of interest, such as bioactive peptides. In addition, the proteolytic enzymes of some LAB are able to hydrolyze the antigenic proteins responsible for food allergy.

The enzymes involved in their proteolytic system may differ amongst different LAB genus, allowing these bacteria to obtain the specific peptides and amino acids required for their growth. Some authors comment that even amongst LAB from the same genus and species, protease expression is strain-specific. For this reason, when intending to use proteolytic LAB to hydrolyze allergenic food proteins, an interesting strategy to enhance the chances of success would be to use strains isolated from the same sources to be fermented. The proteolytic system of LAB has been extensively studied in lactococci and lactobacilli species, both presenting similar mechanisms. In addition, the functioning of this system is frequently described using the model based on the hydrolysis of milk proteins. Although LAB belonging to other genus produce different proteinases and peptidases, the general functioning of their proteolytic system is similar to that of lactococci and lactobacilli species.

In general bases, LAB proteolytic system is composed of extracellular-secreted (E) and surface-associated proteinases (cell envelope proteases, CEP), peptide transporters and intracellular peptidases. It consists of a series of chained reactions, starting with the action of CEPs, which break the proteins into dipeptides, tripeptides, and oligopeptides and are pointed by some authors as the most important enzymes of the system. Five different types of CEPs have already been characterized from different LAB, including PrtP from *Lactococcus lactis* and *Lactobacillus paracasei*, PrtH from *Lactobacillus helveticus*, PrtR from *Lactobacillus rhamnosus*, PrtS from *Streptococcus thermophilus* and PrtB from *Lactobacillus bulgaricus*. In lactococci and lactobacilli, CEPs are serine proteinases distinguished in PI and PII, according to their substrate specificity for αS1, β, and κ-caseins. PI-type primarily degrades β-casein, presenting a strong activity towards this protein fraction. These enzymes also degrade k-casein, but to a lesser extent and present no activity against αS1-casein. PIII-type hydrolyzes all casein fractions (α, β, and κ) equally. The peptides derived from CEP hydrolysis are carried into the cells by peptide transporters, in the second stage of the LAB proteolytic system. The type of transporter involved in this step depends on the size of the molecule been carried (Savijoki et al. 2006, Worsztynowicz et al. 2019).

In the case of amino acids, at least 10 different transport systems, with high specificity, have been identified in lactococci and lactobacilli. These systems can be regulated by ATP hydrolysis (for Glu/Gln, Asn, and Pro/Gly), by proton motive force (PMF for Leu/Val/Ile, Ala/Gly, Ser/Thr, and Met) or by passive transport based on the concentration gradient (for Arg/Orn). The transport of di- and tripeptides is similar to that of essential amino acids. The mechanism is regulated by PMF-driven DtpT and ATP-driven Dpp enzymes, which require proton generation and ATP hydrolysis, respectively. Dpp enzymes allow the transport of di- and tripeptides containing relatively hydrophobic branched-chain amino acids, with higher affinity to tripeptides, while DtpT enzymes present a preference for more hydrophilic and charged di- and tripeptides. Oligopeptides are carried through the cell membrane by enzymes belonging to the ABC family. This system consists of five proteins (Opp A, B, C, D, and F), belonging to a super family of highly conserved ATP-dependent cassette transporters. These proteins can carry oligopeptides containing up to at least 18 residues and the transport kinetics are affected by the nature of peptides been carried. Although the Opp system is well characterized for *L. lactis*, some authors mention that these transporters are similar for other LAB, such as *Lactobacillus* spp. and *Streptococcus* spp. (Savijoki et al. 2006, Rodríguez-Serrano et al. 2018).

The third stage of LAB proteolytic system takes place once the peptides have been carried into the cell. This step involves the action of various intracellular peptidases and aminopeptidases, which further hydrolyze the oligopeptides into small peptides and amino acids. These enzymes present different and partly overlapping specificities. The first enzymes in action at this stage are the intracellular endopeptidases, general aminopeptidases (PepN and PepC), and the X-prolyl dipeptidyl aminopeptidase (PepX). Several endopeptidases were characterized from LAB, the majority belonging to the group of metallopeptidases, but also some cysteine-peptidases and serine peptidases. A thorough description of these enzymes and their functioning can be found in the studies carried out by Kunji et al. 1996, Savijoki et al. 2006, Liu et al. 2010.

As mentioned before, the proteolytic enzymes produced by some LAB can degrade antigenic proteins, causing modifications on their IgE-binding sites and preventing allergy set off. However, their potential application in the manufacture of hypoallergenic products depends on the extension of the hydrolysis and on their ability to modify the specific antigenic sequences from the different proteins. Although some authors point lactic fermentation as a promising strategy to reduce food allergy, to the best of our knowledge, to date no study has reported a total suppression of symptoms by fermentation with proteolytic LAB. Several studies report great proteolysis degrees of food allergens, but the hydrolyzed products still present remaining IgE-binding epitopes, resulting only in a reduction of the antigenic potential of the fermented products rather than a complete allergy suppression. These studies are presented and discussed in more details later in this chapter, in the section addressing specific food allergens.

Probiotics

Many authors discuss the beneficial effects of fermented foods and probiotic LAB cultures in food allergies alleviation and tolerance acquisition. These studies suggest that probiotics may improve the gut microbiota, contribute to the homeostasis of the gastrointestinal tract and modulate the host's immune system.

Imbalance of the gut microbiota can be caused by different factors such as diet, antimicrobial therapy, and gastrointestinal diseases. This scenario, known as dysbiosis, contributes to increasing the occurrence of inflammatory manifestations like those observed in allergic reactions. It is known that some probiotic strains are able to adhere and colonize the gut mucosa, eliminating pathogens by competition and production of antimicrobial compounds, helping to keep a healthy gut microbiota and protecting the gastrointestinal tract from pathogens. Besides some strains can also contribute to enhance the integrity of the intestinal barrier and produce beneficial metabolites, such as short-chain fatty acids (SCFA), helping to keep the homeostasis of the gastrointestinal tract. The ability of some probiotics to modulate the intestinal microbiota, its function and immune responses stimulate their use as adjuvants in the treatment of food allergies (Paparo et al. 2019a, Yousefi et al. 2019).

SCFA are important beneficial compounds produced by LAB and have a pivotal role in the maintenance of the intestine homeostasis. The health benefits related to these compounds include their ability to alleviate colitis and diarrhea episodes, as well as to modulate the inflammatory immune response of the gut, by decreasing the production of pro-inflammatory cytokines. Besides, SCFA produced by probiotics and by the commensal intestinal microbiota can be used as an energy source for enterocytes, which contributes to keeping the integrity of the intestinal barrier. Since the imbalance in the gut microbiota may lead to a low production of organic acids, the use of specific SCFA, such as butyrate, for the treatment of food allergy has already been proposed (Canani et al. 2019).

As for the immune-stimulating role of probiotics, it seems to be related to their ability to increase the secretion of IgA antibodies and to maintain balance between Th1 and Th2 immune responses, by promoting the differentiation of Th1 cells, both resulting in a reduction of the inflammatory response and presenting a positive effect in food allergies decrease (Żukiewicz-Sobczak et al. 2014).

Clinical studies have investigated the positive impact of selected probiotic strains in the treatment of food allergy and reported that their effects in regulating the allergic immune response seem to be strain-specific. These studies are presented and discussed in more details later in this chapter, in the section addressing specific food allergens.

Potential Applications on Specific Allergenic Foods

Cow milk allergy

Bovine milk is one of the first foods introduced into the human diet and its proteins represent the first source of antigens consumed in large amounts during infancy. As a result, cow milk allergy (CMA) is recognized as the most common cause of food allergy in infants (below three years of age) and milk proteins are considered as major food allergens affecting young children. For this population, the prevalence of CMA is about 2 to 3%. Although it is usually outgrown with age, 15% of the patients remain allergic for life. As for the allergy type, the majority of CMA manifestations are IgE-mediated, but non-IgE mediated reactions are also reported. The symptoms vary amongst patients and may include gastrointestinal problems such as nausea and diarrhea, dermatological problems, asthma, rhinitis, inflammations of the gastrointestinal tract and, in more severe cases, anaphylaxis (Bu et al. 2013, Kordesedehi et al. 2018).

The most antigenic proteins in cow's milk are β-lactoglobulin (*Bos d5*), α-lactalbumin (*Bos d4*) and αS1-, αS2-, and β-caseins (*Bos d8*). Amongst them, β-lactoglobulin accounts for 80% of all CMA cases and is identified as the principal milk allergen. Caseins also play a relevant role in CMA development and are related to allergy persistence and the occurrence of more severe symptoms; amongst the casein fractions, αS1 is the most antigenic one. The differences in protein composition between cow's and human's milk are pointed as one of the reasons for CMA development in infants, for instance β-lactoglobulin is not present in human milk and is not hydrolyzed by pepsin, and maintains the IgE binding and T-cell stimulating properties after digestion (Claeys et al. 2013, Kazemi et al. 2018).

As it happens to other food allergies, the main strategy applied to avoid the occurrence of CMA is the dietary exclusion, but the complete elimination of cow milk proteins from the diet is difficult and can lead to nutrient deficiencies in infants and children, the population most affected by CMA. Many technological approaches have been studied in the attempt to propose new strategies to reduce CMA, amongst them fermentation is suggested by many authors as a promising treatment with good potential to be applied in the manufacture of hypoallergenic dairy products.

LAB are part of the natural microbiota of milk. LAB proteolytic metabolism plays an important role in their ability to generate the essential amino acids required for growth since milk is a matrix poor on free amino acids. In addition, the proteolytic enzymes produced by some LAB strains are able to hydrolyze the main milk allergens, destroying or modifying the epitopes responsible for IgE-binding and allergy trigger, leading to a reduction in the antigenic potential of the modified proteins (Pescuma et al. 2015, Kordesedehi et al. 2018).

In the complex proteolytic system of LAB, CEPs are the first enzymes active in the hydrolysis of milk proteins, releasing oligopeptides that are carried into the cells by peptide transporters. These peptides are then hydrolyzed by intracellular peptidases and further metabolized. The extent of proteinase biosynthesis in LAB is influenced by the matrix in which the bacteria growth, complex media containing great amounts of free amino acids may inhibit the induction of proteinase production while matrices poor on these nutrients may boost protease biosynthesis (Beganovic et al. 2013, Kotlar et al. 2015).

Generally, LAB proteolytic enzymes are more active against the casein fraction of milk than against whey proteins. This fact is due to the differences in the structural conformation of both fractions. Caseins are poorly folded proteins and the lack of disulfide bonds on its structure indicates the presence of non-conformational epitopes. In contrast, β-lactoglobulin is a globular protein stabilized by two disulfide bonds, presenting secondary and tertiary structures and conformational epitopes, been harder to hydrolyze than caseins. Heating treatment of β-lactoglobulin, previous to fermentation, can cause the unfolding of its conformational structure, making it easier to hydrolyze (Worsztynowicz et al. 2019).

The proteolytic activity of LAB enzymes against milk allergens is dependent on the strain used and on the process conditions. Different strains have been reported to produce enzymes capable of hydrolyzing milk antigens. Although *Enterococcus* spp. is frequently reported as one of the most proteolytic LAB found in dairy products, possessing a very efficient proteolytic system and a strong activity against both caseins and β-lactoglobulin, its application in foods for technological purposes is a controversial issue, since *Enterococcus* spp. are related to nosocomial infections and reported to harbor several virulence factors (Kim et al. 2016). LAB species belonging to other genera, such as *Lactobacillus* spp., *Lactococcus* spp. and *Streptococcus* spp. have also been reported to produce proteolytic enzymes with activity against milk allergens. The interest in their application in foods relies on their relevant role as fermentation starters and/or on their association with beneficial health effects, some being recognized as probiotics.

Another interesting approach to discuss the role of LAB on CMA alleviation relates to the consumption of probiotics, which can display an immunoregulatory effect and help to reduce allergy manifestations. The trigger of CMA symptoms seems to be related to an overexpression of Th2-type immune responses, causing interference in the Th1/Th2 balance. Consumption of fermented foods containing LAB can enhance the production of Type I and Type II interferons at the systemic level. This immunomodulatory effect regulates the stimulation of Th2 cells, and inhibits the trigger of allergic responses. However, the immune-stimulating effect of probiotics and their positive effect on the inhibition of allergy onset seems to be strain-dependent (Cross et al. 2001, Wang et al. 2018).

Many authors have investigated the potential application of proteolytic LAB in the manufacture of hypoallergenic dairy products. Some authors have also studied the role of probiotics on the modulation of the gut microbiota and on the stimulation of the host's immune system and their implications in CMA reduction and tolerance acquisition. Table 1 presents some examples of recent studies on these subjects.

Table 1: Examples of recent studies addressing the application of lactic acid bacteria to reduce cow milk allergy.

Strain used	Approach	Main results	Reference
Lactococcus lactis	Protein hydrolysis, using purified fraction of β-lactoglobulin. IgE binding of hydrolysates, using ELISA with serum of allergic patients.	Complete hydrolysis of β-lactoglobulin, confirmed by both SDS-PAGE and HPLC analysis, leading to a reduction in IgE binding by up to a 25-fold ratio.	Kazemi et al. 2018
Enterococcus faecium	Protein hydrolysis, using purified fractions of Na-caseinate and αS1-casein. IgE binding of αS1-casein hydrolysates, using ELISA with serum of allergic patients.	Strong hydrolysis of both Na-caseinate and αS1-casein, confirmed by SDS-PAGE and HPLC analysis. Reduction in 50% of the IgE binding ability of αS1-casein hydrolysates.	Kordesedehi et al. 2018
Enterococcus faecalis	Protein hydrolysis, using bovine milk and purified fractions of Na-caseinate, whey proteins and β-lactoglobulin as substrates.	Complete hydrolysis of casein fractions, observed in both milk and purified Na-caseinate substrates, confirmed by SDS-PAGE and HPLC analysis. Partial hydrolysis of β-lactoglobulin in all evaluated substrates.	Biscola et al. 2018
Enterococcus faecalis *Lactobacillus paracasei*	Protein hydrolysis, using bovine milk and purified fractions of Na-caseinate and whey proteins as substrates.	In all evaluated substrates, *E. faecalis* presented higher proteolytic activity than *L. paracasei*, especially on α-casein, β-casein, and β-lactoglobulin. Both strains were more active against caseins than whey proteins; and β-lactoglobulin was the most resistant to hydrolysis. In spite of the substrate and proteolysis condition, only partial hydrolysis was achieved.	Tulini et al. 2016
Lactobacillus delbrueckii subsp. *bulgaricus*	Effect of pre-hydrolysis of purified β-lactoglobulin on the digestion of this protein and on the degradation of its main antigenic sequences. Tests performed using an *in vitro* simulated gastrointestinal system.	Pre-hydrolysis of β-lactoglobulin by bacterial enzymes increased the protein digestion and contributed to the degradation of its antigenic sequences, resulting in the release of peptides possessing antioxidant, antimicrobial and immune-modulating properties.	Pescuma et al. 2015

Table 1 contd. ...

... *Table 1 contd.*

Strain used	Approach	Main results	Reference
Lactobacillus plantarum and *Lactobacillus rhamnosus* GG	Investigation of immunomodulatory effect, intestinal barrier function, and intestinal microbiota modulation, using oral administration in mouse model.	Induction of Th1 or Treg differentiation and inhibition of Th2-biased response. Regulation of Th1/Th2 immune balance. Reduction of serum IgE levels, resulting in the suppression of allergic responses. Enhancing of intestinal barrier function. Regulation of alterations in intestinal microbiota caused by allergies.	Fu et al. 2019
Lactobacillus acidophilus	Influence on miRNA expression, assessed in *in vitro* and *in vivo* models.	Suppression of hypersensitivity responses and Th17 cell differentiation. Inhibition of the expression of allergy-related miRNAs.	Wang et al. 2018
Lactobacillus rhamnosus GG	Immune-regulatory effect against β-lactoglobulin-induced reactions, evaluated in mice model.	Administration of an extensively hydrolyzed casein formula supplemented with LGG resulted in a decrease in IL-4, IL-5, IL-13 and specific IgE production, followed by a significant reduction in β-lactoglobulin-induced allergic reactions.	Aitoro et al. 2017
Lactobacillus rhamnosus GG	Immune-regulatory effect, evaluated in CMA sensitive infants, using oral administration in randomized trials.	Administration of an extensively hydrolyzed casein formula supplemented with probiotic LGG resulted in an acceleration of immune tolerance acquisition in infants suffering from both IgE-mediated and non-IgE-mediated CMA.	Berni Canani et al. 2017
Lactobacillus rhamnosus, Lactobacillus reuteri, Lactobacillus casei, Lactobacillus acidophilus, etc.	Investigation of the effectiveness of probiotic administration in treating food allergies among pediatric patients, using meta-analysis of available literature data.	Nine trials involving 895 CMA patients were included in the study. The analysis revealed a moderate certainty for the efficacy of probiotic oral administration to alleviate symptoms and induce tolerance.	Tan-Lim and Esteban-Ipac 2018

Soy allergy

Soybean (*Glycine max*) is an abundant and inexpensive source of proteins with high nutritional value and interesting functional properties. In addition, due to the presence of several bioactive compounds, the consumption of this grain has been related to health benefits such as cholesterol reduction, prevention of gastrointestinal and cardiovascular diseases, reduction in the risk of diabetes, obesity and breast tumor development, etc. These facts have contributed to raising the interest in soybean proteins and led to an increase in its consumption and incorporation in many formulations. However, the low digestibility and the allergenic potential of soybean proteins may trigger adverse reactions in susceptible individuals (Aguirre et al. 2014, Biscola et al. 2017).

The estimated prevalence of soybean allergy in the general population is relatively low (about 0.5%). However, this incidence is higher (up to 6%) for young children under three years of age, the most affected population for which soybean proteins are considered main allergens. Besides, some authors suggest that the increase in the consumption of soybean and soy-based products raises the exposure to the antigens present in these products and may lead to an increase in allergy prevalence (Aguirre et al. 2014).

Soybean allergy belongs to the category of the IgE-mediated manifestations and at least 16 of its proteins have already been described as immunoreactive to specific IgE, including hydrophobic protein from soybean (*Gly m1*), defensin (*Gly m2*), profilin (*Gly m3*), PR-10 protein (*Gly m4*), seed biotinylated protein (*Gly m7*), *β*-conglycinin (*Gly m5*) and glycinin (*Gly m6*), all of which are officially registered as food allergens. Amongst them, β-conglycinin and glycinin are the most abundant fractions, representing from 70% to 80% of the total protein content. These two fractions are also recognized as the main allergens of soybean. β-conglycinin belongs to the 7S globulin family and is a trimeric structure composed of three subunits (α', α and β). Glycinin belongs to the 11S globulin family and consists of five subunits, each of them composed of a basic and an acidic polypeptide, linked by a disulfide bond. The majority of antigenic epitopes present in soy proteins are found in the "α" subunit of β-conglycinin and in the acidic portion of glycinin subunits. Some studies also point conglycinin (2S globulin) as a soybean major allergen. In this protein fraction, the Kunitz trypsin inhibitor is the most antigenic component (Aguirre et al. 2014, Sung et al. 2014, Yang et al. 2018).

To date the only efficient method to completely avoid soybean allergic reactions is the dietary exclusion, which is difficult and may present nutritional disadvantages. Avoiding soybean allergens is not an easy task since these proteins are frequently present as 'hidden allergens' in various processed foods. Besides, the estimated threshold level for allergy triggering is usually low (ranging from 0.0013 to 500 mg of soy protein) and even the exposure to small amounts of antigens may lead to the occurrence of adverse reactions in susceptible individuals. Therefore, some technological approaches have been proposed in the attempt to find new strategies to reduce soybean allergy, focusing in the modification of the antigenic epitopes and in the reduction of their capability to bind to specific IgE. Amongst them, microbial fermentation is pointed as a promising process with potential to reduce the immunoreactivity of soybean antigens. Some aspects of protein structure such as solubility, digestibility, and stability are associated with allergenic effects. The hydrolysis of soy protein increases both solubility and digestibility, improving the intestinal absorption of protein hydrolysates and presenting a potential to reduce their antigenicity (Song et al. 2008, Sung et al. 2014, Meinlschmidt et al. 2016a).

Although some authors report that LAB fermentation of soybean or soy-based products leads to protein hydrolysis and suggest a potential to reduce the allergenicity, other studies revealed that this approach may not be efficient to completely avoid allergic reactions. The success of this strategy depends on the LAB strain and the fermentation conditions used. In the case of partial hydrolysis of proteins, the hydrolysates may still harbor antigenic epitopes capable to bind specific IgE and cause allergy set off. In addition, partial hydrolysis can cause protein unfolding and result in the exposure of hidden epitopes, leading to an increase of the allergenic potential of hydrolysates. Another adverse effect of soybean proteolysis may be the generation of bitter-tasting peptides, which hinders the sensory quality of the final product (Yang et al. 2018).

The application of proteolytic LAB to reduce soybean immunoreactivity seems to be an interesting research field, but the available studies on this matter are still scarce and more scientific data is necessary before drawing conclusions on the real potential of this technological process as a strategy to avoid allergy. Examples of studies on soy fermentation to decrease allergy can be seen in Table 2. To the best of our knowledge, up to date, there are no literature studies focusing specifically on the application of probiotic LAB cultures to reduce soybean allergy.

Gluten-related disorders

Gluten is an important protein complex found in wheat and other related grains, such as barley (hordein) and rye (secalin). It is composed of two main proteins named gliadin (α-gliadins, γ-gliadins, and ω-gliadins) and glutenin [divided into high-molecular-weight (HMW) and low-molecular-weight (LMW) subunits]. Albumins and globulins are also present in gluten composition, but in lower amounts. Gluten is present in a great variety of food preparations and contributes to improving the technological properties of bakery products by enhancing their structure and function. However, its consumption may trigger adverse reactions in sensitive individuals. It is associated with three important disorders: allergy to wheat proteins; celiac disease; and non-celiac gluten sensitivity. The first one is an IgE-mediated reaction while the other two fit in the category of non-IgE-mediated manifestations (Elli et al. 2015, Stefańska et al. 2016).

Both gliadin and glutenin are complex structures belonging to the prolamins family. These gluten fractions are recognized as the major agents responsible for the adverse reactions related to gluten intake. The high concentration on proline and glutamine contributes to the low digestibility of these proteins. Pepsin and trypsin are not able to hydrolyze the peptides linked to the C-terminal portion of these amino acids. Besides, the cyclic structure of proline enhances the resistance to hydrolysis (Kumar et al. 2017, Owens et al. 2017).

The only currently available treatment to avoid the problems related to gluten intake is a gluten-free diet. However, this strategy is difficult to follow and may present nutritional disadvantages. In this sense, many studies have been carried out in an attempt to find new technological approaches to prevent gluten disorders. This section discusses those related to the potential application of LAB (both proteolytic and probiotic strains) to reduce the three above mentioned diseases.

Wheat allergy is an adverse immunologic reaction to proteins contained in wheat and related grains. Depending on the route of exposure, it presents different clinical manifestations. Food allergy symptoms affect the skin, the gastrointestinal tract or

Table 2: Examples of studies focusing on the application of proteolytic lactic acid bacteria to reduce soy allergy.

Strain used	Approach	Main results	Reference
Mixture of *Lactobacillus casei*, yeast, and *Bacillus subtilis*	Solid-state fermentation of soybean meal. IgE binding ability and allergenic potential of hydrolysates were assessed by ELISA, oral sensitization, and challenge in a BALB/c mice model, respectively.	Fermentation caused the degradation of several allergenic sequences from soy protein. Hydrolysates presented reduced IgE binding ability, lowered mMCP-1 and IgE levels and induced less damage to the intestine of BALB/c mice.	Yang et al. 2018
Enterococcus faecalis	Protein hydrolysis, using purified fractions of β-conglycinin (7S) and glycinin (11S) as substrates. Hydrolysis confirmation by SDS-PAGE and HPLC. Immunoreactivity of hydrolysates observed *in vitro*, using ELISA.	The microbial proteolytic enzymes caused complete hydrolysis of the β-subunit from β-conglycinin and the acidic polypeptide from glycinin. Hydrolysates were less immunoreactive towards specific IgE (more than one logarithmic unit reduction).	Biscola et al. 2017
Lactobacillus helveticus	Induced liquid-state fermentation of soy protein isolate. Proteolysis was observed in SDS-PAGE and the immunoreactivity of hydrolysates was assessed by ELISA and western blot, using sera from soy-sensitive individuals.	*L. helveticus* suppressed the immunoreactivity of fermented soluble protein fractions (up to 100% of reduction). Besides, almost no IgE binding was found in western blot analysis.	Meinlschmidt et al. 2016b
Lactobacillus plantarum, *Bifidobacterium lactis*	Induced fermentation of soybean meal. Immunoreactivity of fermented samples was evaluated by ELISA and western blot, using sera from soy-sensitive individuals. Authors also evaluated the immunoreactivity of commercially available soy-based products (tempeh, miso, and yogurt).	Microbial fermentation reduced the IgE binding ability of hydrolyzed proteins from soybean meal and contributed to increasing the amino acid content. Tested commercial products presented very low immunoreactivity.	Song et al. 2008

the respiratory tract. Other reactions are linked to occupational asthma and rhinitis; wheat-dependent exercise-induced anaphylaxis; and contact urticaria. The clinical manifestations are the result of inflammatory responses, triggered by the activation of specific IgE, and are related with different antigenic proteins, including gliadins and HMW glutenins. This type of allergic reaction affects mainly children and is related to genetic aspects, environmental conditions, diet and modulation of the gut microbiota (Elli et al. 2015, Czaja-Bulsa and Bulsa 2017).

Regarding the potential application of LAB to reduce wheat allergy, *in vitro* studies carried out in cell model, demonstrated that probiotic *Bifidobacterium lactis* was able to inhibit the toxic effects caused by the ingestion of wheat gliadin. Some authors found that *Lactobacillus rhamnosus* GG may contribute to improving gut permeability, presenting potential to prevent the disruption of the gut epithelial barrier caused by this protein. Other results also suggested that the administration of probiotics, such as *Bifidobacterium animalis* subsp. *lactis* BB-12, *Lactobacillus acidophilus*, and *Lactobacillus rhamnosus* GG could suppress the production of specific IgE (Lindfors et al. 2008, Orlando et al. 2014, Paparo et al. 2019b). Still on this matter, De Angelis et al. 2007 observed that commercial probiotic VSL#3, used as starter for dough fermentation, was able to provoke extensive hydrolysis of wheat proteins, cleaving some IgE-binding antigens. Besides, the authors reported that the remaining IgE-binding proteins were completely hydrolyzed by pepsin and pancreatin after simulated digestion.

As for celiac disease (CD), it is a chronic inflammatory enteropathy that affects the small intestine and occurs when genetically predisposed individuals are exposed to gluten, which compromises the integrity of their intestinal epithelium (Herrán et al. 2017, Leonard et al. 2017). About 40% of humans have the HLA-DQ2 and HLA-DQ8 genotypes (present in 90% and 10% of celiac individuals, respectively). However, only 2% to 3% of these individuals develop CD. Its prevalence in the general population is about 1%, and this number can vary in different countries. In Northern Europe, North Africa, Middle East, and India, the occurrence of CD has become an increasingly common problem (Leonard et al. 2017, Watkins and Zawahir 2017).

Celiac individuals develop an immune reaction mediated by T-cells against tissue transglutaminase, which is an enzyme located in the extracellular matrix. The reaction causes mucosal damage and, in some cases, intestinal villous atrophy. In general, peptides originating from the incomplete digestion of gliadin, such as the 33-mer peptide, reach the small intestine intact and cross the intestinal epithelium barrier, attaining the lamina propria, where they undergo a deamidation process. These peptides trigger an innate immune response, resulting in the activation of the adaptive immune response. Such a process stimulates the production of proinflammatory cytokines, which cause the inflammatory process of the small intestine mucosa. The inflammatory process compromises the cellular junctions from the intestinal epithelium, resulting in increased permeability of the gut membrane due to complete histological disorganization (Elli et al. 2015).

There is a strong relationship between intestinal dysbiosis and celiac disease. A normal gut microbiota plays a key role in maintaining intestinal homeostasis and promoting health. CD is associated with a change in the microbiota, mainly with an increase in the number of Gram-negative bacteria and a decrease in the number of Gram-positive bacteria. Changes in the permeability of the intestinal epithelium allow the invasion and action of pathogenic microorganisms, which may contribute to the intensification of inflammatory conditions. In addition, celiac patients have smaller populations of lactobacilli and bifidobacteria, in comparison to healthy individuals, and these microorganisms are recognized for their ability to promote health benefits to the host (Coqueiro et al. 2017, Girbovan et al. 2017, Sommer et al. 2017).

Probiotic microorganisms are known to contribute to the hosts' health by exerting important functions, such as promoting the modulation of intestinal microbiota, helping in the maintenance of a healthy gastrointestinal flow, decreasing the permeability of the intestinal membrane, displaying anti-inflammatory and immunoregulatory actions, and producing beneficial metabolites that contribute to intestinal homeostasis. In this sense, the use of probiotics emerges as a potential therapeutic adjuvant for the treatment of CD, helping to improve the immunological condition of sensitive individuals, reducing the inflammatory process, reestablishing the integrity of the intestinal barrier, and alleviating CD clinical symptoms (Hill et al. 2014, Francavilla et al. 2017).

In recent years, another approach involving the use of LAB to reduce CD manifestations has gotten the attention of the scientific community. It is the use of probiotic and functional LAB strains capable of metabolizing, *in vivo*, the immunogenic gluten peptides derived from incomplete protein digestion, by producing enzymes active against these antigenic molecules.

Several studies found in the literature have described different LAB strains capable of degrading immunogenic gluten peptides. However, those are preliminary studies and more information is needed before drawing conclusions on the real potential application of this strategy in the treatment of CD (Gerez et al. 2012, Duar et al. 2014, Herrán et al. 2017, Rashmi and Gayathri 2017).

As observed for other food antigens, the use of proteolytic LAB may also be an interesting strategy for the development of products with low concentrations of gluten. These hypoallergenic foods constitute a promisor market for celiac people and the use of proteolytic LAB as starter cultures may simplify the production process of gluten-free bakery products and reduce its costs (Stefańska et al. 2016).

Non-Celiac Gluten Sensitivity (NCGS) is the third type of symptomatic response to gluten ingestion. Patients affected by this condition usually report intestinal and extra-intestinal symptoms, arising shortly after the ingestion of gluten-containing foods, and in the absence of CD or wheat allergy. NCGS is also closely related to irritable bowel syndrome and it is considered neither autoimmune nor allergic in nature. This problem is reported to affect up to 6% of the United States general population and, differently from CD and wheat allergy, there are no clinical tests or reliable biomarkers to characterize NCGS. As for symptoms set off, it has been proposed that an innate immune response activated by gluten and several other wheat proteins trigger the development of NCGS. Even though the literature lacks information on the use of LAB to reduce the problems related to NCGS, the use of proteolytic strains as starter cultures for the development of gluten-free products could be an interesting approach to be studied. Besides, this strategy may present economic advantages, since the current market projections predict that sales of gluten-free products will again soar to approach $2 billion in 2020 (Elli et al. 2015, Henggeler et al. 2017, Igbinedion et al. 2017).

Peanut allergy

Over the last decades, the search for a healthier diet has led to a reduction in meat consumption and an increase in vegetarian meals, with peanuts being used as one of the main protein sources. This fact also contributed to increase the exposure to peanut proteins, which are recognized as main food allergens, able to trigger severe reactions in sensitive individuals. This seed is recognized as the third cause of food allergy in the U.S. with a prevalence of about 2.2% in children and 1.8% among adults (Gupta et al. 2019). Prevalence data may differ in other countries since allergy incidence is affected by a variety of factors, such as geographic location; diet exposure effects; dietary habits; differences according to age, race, and ethnicity; etc. (NAS 2016).

Regarding the implicated antigens, the pattern of sensitization to peanut proteins varies among populations in different geographical regions. The three main elicitors of allergic reactions in the U.S. are Ara h 1, Ara h 2, and Ara h 3. In Spain, Ara h 9 is the most antigenic protein, while for Swedish patients Ara h 8 has the highest sensitization rates (Vereda et al. 2011, Ballmer-Weber et al. 2015).

What remains a consensus is that peanut allergy is an IgE-mediated disease which tends to develop in early life and presents high rates of persistence (only 20% of patients overcame the problem in adulthood) and cross-reactivity with other antigenic food proteins. It is related to severe symptoms and the consumption of even minimal doses of antigen can induce anaphylaxis in sensitive patients (Bublin and Breiteneder 2014).

Scientific data on LAB application to reduce peanut allergy are scarce, but this chapter brings a few examples found in the literature. Zhang et al. 2008 studied the therapeutic effects of a fermented soy product (ImmuBalance™) on peanut hypersensitivity and demonstrated that ImmuBalance consumption significantly suppressed anaphylactic symptoms. This effect was related to a suppression in the production of specific IgE and in histamine release. In addition, ImmuBalance showed an immunomodulatory effect, evidenced by a reduction in Th2 and an increase in Th1 responses. Barletta et al. (2013) observed in peanut-hypersensitive mice that *Lactobacillus* spp. and *Bifidobacterium* spp. successfully reduced allergic disease symptoms by increasing IL-10 and decreasing IL-13 levels, which are cytokines respectively associated with Treg cells and Th2 response induction. Ren et al. (2014) suggested that genetically modified *Lactococcus lactis,* producing peanut allergen Ara h 2, modulated allergic immune responses in mice, decreasing peanut-specific IL-4 and IL-10 responses while increasing IFN-γ. Tang et al. (2015) reported that the administration of probiotic strain *Lactobacillus rhamnosus* CGMCC 1.3724 induced sustained tolerance in children allergic to peanut and that the effects remained for 2 to 5 weeks after discontinuation of treatment. Hsiao et al. (2017), using the same strain, reported a long-term sustained tolerance, persisting for up to 4 years after completing treatment.

Chicken egg allergy

Chicken egg is a source of good quality proteins highly consumed all over the world and is used as ingredient in a lot of food products. However, it also recognized as a high antigenic food. The prevalence of chicken egg allergy is higher among children (between 0.5% and 2.5%) and about 66% of patients acquire tolerance by the age of five. Egg's allergy includes both IgE-mediated and non-IgE-mediated reactions. The first is the most common type and its major symptoms are urticaria or angioedema, but reactions related to the gastrointestinal or the respiratory tracts were also reported (Colver et al. 2005, Rona et al. 2007).

There are five major allergenic proteins from domestic chicken egg (*Gallus domesticus*) named as Gal d 1–5. The egg white is the portion that harbors the majority of antigens, presenting higher allergenic potential than egg yolk. Egg white's allergenic proteins are ovalbumin (Gal d 2, 54%), ovomucoid (Gal d 1, 11%), ovotransferrin (Gal d 3, 12%) and lysozyme (Gal d 4, 3.4%). Despite the fact that ovalbumin is the most abundant protein, ovomucoid is the dominant allergen in eggs. This protein is heat stable and resistant to proteolytic digestion. Its strong allergenicity is possibly associated with the presence of strong disulfide bonds that stabilize this highly glycosylated protein. As for the egg yolk, the antigenic molecules are the low-density lipoproteins and the livetins (α-livetin – Gal d 5) (Cooke and Sampson 1997, Julià et al. 2007, Mann and Mann 2008).

Although the only current approach to completely prevent the triggering of egg allergy is avoidance, studies have been carried out to find new strategies to reduce the antigenicity of these proteins. Regarding the role of LAB, the literature lacks information on the potential application of lactic fermentation and no studies assessing the role of LAB proteolytic enzymes to modify egg's antigenic proteins were found. However, the beneficial effects of LAB intake and the contribution of their immune-stimulation potential to suppress allergenic reactions are being extensively investigated. A few of these studies focus on the alleviation of egg's allergy symptoms.

Shida et al. (2002) reported that *Lactobacillus casei* strain Shirota decreased serum ovalbumin-specific IgE and IgG1 responses, and reduced systemic anaphylaxis in a mouse model for food allergy. Kimoto-Nira et al. (2007) observed that the treatment of mice with *Lactococcus lactis* subsp. *lactis* GB50 enhanced Th1 response and reduced the levels of ovomucoid-specific IgE and IgG1 antibodies; the results were not the same for the group treated with *Lactobacillus acidophilus* NCFB1748. In another study, Toomer et al. (2014) observed that early probiotic supplementation (*Lactobacillus acidophilus*, *Lactobacillus casei*, *Bifidobacterium bifidium*, and *Enterococcus faecium*) on ovalbumin-sensitized mice improved the gut microbiota and decreased plasma ovalbumin-IgE levels, enhancing Treg cells and providing host protection from hypersensitivity to food allergens. Still about early supplementation, but in newborn pigs sensitized with ovomucoid, Rupa et al. (2011) reported that *Lactococcus lactis* reduced both the frequency of clinical signs of allergy and the sensitization of piglets, suggesting that probiotics could be used as neonatal immune-regulators.

Conclusions

This chapter discussed relevant aspects of food allergies as well as the potential application of LAB to reduce the antigenicity of food proteins and to display an immune-modulating role in the host, leading to a decrease in allergy manifestations. The focus was the major food allergens, belonging to the Big 8 group and for which literature reports the use of LAB to decrease the allergy.

The majority of studies found in the literature were performed with milk proteins. Regarding the other members of the Big 8 group, it is possible to find data on soy, peanut, wheat, and eggs, even if for some of these matrices the studies are scarce. As for fish, crustacean shellfish and tree nuts, to the best of our knowledge, the literature lacks information on the use of LAB to reduce the allergy manifestations related to these foods, therefore they were not included in this chapter.

The strategies involving LAB application in the attempt to prevent the triggering of allergy symptoms are mainly based on two principles: (i) fermentation with proteolytic strains, leading to antigen cleavage and modification of IgE-binding sites;

and (ii) modulation of the host's microbiota and immune-stimulation through the consumption of probiotic strains. Both approaches are described in the literature as promising strategies, but results shall be interpreted carefully.

Although many studies describe high hydrolysis degrees of food allergens by LAB enzymes, some even reporting reductions in IgE-binding levels, results observed in *in vitro* tests are not always confirmed *in vivo*. These results evidence a potential of application in the production of hypoallergenic products. However, the only reliable results of actual allergy reduction are those obtained by tests performed with allergic patients. In addition, the majority of studies reporting high degrees of protein degradation also report a residual IgE binding ability of hydrolysates, suggesting the presence of remaining antigenic epitopes rather than a total allergy suppression. The same is valid for tests regarding the immune-regulatory effect of probiotics. Studies performed *in vitro* or using animal models can serve as indicators of the immune-stimulating potential of these LAB; however, confirmation of their potential application for allergy reduction or suppression requires thorough evaluations performed with allergic patients. Besides, the immune-regulatory effect of probiotics depends on the used strain and on the response of the host. While some studies present promising results, others report no positive effect on allergy reduction.

Acknowledgments

Albuquerque, MAC was supported by the grant #2018/10540-6, São Paulo Research Foundation (FAPESP).

References

Aguirre, L., E.M. Hebert, M.S. Garro and G.S. Giori. 2014. Proteolytic activity of *Lactobacillus* strains on soybean proteins. LWT - Food Sci. Technol. 59: 780–785.

Aitoro, R., R. Simeoli, A. Amoroso, L. Paparo, R. Nocerino, C. Pirozzi, M. di Costanzo, R. Meli, C. de Caro, G. Picariello, G. Mamone, A. Calignano, C.R. Nagler and R. Berni Canani. 2017. Extensively hydrolyzed casein formula alone or with *Lactobacillis rhamnosus* GG reduces β-lactoglobulin sensitization in mice. Pediatr. Allergy Immunol. 28: 230–237.

Ballmer-Weber, B.K., J. Lidholm, M. Fernandez-Riva, S. Seneviratne, K.M. Hanschmann, L.Vogel, P. Bures, P. Fritsche, C. Summers, A.C. Knulst, T.M. Le, I. Reig, N.G. Papadopoulos, A. Sinaniotis, S. Belohlavkova, T. Popov, T. Kralimarkova, F. deBlay, A. Purohit, M. Clausen, M. Jedrzejczak-Czechowcz, M.L. Kowalski, R. Asero, R. Dubakiene, L. Barreales, E.N. Clare Mills, R. van Ree and S. Vieths. 2015. IgE recognition patterns in peanut allergy are age dependent: perspectives of the EuroPrevall study. Allergy 70: 391–407.

Barletta, B., G. Rossi, E. Schiavi, C. Butteroni, S. Corinti, M. Boirivant and G. Di Felice. 2013. Probiotic VSL#3-induced TGF-β ameliorates food allergy inflammation in a mouse model of peanut sensitization through the induction of regulatory T cells in the gut mucosa. Mol. Nutr. Food Res. 57: 2233–2244.

Beganovic, J., B. Kos, A.L. Pavunc, K. Uroic, P. Dzidara and J. Suskovic. 2013. Proteolytic activity of probiotic strain *Lactobacillus helveticus* M92. Anaerobe 20: 58–64.

Berni Canani, R., M. di Costanzo, G. Bedogni, A. Amoroso, L. Cosenza, C. di Scala, V. Granata and R. Nocerino. 2017. Extensively hydrolyzed casein formula containing *Lactobacillus rhamnosus* GG reduces the occurrence of other allergic manifestations in children with cow's milk allergy: 3-year randomized controlled trial. J. Allergy Clin. Immunol. 139: 1906–1913.

Biscola, V., A. Rodriguez de Olmos, Y. Choiset, H. Rabesona, M.S. Garro, F. Mozzi, J.M. Chobert, M. Drouet, T. Haertlé and B.D.G.M. Franco. 2017. Soymilk fermentation by *Enterococcus*

faecalis VB43 leads to reduction in the immunoreactivity of allergenic proteins β-conglycinin (7S) and glycinin (11S). Benef. Microbes 8: 635–643.
Biscola, V., Y. Choiset, H. Rabesona, J.M. Chobert, T. Haertlé and B.D.G.M. Franco. 2018. Brazilian artisanal ripened cheeses as sources of proteolytic lactic acid bacteria capable to reduce cow milk allergy. J. Appl. Microbiol. 125: 564–574.
Bu, G., Y. Luo, F. Chen, K. Liu and T. Zhu. 2013. Milk processing as a tool to reduce cow's milk allergenicity: a mini-review. Dairy Sci. Technol. 93: 211–223.
Bublin, M. and H. Breiteneder. 2014. Developing therapies for peanut allergy. Int. Arch. Allergy Immunol. 165: 179–194
Canani, R.B., L. Paparo, R. Nocerino, C. Di Scala, G.D. Gatta, Y. Maddalena, A. Buono, C. Bruno, L. Voto and D. Ercolini. 2019. Gut microbiome as target for innovative strategies against food allergy. Front Immunol. 10: 191.
Claeys, W.L., S. Cardoen, G. Daube, J. De-Block, K. Dewettinck, K. Dierick, L. De-Zutter, A. Huyghebaert, H. Imberechts, P. Thiange, Y. Vandenplans and L. Herman. 2013. Raw or heated cow milk consumption: Review of risks and benefits. Food Control. 31: 251–262.
Colver, A.F., H. Nevantaus, C.F. Macdougall and A.J. Cant. 2005. Severe food-allergic reactions in children across the UK and Ireland, 1998–2000. Acta Paediatr. 94: 689–695.
Cooke, S.K. and H.A. Sampson. 1997. Allergenic properties of ovomucoid in man. J. Immunol. 159: 2026–2032.
Coqueiro, A.Y., A. Bonvini, J. Tirapegui and M.M. Rogero. 2017. Probiotics supplementation as an alternative method for celiac disease treatment. Int. J. Probiotics Prebiotics 12: 23–32.
Cross, M.L., L.M. Stevenson and H.S. Gill. 2001. Anti-allergy properties of fermented foods: an important immunoregulatory mechanism of lactic acid bacteria? Int. Immunopharmacol. 1: 891–901.
Czaja-Bulsa, G. and M. Bulsa. 2017. What do we know about IgE-mediated wheat allergy in children? Nutrients 9(1): E35.
De Angelis, M., C.G. Rizzello, E. Scala, C. De Simone, G.A. Farris, F. Turrini and M. Gobbeti. 2007. Probiotic preparation has the capacity to hydrolyze proteins responsible for wheat allergy. J. Food Prot. 70: 135–144.
Duar, R.M., K.J. Clark, P.B. Patil, C. Hernández, S. Brüning, T.E. Burkey, N. Madayiputhiya, S.L. Taylor and J. Walter. 2014. Identification and characterization of intestinal lactobacilli strains capable of degrading immunotoxic peptides present in gluten. J. Appl. Microbiol. 118: 515–527.
Elli, L., F. Branchi, C. Tomba, D. Villalata, L. Norsa, F. Ferretti, L. Roncoroni and M.T. Bardella. 2015. Diagnosis of gluten related disorders: celiac disease, wheat allergy and non-celiac gluten sensitivity. World J. Gastroenterol. 21: 7110–7119.
Francavilla, R.M., De Angelis, C.G. Rizzello, N. Cavallo, F. Dal Bello and M. Gobbetti. 2017. Selected probiotic lactobacilli have the capacity to hydrolyze gluten peptides during simulated gastrointestinal digestion. Appl. Environ. Microbiol. 83(14): e00376–17.
Fu, G., K. Zhao, H. Chen, Y. Wang, L. Nie, H. Wei and C. Wan. 2019. Effect of 3 lactobacilli on immunoregulation and intestinal microbiota in a β-lactoglobulin-induced allergic mouse model. J. Dairy Sci. 102(3): 1943–1958.
Gerez, C.L., A. Dallagnol, G. Rollán and G.F. De Valdez. 2012. A combination of two lactic acid bacteria improves the hydrolysis of gliadin during wheat dough fermentation. Food Microbiol. 32: 427–430.
Girbovan, A., G. Sur, G. Samasca and I. Lupan. 2017. Dysbiosis a risk factor for celiac disease. Med. Microbiol. Immunol. 206: 83–91.
Gupta, R.S., C.M. Warren, B.M. Smith, J. Jiang, J.A. Blumenstock, M.W. Davis, R.P. Schleimer and K.C. Nadeau. 2019. Prevalence and severity of food allergies among US adults. JAMA Netw. Open. 2(1): e185630.
Henggeler, J.C., M. Veríssimo and F. Ramos. 2017. Non-celiac gluten sensitivity: a review of the literature. Trends Food Sci. Technol. 66: 84–92.
Herrán, A.R., J. Pérez-Andrés, A. Caminero, E. Nistal, S. Vivas, J.M.R. De Morales and J. Casqueiro. 2017. Gluten-degrading bacteria are presente in the human small intestine of healthy volunteers and celiac patients. Res. Microbiol. 168: 673–684.

Hill, C., F. Guarner, G. Reid, G.R. Gibson, D.J. Merenstein, B. Pot, L. Morelli, R.B. Canani, H.J. Flint, S. Salminen, P.C. Calder and M.E. Sanders. 2014. The International Scientific Association for Probiotics and Prebiotics consensus statement on the scope and appropriate use of the term probiotic. Nat. Rev. Gastroenterol. Hepatol. 11: 506–514.

Hsiao, K.-C., A.L. Ponsonby, C. Axelrad, S. Pitkin and M.L.K. Tang. 2017. Long-term clinical and immunological effects of probiotic and peanut oral immunotherapy after treatment cessation: 4-year follow-up of a randomised, double-blind, placebo-controlled trial. Lancet Child Adolesc. Health 1: 97–105.

Julià, S., L. Sánchez, M.D. Pérez, M. Lavilla, C. Conesa and M. Calvo. 2007. Effect of heat treatment on hen's egg ovomucoid: an immunochemical and calorimetric study. Food Res. Int. 40: 603–612.

Igbinedion, S.O., J. Ansari, A. Vasikaran, F.N. Gavins, P. Jordan, M. Boktor and J.S. Alexander. 2017. Non-celiac gluten sensitivity: all wheat attack is not celiac. World J. Gastroenterol. 23: 7201–7210.

Kumar, J., M. Kumar, R. Pandey and N.S. Chauhan. 2017. Physiopathology and management of gluten-induced celiac disease. J. Food Sci. 82: 270–277.

Kazemi, R., A. Taheri-Kafrani, A. Motahari and R. Kordesedehi. 2018. Allergenicity reduction of bovine milk β-lactoglobulin by proteolytic activity of *Lactococcus lactis* BMC12C and BMC19H isolated from Iranian dairy products. Int. J. Biol. Macromol. 112: 876–881.

Kim, E.B., G.D. Jin, J.E. Lee and Y.J. Choi. 2016. Genomic features and niche-adaptation of *Enterococcus faecium* strains from Korean soybean-fermented foods. PLoS ONE 11: e0153279.

Kimoto-Nira, H., K. Mizumachi, M. Nomura, M. Kobayashi, Y. Fujita, T. Okamoto, I. Suzuki, N.M. Tsuji, J. Kurisaki and S. Ohmomo. 2007. *Lactococcus* sp. as potential probiotic lactic acid bacteria. Jpn. Agric. Res. Q. 41: 181–189.

Kordesedehi, R., A. Taheri-Kafrani, M. Rabbani-Khorasgani, R. Kazemi, D. Mutangadura and T. Haertlé. 2018. Modification of IgE binding to αS1-casein by proteolytic activity of *Enterococcus faecium* isolated from Iranian camel milk samples. J. Biotechnol. 20: 276–277.

Kotlar, C., A. Ponce and S. Roura. 2015. Characterization of a novel protease from *Bacillus cereus* and evaluation of an eco-friendly hydrolysis of a brewery byproduct. J. Inst. Brew. 121: 558–564.

Kunji, E.R., I. Mierau, A. Hagting, B. Poolman and W.N. Konings. 1996. The proteolytic systems of lactic acid bacteria. Antonie Van Leeuwenhoek 70: 187–221.

Leonard, M.M., A. Sapone, C. Catassi and A. Fasano. 2017. Celiac disease and nonceliac gluten sensitivity: a review. JAMA 318: 647–656.

Lindifors, K., T. Blomqvist, K. Juuti-Uusitalo, S. Stenman, J. Venäläinen, M. Mäki and K. Kaukinen. 2008. Live probiotic *Bifidobacterium lactis* bacteria inhibit the toxic effects of induced by wheat gliadin in epithelial cell culture. Clin. Exp. Immunol. 152: 552–558.

Liu, M., J.R. Bayjanov, B. Renckens, A. Nauta and R.J. Siezen. 2010. The proteolytic system of lactic acid bacteria revisited: a genomic comparison. BMC Genomics 11: 36.

Loh, W. and M.L.K. Tang. 2018. The epidemiology of food allergy in the global context. Int. J. Environ. Res. 15: e2043.

Meinlschmidt, P., U. Schweiggert-Weisz and P. Eisner. 2016a. Soy protein hydrolysates fermentation: effect of debittering and degradation of major soy allergens. LWT - Food Sci. Technol. 71: 202–212.

Meinlschmidt, P., E. Ueberham, J. Lehmann, U. Schweiggert-Weisz and P. Eisner. 2016b. Immunoreactivity, sensory and physicochemical properties of fermented soy protein isolate. Food Chem. 205: 229–238.

[NAS] National Academies of Sciences, Engineering and Medicine. 2016. Finding a Path to Safety in Food Allergy: Assessment of Global Burden, Causes, Prevention, Management, and Public Policy. National Academy Press, Washington (DC): 575 pp.

Okolie, C.L., A.N.A. Aryee and C.C. Udenigwe. 2018. Detection and deactivation of allergens in food. pp. 367–387. *In*: Yada, R.Y. (ed.). Proteins in Food Processing 2nd edition. Woodhead Publishing/Elsevier, Cambridge, UK.

Orlando, A., M. Linsalata, M. Notarnicola, V. Tutino and F. Russo. 2014. *Lactobacillus* GG restoration of the gliadin induced epitelial barrier disruption: the role of cellular polyamines. BMC Microbiol. 14: 19.

Owens, S.R., C.J. Lapedis and J.K. Greenson. 2017. Coeliac disease and their intraepithelial lymphocytic disorders of the upper gastrointestinal tract. Diagn. Histopathol. 23: 62–72.

Paparo, L., R. Nocerino, C. Di Scala, G.D. Gatta, M. Di Constanzo, A. Buono, C. Bruno and R.B. Canani. 2019a. Targeting food allergy with probiotics. Adv. Exp. Med. Biol. In press.

Pescuma, M., E.M. Hebert, T. Haertlé, J M. Chobert, F. Mozzi and G.F. Valdez. 2015. *Lactobacillus delbrueckii* ssp. *bulgaricus* CRL454 cleaves allergenic peptides of β-lactoglobulin. Food Chem. 170: 407–414.

Rashmi, B.S. and D. Gayathri. 2017. Molecular characterization of gluten hydrolyzing *Bacillus* sp. and their efficacy and biotherapeutic potential as probiotic using Caco-2 cell line. J. Appl. Microbiol. 123: 759–772.

Ren, C., Q. Zhang, G. Wang, C. Ai, M. Hu, X. Liu, F. Tian, J. Zhao, Y. Chen, M. Wang, H. Zhang and W. Chen. 2014. Modulation of peanut-induced allergic immune responses by oral lactic acid bacteria-based vaccines in mice. Appl. Microbiol. Biotechnol. 98: 6353–6364.

Rodríguez-Serrano, G.M., M. García-Garibay, A.E. Cruz-Guerrero, L. Gómez-Ruiz, A. Ayala-Niño, A. Castañeda-Ovando and L.G. González-Olivares. 2018. Proteolytic system of *Streptococcus thermophilus*. J. Microbiol. Biotechnol. 28: 1581–1588.

Rona, R.J., T. Keil, C. Summers, D. Gislason, L. Zuidmeer, E. Sodergren, S.T. Sigurdardottir, T. Lindner, K. Goldhahn, J. Dahlstrom, D. McBride and C. Madsen. 2007. The prevalence of food allergy: A meta-analysis. J. Allergy Clin. Immunol. 120: 638–646.

Rupa, P., J. Schmied and B.N. Wilkie. 2011. Prophylaxis of experimentally induced ovomucoid allergy in neonatal pigs using *Lactococcus lactis*. Vet. Immunol. Immunopathol. 140: 23–29.

Savijoki, K., H. Ingmer and P. Varmanen. 2006. Proteolytic systems of lactic acid bacteria. Appl. Microbiol. Biotechnol. 71: 394–406.

Shida, K., R. Takahashi, E. Iwadate, K. Takamizawa, H. Yasui, T. Sato, S. Habu, S. Hachimuri and S. Kaminogawa. 2002. *Lactobacillus casei* strain Shirota suppresses serum immunoglobulin E and Immunoglobulin G1 responses and systemic anaphylaxis in food allergy model. Clin. Exp. Allergy 32: 563–572.

Sommer, F., J.M. Anderson, R. Bharti, J. Raes and P. Rosenstiel. 2017. The resilience of the intestinal microbiota influences health and disease. Nat. Rev. Microbiol. 15: 630–638.

Song, Y.S., J. Frias, C. Martinez-Villaluenga, C. Vidal-Valdeverde and E. Gonzales de Mejia. 2008. Immunoreactivity reduction of soybean meal by fermentation, effect on amino acid composition and antigenicity of commercial soy products. Food Chem. 108: 571–581.

Song, Y.S., V.G. Pérez, J.E. Pettigrew, C. Martinez-Villaluenga and E. Gonzales de Mejia. 2010. Fermentation of soybean meal and its inclusion in diets for newly weaned pigs reduced diarrhea and measures of immunoreactivity in the plasma. Anim. Feed Sci. Technol. 159: 41–49.

Stefánska, I., K. Piasecka-Jóźwiak, D. Kotyrba, M. Kolenda and K.M. Stecka. 2016. Selection of lactic acid bacteria strains fot the hydrolysis of allergenic proteins of wheat flour. J. Sci. Food Agric. 96: 3897–3905.

Sung, D., K.M. Ahn, S.Y. Lim and S. Oh. 2014. Allergenicity of an enzymatic hydrolysate of soybean 2S protein. J. Sci. Food. Agric. 94: 2482–2487.

Tang, M.L., A.L. Ponsonby, F. Orsini, D. Tey, M. Robinson, E.L. Su, P. Licciardi, W. Burks and S. Donath. 2015. Administration of a probiotic with peanut oral immunotherapy: a randomized trial. J. Allergy Clin. Immunol. 135: 737–744.

Tan-Lim, C.S.C. and N.A.R. Esteban-Ipac. 2018. Probiotics as treatment for food allergies among pediatric patients: a meta-analysis. World Allergy Organ. J. 11: e25.

Toomer, O.T., M. Ferguson, M. Pereira, A. Do, E. Bigley, D. Gaines and K. Williams. 2014. Maternal and postnatal dietary probiotic supplementation enhances splenic regulatory T helper cell population and reduces ovalbumin allergen-induced hypersensitivity responses in mice. Immunobiology 219: 367–376.

Tulini, F.L., B. Vanessa, Y. Choiset, N. Hymery, G. Le Blay, E.C.P. de Martinis, J.M. Chobert and T. Haertlé. 2016. Evaluation of the proteolytic activity of *Enterococcus faecalis* FT132 and *Lactobacillus paracasei* FT700, isolated from dairy products in Brazil, using milk proteins as substrates. Eur. Food Res. Technol. 241: 385–392.

Vereda, A., M. van Hage, S. Ahlstedt, M.D. Ibanez, J. Cuesta-Herranz, J. van Odijk, M. Wickman and H.A. Sampson. 2011. Peanut allergy: clinical and immunologic differences among patients from 3 different geographic regions. J. Allergy Clin. Immunol. 127: 603–607.

Verhoeckx, K.C.M., Y.M. Vissers, J.L. Baumert, R. Faludi, M. Feys, S. Flanagan, C. Herouet-Guicheney, T. Holzhauser, R. Shimojo, N. van der Bolt, H. Wichers and I. Kimber. 2015. Food processing and allergenicity. Food Chem. Toxicol. 80: 223–240.

Wang, J.J., S.H. Li, A.L. Li, Q.M. Zhang, W.W. Ni, M.N. Li, X.C. Meng, C. Li, S.L. Jiang, J.C. Pan and Y.Y. Li. 2018. Effect of *Lactobacillus acidophilus* KLDS 1.0738H on miRNA expression in *in vitro* and *in vivo* models of β-lactoglobulin allergy. Biosci. Biotechnol. Biochem. 82: 1955–1963.

Watkins, R.D. and S. Zawahir. 2017. Celiac disease and nonceliac gluten sensitivity. Pediatr. Clin. N. Am. 64: 563–576.

Worsztynowicz, P., A. Olejnik-Schmidt, W. Białas and W. Grajek. 2019. Identification and partial characterization of proteolytic activity of *Enterococcus faecalis* relevant to their application in the dairy industry. Acta Biochim. Pol. 66: e2714.

Yang, A., L. Zuo, Y. Cheng, Z. Wu, X. Li, P. Tong and H. Chen. 2018. Degradation of major allergens and allergenicity reduction of soybean meal through solid-state fermentation with microorganisms. Food Funct. 9: 1899–1909.

Yousefi, B., M. Eslami, A. Ghasemian, P. Kokhaei, A.S. Farrokhi and N. Darabi. 2019. Probiotics importance and their immunomodulatory properties. J. Cell Physiol. 234: 8008–8018.

Zhang, T., W. Pan, M. Takebe, B. Schofield, H. Sampson and X.-M. Li. 2008. Therapeutic effects of a fermented soy product on peanut hypersensitivity is associated with modulation of T-helper type 1 and T-helper type 2 responses. Clin. Exp. Allergy 38: 1808–1818.

Żukiewicz-Sobczak, W., P. Wróblewskal, P. Adamczuk and W. Silny. 2014. Probiotic lactic acid bacteria and their potential in the prevention and treatment of allergic diseases. Centr. Eur. J. Immunol. 39: 104–108.

5

Lactic Acid Bacteria Bacteriocins and their Impact on Human Health

Svetoslav D Todorov[1,*] and *Michael L Chikindas*[2,3]

Bacteriocins, Useful Tools not only for Biopreservation

Human society's evolution is driven by many factors with economy and science development being amongst most essential variables. Conversely, changes of the society can catalyze economical growth and influence scientific research. When these processes work in harmony; economy, society, and science form a stable "continuum" with a "vector force" called "progress". The early 19th century's industrial revolution was critical for development of numerous branches of applied and fundamental sciences. Science became more popular, accessible and not only for the elite with chosen privilege. Together with the development of the society, popularization of science and communications were reflected in the triggering of numerous discoveries. The late 20th century's rapid development of hardware computer technologies was yet another catalyst for scientific progress. Sophisticated databases facilitated research and directly influenced productivity of science, making it a vitally important field of industry.

[1] Universidade de São Paulo, Faculdade de Ciências Farmacêuticas, Departamento de Alimentos e Nutrição Experimental, Laboratório de Microbiologia de Alimentos, Av. Prof. Lineu Prestes 580 Bloco 14, 05508-900, São Paulo, SP, Brazil.

[2] Health Promoting Naturals Laboratory, School of Environmental and Biological Sciences, Rutgers State University, New Brunswick, New Jersey 08904, USA.

[3] Academy of Biology and Biotechnology, Southern Federal University, Rostov-on-Don, 344090 Russia.

Email: tchikind@sebs.rutgers.edu

* Corresponding author: slavi310570@abv.bg

Related to all this, human health is no longer a task exclusively managed by medical physicians. Preventive medicine, natural/alternative, and complementary approaches, healthy products, etc., are just a few keywords of the modern medicine. Moreover, an idea for One Health is a general social movement aimed at putting several pieces of the puzzle together, with one general goal, the wellbeing of the society.

We should acknowledge that historically, microbiology, food, nutrition, and human health were always interconnected as inseparable parts of the same, although not always well-balanced structure. Hippocrates (400 BC) suggested perhaps the most captivating link between medicine and food: "Let thy food be thy medicine and medicine be thy food." From the modern science's perspective, we can acknowledge and appreciate the role of various starters and beneficial microbial cultures in preparation, safety, gastronomical characteristics and beneficial properties of numerous fermented food products. These food fermenting microorganisms and their metabolites are essential components of complex benefits provided to consumers with microbially-processed food. Therefore, it is our objective, in the light of disease prevention and wellbeing modulation to utilize these products for the benefit of the human health. Saying that, traditional (people's) medicine can be a priceless source of information on what microbiology may offer to be in good balance with the nature.

Lactic acid bacteria (LAB) are important contributors to many foods: they are essential components of various fermentation processes converting substrates into food products, reducing allergenicity (Hajavi et al. 2018) and modulating the immune system, when consumed, dead or alive (Dargahi et al. 2019, Tsai et al. 2019); as important food safety and quality components, they are outcompeting spoilage and pathogenic microorganisms or controlling them with various naturally produced antimicrobial agents (Todorov et al. 2019). Some LAB are a part of our gastrointestinal tract (GIT) microbiota modulating the host's metabolism (Leu et al. 2018, Quigley 2019) and even influencing our behavior (Liu et al. 2018, Sarkar and Banerrjee 2019). Antimicrobial peptides-producing LAB are playing essential role in modulation of the GIT microbiota (Saarela 2019, Todorov et al. 2019). Antimicrobial proteins of bacterial origin were discovered in the early 20th century, with colicins, bacteriocins produced by *Escherichia* (*E.*) *coli*, effective in inhibition of other Gram-negative organisms (for recent review, see Todorov et al. 2019), being the first reported in this group of molecules. Shortly after, in 1933, nisin A, a bacteriocin produced by *Lactococcus lactis* was described (for review, see Todorov et al. 2019) and based on its well-demonstrated ability to inhibit several Gram-positive foodborne pathogens it is presently used as a food preservative in variety of products. What is particularly attractive about LAB and the proteinaceous antimicrobials they produce is that food fermenting LAB are safe for human consumption and the antimicrobial proteins they produce ribosomally (also known as bacteriocins) so far have no reports indicating any negative safety concerns.

Definitions and classifications of bacteriocins have been proposed and discussed in many papers and have changed over time with growing understanding of the nature, activities, and biological significance of these substances in the natural environment of their producers. Perhaps the essence of definition is that bacteriocins are proteinaceous antimicrobials, ribosomally encoded, produced by bacteria that primarily inhibit the growth of other closely-related bacteria (Chikindas et al. 2018), while some authors mention signaling function, in addition to antimicrobial, as an important feature observed in many bacteriocins (Cotter et al. 2005). It is only logical to assume that in the natural microbial environment small quantities of bacteriocins are involved in intra- and inter-species interactions, while in the higher concentrations can present their antimicrobial, spermicidal, anti-viral, and anti-cancer activities (reviewed by Chikindas et al. 2018). Bacteriocins-producing LAB are self-protected by the immunity mechanisms specific to their bacteriocins. The antibacterial mode of action of bacteriocin can be bactericidal or bacteriostatic, and under certain conditions, bacteriolytic (Coter et al. 2005). Bacteriocinogenicity (the production of bacteriocins) is considered a ubiquitous characteristic of virtually all bacteria and is presumed to play a central role in the microbial population fitness. Research criteria for study of bacteriocins have been well standardized (Schillinger and Holzapfel 1996), and various classifications of the bacteriocins of Gram-positive bacteria have been proposed (Klaenhammer 1988, De Vuyst and Vandamme 1994, Coter et al. 2005, Heng et al. 2007), with, perhaps the most comprehensive classification of LAB bacteriocins offered by Alvarez-Sieiro et al. (2016). Most bacteriocins are small peptides. Some of them are post-translationally modified, as in the case of the lantibiotics (Heng et al. 2007), and others are complex molecules that also incorporate non-protein moieties responsible for their antibacterial activity (class IV of classification by Klaenhammer 1988). Although the majority of the LAB bacteriocins are linear molecules, exceptions occur such as the cyclic enterocin AS48, garvicin ML, uberolysin A, carnocyclin A, etc. (Gabrielsen et al. 2014). The rest of the LAB bacteriocins (e.g., classes II and III) are classified based on their molecular size, as smaller or bigger than 10 kDa, with some subclasses based on their specificity of molecular structure and mode of action (Heng et al. 2007).

Bacteriocins' Roles: Not Just Simple Antimicrobials

Initially, bacteriocins were studied as antimicrobial peptides with very specific spectrum of activity (Klaenhammer 1988, De Vuyst and Vandamme 1994). However, the last decades of bacteriocin studies have shown that these antimicrobial peptides are involved in signaling mechanisms and quorum sensing processes (for review see Chikindas et al. 2018). There are emerging reports pointing at LAB bacteriocins as very useful antimicrobial agents for control of multidrug resistant microorganisms, including methicillin resistant *Staphylococcus aureus* or vancomycin resistant *Enterococcus* spp. (Brussow 2017, Lewis and Pamer 2017, Phumisantiphong et al. 2017, Perales-Adán et al. 2018). LAB bacteriocins were described as effective anti-viral agents and some aspects of this specific mode of action were elucidated (Torres et al. 2013, Wachsman et al. 2003, Quintana et al. 2014). Also, their effect on prokaryotic cells, including *Candida albicans* was reported (Smaoui et al. 2010, Vidhyasagar and Jeevaratnam 2013). As for difficulty to control microorganisms such as *Mycobacterium* spp., their inhibition by LAB bacteriocins was shown as well (Montville et al. 1999, Donaghy 2010, Carroll and O'Mahony 2011). When utilized as antimicrobials, bacteriocins can be used as purified to homogeneity (e.g., medical applications) or partially purified (e.g., food preservation) preparations, or as in the latter case, they can be produced by the food fermenting starter cultures. Several investigators suggested a sophisticated approach of providing consumers with bacteriocin-producing probiotic LAB (Carroll et al. 2010, Donaghy 2010, Carroll and O'Mahony 2011).

Signaling Properties of Bacteriocins

LAB can produce numerous antimicrobial compounds extracellularly as a result of their primary and secondary metabolism. Some of these antimicrobials can be involved in intra- or/and inter-species quorum-sensing related processes (Asfahl and Schuster 2016, Westhoff et al. 2017). Due to their diffusivity, these antimicrobial peptides may be engaged in recognition of and/or communication with the potential competitors at the distances closer than marked for mentioned substances.

Some of the previously mentioned molecules can have similarities to the Class II bacteriocins. Class II bacteriocins belong to the Gly-Gly-type peptide family (Cook and Federle 2014). These peptides contain a double glycine motif in their conserved leader sequence (LSX2ELX2IXGG) (Havarstein et al. 1994). Gly-Gly peptides have shown to be actively related in quorum sensing processes and transmit a signal internally via phosphorylation of cognate response regulators (Cook and Federle 2014). Regulation of bacteriocin production is a complex process, including self-regulation and self-induction of the bacteriocin production (Cotter et al. 2005). Moreover, signaling molecules can be different from the bacteriocins and bacteriocin-like polypeptides; these may include cell surface proteins from potential bacteriocin-targeted organisms. Recent studies (Westhoff et al. 2017) have shown that bacteria can modify their metabolome and their antimicrobial production ability when they are co-cultured with or in close physical proximity to competitors (Abrudan et al. 2015, Gutiérrez-Cortés et al. 2018). Because of this, such co-cultures offer promising avenues for new drug discovery (Wu et al. 2015).

It is still unknown if the produced antimicrobials can be the signals for the production of the same or other bacteriocins and what are the other functions of the bacteriocins. Chikindas et al. (2018) reviewed the other important roles of bacteriocins highlighting their role in intra-species communication and quorum sensing process as well as in inter-species interactions and quorum quenching and/or quorum sensing. The hypothesis raised questions regarding if some of the polypeptides produced by the LAB with regulatory function, can share similarities with antimicrobial peptides and if they can exhibit some antimicrobial activity as well (Rodali et al. 2013, Kjos et al. 2014).

Bacteriocins and/or Probiotics

The idea to explore probiotics with bacteriocinogenic properties is an ambitious and challenging task. Major points that need to be covered are not just optimal spectrum of activity (controlling some groups of bacteria while keeping alive friendly microorganisms), but also the sufficient production yield during the desired time interval. Even with all the knowledge of optimization of fermentation processes for higher titer production and growth of bacteriocin producers (temperature, agitation, pH, medium components, etc.) (Abbasiliasi et al. 2017), we still know very little about bacteriocins production and their gene expression in the human GIT, which is a dynamic system where conditions vary dramatically and depend on physiology, age, gender, diet, stress, medications, etc. We are in a very beginning stage of investigation of the role of the GIT specific environment on survival, growth and production of bacteriocins by the targeted microorganisms.

The role of bacteriocins inside the eukaryotic GIT is still unclear, and results and potential applications of live cultures that can produce bacteriocins inside the GIT are, although interesting, are still questionable and speculative. It is still unclear if the GIT conditions are appropriate for the most effective production of bacteriocins at high yield and whether the high yield is actually needed for the bacteriocins to act at the targeted site. The "sufficient" production of bacteriocins in the eukaryotic GIT environment is likely driven by a purpose, which takes us to the most important question: what is the natural role of a bacteriocin in the cell-producer's natural environment? Is this a killing weapon or a signaling tool for communication with friendly neighboring microorganisms, repelling intruding bacteria, and perhaps even for communicating with the eukaryotic host in the case of bacteriocin-producers being a part of a GIT microbiota. What will be the influence of other bacteria in the GIT and how the produced bacteriocin will affect microbial balance in the GIT? Some authors are skeptical about possible application of life bacteriocin producing LAB cultures for control of pathogenic bacteria and proposing application of semi-purified bacteriocin as pharmaceutical applications (Ahmad et al. 2017). At the same time, some studies, while not firmly decisive, shed light on possible use of bacteriocin-producing LAB for the correction of the host microbiota. For instance, Murphy et al. (2013) and Clarke et al. (2013) reported on the possible role of a bacteriocin producing *Lactobacillus salivarius* UCC118 in altering the composition of the gut microbiota in diet-induced obese mice by a mechanism involving bacteriocin production. This and other promising applications of probiotics with bacteriocin production capacity as a positive functional trait were recently reviewed by Hegarty et al. (2016) with applications aimed at the control of the host microbiota in obesity, diabetes and even cancer being mentioned. Moreover, some *in vivo* models showed that bacteriocin production can be a part of a larger "scheme" of antimicrobial compounds production by LAB in the GIT.

Ganzle et al. (1999) proposed application of bacteriocin-producing lactobacilli for control of *E. coli* and *Listeria* spp. studied in a dynamic model simulating the stomach and the small intestine. Authors explored the behavior of *Lactobacillus curvatus* LTH 1174 bacteriocin producing strain and its bacteriocin negative mutant in combination with *E. coli* and *Listeria innocua* during transit trough a dynamic GIT model. As it was expected, *Lactobacillus curvatus* was rapidly killed in the gastric acidity environment at pH < 2.0 and only less than 0.01% of the cells were delivered to the small intestinal compartments, were they recovered from the ileal compartment of the studied model (Ganzle et al. 1999), already showing potential problems for the *in vivo* application of unprotected bacteriocins producers. However, the authors have shown that meat extracts can have a protective effect on the lethal action of bile against *Lactobacillus curvatus*. Moreover, Ganzle et al. (1999) pointed at a very important observation indicating that low pH, lactic acid, and curvacin A, produced by the investigated *Lactobacillus curvatus* strain had a cumulative or even synergistic effect in inhibition of *E. coli* in the mentioned model. In addition, applying a dynamic model system by Ganzle et al. (1999) was a very important point, since different behavior of *E. coli* was observed when experiments were performed in anaerobic, strongly or moderate agitated conditions. Experiment with bacteriocin-negative mutant of *Lactobacillus curvatus* confirmed importance of curvacin A in control of *E. coli* in the studied model (Ganzle et al. 1999).

In addition, Ganzle et al. (1999) evaluated *Listeria innocua* as a model for bacteriocinogenic effect of curvacin A. Both *Listeria inncua* and *Lactobacillus curvatus,* survived at similar rates in the GIT model when applied as single cultures. However, when applied in co-cultures, bacteriocinogenic *Lactobacillus curvatus* was able to produce curvacin A leading to killing effectively co-cultured listeria, even after just 10 minutes of co-incubation. Using bacteriocin-negative mutant as a control, it was confirmed that the principal agent responsible for the reduction of *Listeria innocua* in the studied model was curvacin A, a bacteriocin produced by *Lactobacillus curvatus*. In addition, curvacin A was not degraded in the gastric compartment and was detected in the ileal compartment during the first 180 minutes upon addition of the meal (Ganzle et al. 1999).

Recently, Naimi et al. (2018) evaluated stability and bacteriocin activity of microcin J25 produced by *E. coli* in TIM-1 model, simulating human GIT conditions. Microcin J25 is the bacteriocin active against *Salmonella*, *Shigella*, and pathogenic *E. coli* strains. Microcin J25 is a 21 amino acid peptide, with remarkable stability to heat and extreme pH values. It is also resistant to several proteases, resistance related to bacteriocins specific lasso structure (Naimi et al. 2018). Authors have used LC-MS/MS analysis and subsequent molecular networking analysis on the Global Natural Products Social Molecular Networking platform (GNPS), and analysis of the peptide degradation in the presence of proteolytic enzymes mimicking the GIT conditions with the aim to delineate the fate of microcin J25 through identification of the main degradation products. Microcin J25 was relatively stable under studied gastric conditions, but degraded rapidly in the compartment mimicking the duodenum, notably in the presence of pancreatin. Based on Naimi et al. (2018), among pancreatin components, elastase I appeared primarily responsible for microcin J25 breakdown while α-chymotrypsin was less effective. Authors have provided detailed analysis of the breakdown of microcin J25 in different parts of the GIT and predicted levels of activity during the passage through the GIT.

Granger et al. (2008) explored the effect of the GIT conditions on the survival of *Enterococcus mundtii* and bacteriocin production based on RT-PCR analysis. *Enterococus mundtii* was cultured in MRS broth at different pH and supplemented with bile pancreatic enzymes, bile and contents of the stomach and small intestine of pigs, respectively. Authors have evaluated the bacteriocin expression based on RT-PCR analysis; however, neglecting a very important point that in *in vivo* conditions the bacteriocin producer will be exposed to much more complex conditions and not in optimal for LAB growing medium such as MRS. Under the studied experimental conditions, the authors pointed that *Enterococcus mundtii* can survive passage through the stomach (pH 2.5) and may produce low levels of the bacteriocin in the duodenum; however, the question remains: will this be the same in a real-life scenario. In the next step of the study, the microbial growth can be simulated in the small intestine (pH 3.5) and in the colon (pH 5.5) mimicking environments. Moreover, authors claimed that the bacteriocin activity is not expected to increase in the duodenum due to the presence of proteolytic enzymes in the pancreatic juice. Higher yields of the bacteriocin may be produced in the lower parts of the intestinal tract (jejunum, ileum and colon).

However, bacteriocin production seems to be more complex than anticipated and it may not be directly related to the number of the producer cells (Grander et al. 2008). Moreover, some authors are frequently misinterpreting the detected bacteriocin activity as the level of production and/or the level of effectively recorded bacteriocin after interaction of the expressed bacteriocin with the bacteriocin-producer cells and other components of the investigated system. It has been shown for the bacteriocin producer strain *Enterococcus mundtii* ST4SA that the bacteriocin can interact with different test microorganisms and this interaction depends on the temperature, pH, presence of different chemicals, etc. (Todorov and Dicks 2009).

While the studied strain of *E. mundtii* can grow well at low pH, at the same time it has been shown that it cannot produce the bacteriocin at initial pH of the growth medium lower than 4.0 (Todorov and Dicks 2009). Taking this into consideration and the fact of the complexity of the GIT environmental conditions, the presence of different proteolytic enzymes in specific parts of the GIT, chances to have bacteriocin production and effect of eventually produced bacteriocin is very low. Related to all these essential challenges, possible strategies aimed at specific effects of bacteriocins in definite GIT conditions needs to be designed, taking into consideration all essential parameters/variables that can influence this process.

The pharmaceutical industry is offering various approaches for solving problems with stability and delivery of the bacteriocins as a part of pharmaceutical preparation. Modifications of proteins and peptides by hydrophobic fatty acid residues or amphiphilic block copolymers have been shown to be promising strategies to deliver proteins (and peptides) across numerous tissue barriers (Carmona-Ribeiro and Carrasco 2014). C-terminal immobilization distinctly reduced the activity of potent antimicrobial substances of proteinaceous origin trough the correlation between the antimicrobial activity and structural properties such as amphipathicity and the structure-related activity profile of the investigated peptides, which should not change with immobilization. For example, nisin immobilized on the micro-bead surface, when submitted to proteolysis, conserved its bacteriocin activity, and strongly inhibited the growth of target microbial cells; however, free nisin totally lost its activity. As an alternative to internal immobilization, activated alginate was first beaded by ionotropic gelation and then the protein immobilized on to the bead surface with importance of the association of covalent coupling and ionotropic inclusion for internal or external immobilization of nisin (Le-Tien et al. 2004). Antimicrobial peptides can be formulated with lipids or polymers for protection against peptidase and to assure proper delivery. Anionic liposomes can be utilized for effective encapsulation of cationic, α-helical antimicrobial peptides and their delivery by nebulization to the lungs (Carmona-Ribeiro and Carrasco 2014).

Applications in Medicine

By definition (Klaenhamer 1988, Cotter et al. 2005, Alvarez-Sieiro et al. 2016) bacteriocins are active against closely related bacterial species. Based on this postulate, bacteriocins produced by LAB are expected to inhibit or to kill other Gram-positive bacterial species. However, evidences from the last two decades indicate that some bacteriocins from LAB can be effective against some Gram-negative bacteria, yeasts, viruses, and even difficult to kill *Mycobacterium* spp. (Montville et al. 1999, Gupta et al. 2016, Todorov et al. 2019).

The idea of exploring applications of bacteriocins in human and veterinary medicine is becoming more and more attractive, since it was shown that bacteriocins can act in synergistic manner with antibiotics and other antimicrobials against different resistant bacterial species, and can be even equally effective or even work better than antibiotics (for review see Todorov et al. 2019). Some studies are reporting effectiveness of combination of bacteriocins and antibiotics in inhibition of various human pathogens, including high risk of mortality *Listeria monocytogenes* (Minahk et al. 2004, Todorov and Dicks 2009, Salvucci et al. 2010, Todorov et al. 2010, Todorov 2010).

The anti-*Listeria* efficacy of enterocin CRL35, when combined with monensin, bacitracin or gramicidin (the cell wall membrane-acting antibiotics), and mutanolysin and lysozyme (muralytic enzymes) was evaluated by Salvucci et al. (2010) and showed possible synergistic effects. Salvucci et al. (2010) suggested that enterocin CRL35 is a promising agent having applications not only for maintenance of food quality and safety, but also for medical applications in combination with conventional therapeutic agents.

A bacteriocin from *Enterococcus faecium* ST5Ha was evaluated in combination with sub-lethal doses of ciprofloxacin and showed enhancement of activity against *Listeria ivanovii* subsp. *ivanovii* ATCC19119 (Todorov et al. 2010). Most probably, the combination of bacteriocin ST5Ha with the antibiotic positively influenced the effectiveness of ciprofloxacin by dissipating the membrane proton gradient and resulted in better inhibition of *Listeria ivanovii* subsp. *ivanovii* ATCC19119 as compared to the use of the tested antibacterial agents alone (Todorov et al. 2010).

Synergism between ciprofloxacin and bacteriocin ST44AM from *Pediococcus pentosaceus* was also reported (Todorov and Dicks 2009). Similar interactions occurred between ciprofloxacin and five other bacteriocins produced by *Lactobacillus plantarum* ST69BZ, *Enterococcus faecium* ST62BZ and *Leuconostoc lactis* strains ST63BZ, ST611BZ and ST612BZ. There may be different mechanisms explaining synergy between antibiotics and bacteriocins; however, we speculate that pore formation by the bacteriocins facilitates the antibiotic killing action against sensitive bacterial cells, especially in the case of antibiotics acting on the inner components of the targeted cell. These results indicate that cationic peptides increase the effectiveness of ciprofloxacin through a similar mechanism.

Moreover, Hanchi et al. (2017) proposed combined application of duracin 61A with reuterin or duracin 61A with pediocin PA-1 as synergetic agents for control of *Clostridium difficile*. Authors reported of reduction in minimum inhibitory concentration (MIC) of duracin from 16 mg/l to 0.12 and 0.06 mg/l respectively when the bacteriocin is combined with reuterin or pediocin.

Design and implementation of delivery systems that are expected to improve stability and effectiveness of antimicrobial action by bacteriocins are still a significant challenge in development of technologically-suitable formulations. Some options for bacteriocin delivery systems include silver or carbohydrate-based nanoparticles, nanofiber scaffolds, nanospheres, implant impregnation, catheter coating, hydrogel, oral tablets, gum technology, livestock feed supplementation, aquaculture dry spray, addition of the bacteriocin or bacteriocin producer strain to food products, and incorporation of bacteriocins into food packaging (Favaro et al. 2015). Bacteriocins and other antimicrobial peptides can be considered as a highly-promising opportunity for development as antibiotic alternatives or synergistic antimicrobial pharmaceutical preparations suitable for application in the prevention and/or treatment of microbial diseases. Pilot studies with bacteriocins have already proven efficacy in laboratory controlled conditions, and some were already applied in industrial processes (Todorov et al. 2019). It seems that *in vivo* studies are now imperative before bacteriocins can be more regularly incorporated into the direct control of bacterial and viral infections in human and veterinary medicine (Todorov et al. 2019).

Some bacteriocins can be considered as potential candidates for cancer therapy (Baumal et al. 1982, Farkas-Himsley and Yu 1985) or as diagnostic agents for certain types of cancer (Kaur and Kaur 2015, Loiseau et al. 2016). At the present, most of these anticancer applications remain at an experimental stage and outside of the mainstream of cancer research (Saito et al. 1979, Musclow et al. 1987, Farkas-Himsley et al. 1995, Cruz-Chamorro et al. 2006, Sand et al. 2007, Sand et al. 2010, Sand et al. 2013). At the moment, this area of application of bacteriocins is generating more questions than providing concrete solutions. However, the accumulating evidence allows us to speculate on potential efficacy of bacteriocins that are also capable of either killing or inhibiting mammalian tumor cells. Bacteriocins, including well studied pediocin PA-1, nisin A or plantaricin A have relatively high cytotoxicity toward cancer cells as compared to the healthy cell lines. This selective toxicity is crucial for development of potential candidates for further investigation and clinical trials of application of bacteriocins as anticancer agents (Kaur and Kaur 2015).

Mycobacterium tuberculosis, being one of the most robust and antibiotic resistant microorganisms, still remains one of the most serious challenges for modern medicine (Carroll and O'Mahony 2011). Tuberculosis was always related in part to the low quality of life and widespread malnutrition (Carroll and O'Mahony 2011). In the last decades, tuberculosis once again attracted the attention of the medical professionals due to emerging problems with antibiotic resistance (for review see Todorov et al. 2019). Some factors influencing this process are related to the uncontrolled use of antibiotics. Therefore, there is a growing demand for alternative treatments and prophylactic strategies. Based on traditional medicine, LAB have emerged as promising tools in this resurgent crusade against *Mycobacterium tuberculosis,* including the Multidrug-Resistant and Extensively Drug Resistant *Mycobacterium tuberculosis* (Carroll and O'Mahony 2011). Bacteriocins were proposed as possible tools in these processes (Carroll and O'Mahony 2011), suggesting a potential mode of action of bacteriocins in control of *Mycobacterium* spp., such as involving lipid II and the relevance of modifications of lipid II occurring in *Mycobacterium* spp. (Carroll and O'Mahony 2011). Carroll et al. (2010), studied modifications in primary structure of known bacteriocins in order to generate more effective antimicrobials against *Mycobacterium* spp. These modified/engineered bacteriocins effectively inhibited *Mycobacterium* spp., including *Mycobacterium tuberculosis* H37Ra, *Mycobacterium kansasii* CIT11/06, *Mycobacterium avium* subsp. *hominissuis* CIT05/03 and *Mycobacterium avium* subsp. *paratuberculosis* MAP (ATCC19698) (Carroll et al. 2010). According to Carroll et al. (2010), in order to improve the bacteriocin's efficacy, these modifications need to be related to amino-acids in positions 21 and 22 of nisin A. The creation, molecular characterization and clinical evaluation of nisin T, nisin S and nisin V may be considered as the first step in the production of species or even sub-species specific bio-engineered anti-mycobacterial peptides (Carroll et al. 2010).

Mycobacterium species of animal origin, closely related to *Mycobacterium tuberculosis*, can also be responsible for the spread of tuberculosis to and between humans (Wedlock et al. 2002, Thoen et al. 2006). The general animals-related pathogen *Mycobacterium bovis* can cause tuberculosis in a variety of livestock species, as well as in humans and wild animals (O'Reilly and Daborn 1995). Various bacteriocins produced by different strains of *Lactobacillus plantarum*, *Enterococcus faecium*, *Lactobacillus sakei* were investigated on inhibition of *Mycobacterium tuberculosis* (Todorov et al. 2013). Bacteriocins can be applied as accompanying/complementary therapy or as a preventive measure in decreasing *Mycobacterium tuberculosis* infection, especially in countries with high infection rates. These bacteriocin-producing strains in fermented meat or milk products may contribute to the prevention of the transfer of the pathogen via the food chain. The consumption of probiotic products having anti-tuberculosis activity is considered as an advantageous treatment/prophylactic measure for workers exposed to a high risk of *Mycobacterium tuberculosis* contamination (Todorov et al. 2013).

Conclusion

We hope those exploring LAB and their bacteriocins are captivated by the quote that started this essay: "Let thy food be thy medicine and medicine be thy food" (Hippocrates, 400 BC). Are bacteriocins and probiotics contributing to the fulfillment of Hippocrates' dream? The food fermenting microorganisms are an essential part of our nutritional and wellbeing life style. Metabolites from starter cultures, beneficial and probiotics microorganisms are essential components of complex benefits provided to the consumers with microbially-processed food. However, we are evolving into the next phase of the healthier lifestyle, where the knowledge of microbially produced health-promoting substances contributes to the design of new formulations utilized in medical applications for treatment, disease prevention and wellbeing modulation (Dicks et al. 2011). Traditional (people's) medicine can be a priceless source of information on what microbiology may offer. Antimicrobial peptides can be a powerful weapon in the centuries-long crusade against pathogenic bacteria and even some types of cancer. One of the old traditional Bulgarian proverbs says: "For every pain there is a medical herb". Perhaps, we can add to this: "... and a lactic acid bacteria."

Acknowledgements

S.T. was supported by the University of Sao Paulo Program for Visiting Professors (2016.1.920.93), USP, Sao Paulo, SP, Brazil. MC was supported in part by the Ministry of Science of the Russian Federation (Project Number 19.6015.2017/8.9).

References

Abbasiliasi, S., J.S. Tan, T.A.T. Ibrahim, F. Bashokouh, N.R. Ramakrishnan, S. Mustafa and A.B. Ariff. 2017. Fermentation factors influencing the production of bacteriocins by lactic acid bacteria: a review. RSC Adv. 7: 29395–29420.

Abrudan, M.I., F. Smakman, A.J. Grimbergen, S. Westhoff, E.L. Miller, G.P. van Wezel and D.E. Rozen. 2015. Socially mediated induction and suppression of antibiosis during bacterial coexistence. Proc. Nat. Acad. Sci. USA 112: 11054–11059.

Ahmad, V., Q.M.S. Jamal, M.U. Siddiqui, A.K. Shukla, M.A. Alzohairy, M.A. Al Karaawi and M.A. Kamal. 2017. Methods of screening-purification and antimicrobial potentialities of bacteriocins in health care. Curr. Drug Methabol. 18: 814–830.

Alvarez-Sieiro, P., M. Montalban-Lopez, D. Um and O.P. Kuipers. 2016. Bacteriocins of lactic acid bacteria: extending the family. Appl. Microbiol. Biotechnol. 100: 2939–2951.

Asfahl, K.L. and M. Schuster. 2016. Social interactions in bacterial cell–cell signaling. FEMS Microbiol. Rev. 41: 92–107.

Baumal, R., E. Musclow and H. Farkas-Himsley. 1982. Variants of an interspecies hydroma with altered tumorigenicity and protective ability against mouse myeloma tumors. Cancer Res. 42(5): 1904–1908.

Brussow, H. 2017. Infection therapy: the problem of drug resistance—and possible solutions. Microb. Biotechnol. 10: 1041–1046.

Carmona-Ribeiro, A.M. and L.D.M. Carrasco. 2014. Novel formulation for antimicrobial peptides. Int. J. Mol. Sci. 15: 18040–18083.

Carroll, J., D. Field, P.M. O'Connor, P.D. Cotter, A. Coffer, C. Hill, R.P. Ross and J. O'Mahony. 2010. Gene encoded antimicrobial peptides, a template for the design of novel anti-micobacterial drugs. Bioengin. Bugs 1(6): 408–412.

Carroll, J. and J. O'Mahony. 2011. Anti-mycobacterial peptides: Made to order with delivery included. Bioengin. Bugs 2(5): 241–246.

Chikindas, M.L., R. Weeks, D. Drider, V.A. Chistyakov and L.M.T. Dicks. 2018. Function and emerging applications of bacteriocins. Curr. Opinion Biotechnol. 49: 23–28.

Clarke, S.F., E.F. Murphy, O. O'Sullivan, R.P. Ross, P.W. O'Toole, F. Shanahan and P.D. Cotter. 2013. Targeting the microbiota to address diet-induced obesity: a time dependent challenge. PLoS One 8(6): e65790.

Cook, L.C. and M.J. Federle. 2014. Peptide pheromone signaling in *Streptococcus* and *Enterococcus*. FEMS Microbiol. Rev. 38: 473–492.

Cotter, P.D., C. Hill and R.P. Ross. 2005. Food microbiology: Bacteriocins: Developing innate immunity for food. Nat. Rev. Microbiol. 3(10): 777–788.

Cruz-Chamorro, L., M.A. Puertollano, E. Puertollano, G.A. de Cienfuegos and M.A. de Pablo. 2006. *In vitro* biological activities of magainin alone or in combination with nisin. Peptides 27(6): 1201–1209.

Dargahi, N., J. Johnson, O. Donkor, T. Vasiljevic and V. Apostolopoulos. 2019. Immunomodulatory effects of probiotics: Can they be used to treat allergies and autoimmune diseases? Maturitas. 119: 25–38.

De Vuyst, L. and E.J. Vandamme. 1994. Lactic acid bacteria and bacteriocins: their practical importance. pp. 1–12. *In*: De Vuyst, L. and E.J. Vandamme (eds.). Bacteriocins of Lactic Acid Bacteria: Microbiology, Genetics and Applications. Blackie Academic & Professional, Glasgow, Scotland.

Dicks, L.M.T., T.D.J. Heunis, D.A. van Staden, A. Brand, K. Sutyak Noll and M.L. Chikindas. 2011. Medical and personal care applications of bacteriocins produced by lactic acid bacteria. pp. 391–421. *In*: Drider, D. and S. Rebuffat (eds.). Prokaryotic Antimicrobial Peptides. Springer, New York, USA.

Donaghy, J. 2010. Lantibiotics as prospective antimycobacterium agents. Bioengin. Bugs 1(6): 437–439.

Farkas-Himsley, H. and H. Yu. 1985. Purified colicin as cytotoxic agent of neoplasia; Comparative study with crude colicin. Cytobios 42: 193–207.

Farkas-Himsley, H., R. Hill, B. Rosen, S. Arab and C.A. Lingwood. 1995. The bacterial colicin active against tumor cells *in vitro* and *in vivo* in verotoxin A. Proceed. Nat. Acad. Sci. USA 92(15): 6996–7000.

Favaro, L., A.L.B. Penna and S.D. Todorov. 2015. Bacteriocinogenic LAB from cheeses—application in biopreservation? Trends Food Sci. Technol. 41(1): 37–48.

Gabrielsen, C., D.A. Brede, I.F. Nes and D.B. Diep. 2014. Circular bacteriocins: Biosynthesis and mode of action. Appl. Environ. Microbiol. 80(22): 6854–6862.

Ganzle, M.G., C. Hertel, J.M.B.M. Van Der Vossen and W.P. Hammes. 1999. Effect of bacteriocin-producing lactobacilli on the survival of *Escherichia coli* and *Listeria* in a dynamic model of the stomach and the small intestine. Int. J. Food Microbiol. 48: 21–35.

Granger, M., C.A. van Reenen and L.M.T. Dicks. 2008. Effect of gastro-intestinal conditions on the growth of *Enterococcus mundtii* ST4SA, and production of bacteriocin ST4SA recorded by the real-time PCR. Int. J. Food Microbiol. 123: 277–280.

Gupta, A., S.K. Tiwari, V. Netrebov and M.L. Chikindas. 2016. Biochemical properties and mechanism of action of enterocin LD3 purified from *Enterococcus hirae* LD3. Prob. Antimicrob. Prot. 8: 161–169.

Gutiérrez-Cortés, C., H. Suarez, G. Buitrago, L.A. Nero and S.D. Todorov. 2018. Enhanced bacteriocin production by *Pediococcus pentosaceus* 147 in co-culture with *Lactobacillus plantarum* LE27 on cheese whey broth. Front. Microbiol. 9: 2952. doi: 10.3389/fmicb.2018.02952.

Hajavi, J., S.-A. Esmaeili, A.-R. Varasteh, H. Vazini, H. Atabati, F. Mardani, A.A. Momtazi-Borojeni, M. Hashemi, M. Sankian and A. Sahebkar. 2018. The immunomodulatory role of probiotics in allergy therapy. J. Cell Physiol. 234: 2386–2398.

Hanchi, H., R. Hammami, H. Gingras, R. Kourda, M.G. Bergeron, J. Ben Hamida, M. Ouellette and I. Fliss. 2017. Inhibition of MRSA and of *Clostridium difficile* by durancin 61A: Synergy with bacteriocins and antibiotics. Future Microbiol. 12: 205–212.

Havarstein, L.S., H. Holo and I.F. Nes. 1994. The leader peptide of colicin V shares consensus sequences with leader peptides that are common among peptide bacteriocins produced by gram-positive bacteria. Microbiology 140: 2383–2389.

Hegarty, J.W., C.M. Guinane, R.P. Ross, C. Hill and P.D. Cotter. 2016. Bacteriocin production: a relatively unharnessed probiotic trait? F1000Res. 5: 2587.

Heng, N.C.K., P.A. Wescombre, J.P. Burton, R.W. Jack and J.R. Tagg. 2007. The diversity of bacteriocins in gram-positive bacteria. pp. 45–92. *In*: Riley, M.A. and M.A. Chavan (eds.). Bacteriocins: Ecology and Evolution. Springer-Verlag, Berlin Heidelberg, New York.

Kaur, S. and S. Kaur. 2015. Bacteriocins as potential anticancer agents. Front. Pharmacol. 6: 272.

Kjos, M., C. Oppegard and D.B. Diep. 2014. Sensitivity of two-peptide bacteriocinlactocin G in dependent on UppP, an enzyme involved in cell-wall synthesis. Mol. Microbiol. 92: 1177–1187.

Klaenhammer, T. 1988. Bacteriocins of lactic acid bacteria. Biochimie. 70: 337–349.

Le-Tien, C., M. Millette, M. Lacroix and M.-A. Mateescu. 2004. Modified alginate matrices for the immobilization of bioactive agents. Biotechnol. Appl. Biochem. 39: 189–198.

Lew, L.-C., S.-B. Choi, B.-Y. Khoo, S. Sreenivasan, K.-L. Ong and M.-T. Liong. 2018. *Lactobacillus plantarum* DR7 reduces choliesterol via phosphorylation of AMPK that down-regulated the mRNA expression of HNG-CoA reductase. Korean J. Food Sci. Animal Resources 38: 350–361.

Lewis, B.B. and E.G. Pamer. 2017. Microbiota-based therapies for *Clostridium difficile* and antibiotic-resistant enteric infections. Ann. Rev. Microbiol. 71: 157–178.

Liu, Y.-W., M.T. Liong and Y.-C. Tsai. 2018. New perspectives of *Lactobacillus plantarum* as a probiotic: The gut-heart-brain axis. J. Microbiol. 56: 601–613.

Loiseau, C., J. Augenstreich, A. Marchand, E. Harte, M. Garcia, J. Verdon, M. Mesnil and S. Lecomte. 2016. Specific anti-leukemia activity of the peptide warnericin RK and analogues and visualization of their effect on cancer cells by chemical raman imaging. PLoS ONE 11(9): Article number 162007.

Minahk, C.J., F. Dupuy and R.D. Morero. 2004. Enhancement of antibiotic activity by sub-lethal concentrations of enterocin CRL35. J. Antimicrob. Chemother. 53: 240–246.

Montville, T.J., H.-J. Chung, M.L. Chikindas and Y. Chen. 1999. Nisin A depletes intracellular ATP and acts in bactericidal manner against *Mycobacterium smegmatis*. Lett. Appl. Microbiol. 28: 189–193.

Murphy, E.F., P.D. Cotter, A. Hogan, O. O'Sullivan, A. Joyce, F. Fouhy, S.F. Clarke, T.M. Marques, P.W. O'Toole, C. Stanton, E.M. Quigley, C. Daly, P.R. Ross, R.M. O'Doherty and F. Shanahan. 2013. Divergent metabolic outcomes arising from targeted manipulation of the gut microbiota in diet-induced obesity. Gut 62(2): 220–226.

Musclow, C.E., H. Farkas-Himsley, S.S. Weitzman and M. Herridge. 1987. Acute lymphoblastic leukemia of childhood monitored by bacteriocin and flowcytometry. Eur. J. Cancer Clin. Oncol. 23(4): 411–418.

Naimi, S., S. Zirah, R. Hammami, B. Fernandez, S. Rebuffat and I. Fliss. 2018. Fate and biological activity of the antimicrobial lasso peptide microcin J25 under gastrointestinal tract conditions. Front. Microbiol. 9, article number 1764.

O'Reilly, L.M. and C.J. Daborn. 1995. The epidemiology of *Mycobacterium bovis* infections in animals and man: A review. Tubercle. Lung Dis. 76(supplement 1): 1–46.

Perales-Adán, J., S. Rubino, M. Martinez-Bueno, E. Valdivia, M. Montalban-Lopez, R. Cebrian and M. Maqueda. 2018. LAB bacteriocins controlling the food isolated (drug-resistant) *Staphylococci*. Front. Microbiol. 9: 1143.

Phumisantiphong, U., K. Siripanichgon, O. Reamtong and P. Diraphat. 2017. A novel bacteriocin from *Enterococcus faecalis* 478 exhibit a potent activity against vancomycin-resistant enterococci. PLOS One 12: 1–22.

Quigley, E.M.M. 2019. Prebiotics and probiotics in digestive health. Clin. Gastroenterol. Hepatol. 17: 333–344.

Quintana, V.M., N.I. Torres, M.B. Wachsman, P.J. Sinko, V. Castilla and M.L. Chikindas. 2014. Antiherpes simplex virus type 2 activity of the antimicrobial peptide subtilosin. J. Appl. Microbiol. 117: 1253–1259.

Rodali, V.P., V.K. Lingala, A.P. Karlapudi, M. Indira, T.C. Venkateswarulu and B.D. John. 2013. Biosynthesis and potential application of bacteriocins. J. Pure Appl. Microbiol. 7: 2933–2945.

Saarela, M.H. 2019. Safety aspects of next generation probiotics. Curr. Opin. Food Sci. 30: 8–13.

Saito, H., T. Watanabe and H. Tomioka. 1979. Purification, properties and cytotoxic effect of a bacteriocin from *Mycobacterium smegmatis*. Antimicrob. Agents Chemother. 15: 504–509.

Salvucci, E., E.M. Hebert, F. Sesma and L. Saavedra. 2010. Combined effect of syntetic enterocin CRL35 with cell wall, membrane-acting antibiotics and muranolytic enzymes against *Listeria* cells. Lett. Appl. Microbiol. 51(2): 191–195.

Sand, S.L., T.M. Haug, J. Nissen-Meyer and O. Sand. 2007. The bacterial peptide pheromone plantaricin A permeabilizes cancerous, but not normal, rat pituitary cells and differentiates between the outer and inner membrane leaflet. J. Membrane Biol. 216: 61–71.

Sand, S.L., C. Oppegård, S. Ohara, T. Iijima, S. Naderi and H.K. Blomhoff. 2010. Plantaricin A, a peptide pheromone produced by *Lactobacillus plantarum*, permeabilizes the cell membrane of both normal and cancerous lymphocytes and neuronal cells. Peptides 31: 1237–1244.

Sand, S.L., J. Nissen-Meyer, O. Sand and T.M. Haug. 2013. Plantaricin A, a cationic peptide produced by *Lactobacillus plantarum*, permeabilizes eukaryotic cell membranes by a mechanism dependent on negative surface charge linked to glycosylated membrane proteins. Biochim. Biophys. Acta 1828: 249–259.

Sarkar, S.R. and S. Banerrjee. 2019. Gut microbiota in neurodegenerative disorders. J. Neuroimmunol. 328: 98–104.

Schillinger, U. and W.H. Holzapfel. 1996. Guidelines for manuscripts on bacteriocins of lactic acid bacteria. Int. J. Food Microbiol. 33(2-3): iii–v.

Smaoui, S., L. Elleuch, W. Bejar, I. Karray-Rebai, I. Ayadi, B. Jaouadi, F. Mathieu, H. Chouayekh, S. Bejar and L. Mellouli. 2010. Inhibition of fungi and Gram-negative bacteria by bacteriocin BacTN635 produced by *Lactobacillus plantarum* sp. TN635. Appl. Biochem. Biotechnol. 162(4): 1132–1146.

Thoen, C., P. LoBue and I. De Kantor. 2006. The importance of *Mycobacterium bovis* as a zoonosis. Vet. Microbiol. 112(2-4 Special issue): 339–345.

Todorov, S.D. and L.M.T. Dicks. 2009. Bacteriocin production by *Pediococcus pentosaceus* isolated from marula (*Scerocarya birrea*). Int. J. Food Microbiol. 132(2-3): 117–126.

Todorov, S.D. and L.M.T. Dicks. 2009. Effect of modified MRS medium on production and purification of antimicrobial peptide ST4SA produced by *Enterococcus mundtii*. Anaerobe. 15(3): 65–73.

Todorov, S.D. 2010. Diversity of bacteriocinogenic lactic acid bacteria isolated from boza, a cereal-based fermented beverage from Bulgaria. Food Control. 21(7): 1011–1021.

Todorov, S.D., M. Wachsman, E. Tomé, X. Dousset, M.T. Destro, L.M.T. Dicks, B.D.G.M. Franco, M. Vaz-Velho and D. Drider. 2010. Characterisation of an antiviral pediocin-like bacteriocin produced by *Enterococcus faecium*. Food Microbiol. 27(7): 869–879.

Todorov, S.D., B.D.G.M. Franco and I.J. Wiid. 2013. *In vitro* study of beneficial properties and safety of lactic acid bacteria isolated from Portuguese fermented meat products. Beneficial Microbs. 5(3): 351–366.

Todorov, S.D., B.D.G.M. Franco and J.R. Tagg. 2019. Bacteriocins of Gram positive bacteria having activity spectra extending beyond closely-related species. Beneficial Microbes. 10(3): 315–328.

Torres, N.I., K.S. Noll, S. Xu, J. Li, Q. Huang, P.J. Sinko, M.B. Wachsman and M.L. Chikindas. 2013. Safety, formulation and *in vitro* antiviral activity of the antimicrobial peptide subtilosin against herpes simplex virus type 1. Prob. Antimicrob. Prot. 5: 26–35.

Tsai, Y.-L., T.-L. Lin, C.-J. Chang, T.-R. Wu, W.-F. Lai, C.-C. Lu and H.-C. Lai. 2019. Probiotics, prebiotics and amelioration of diseases. J. Biomed. Sci. 26: 3.

Vidhyasagar, V. and K. Jeevaratnam. 2013. Bacteriocin activity against various pathogens produced by *Pediococcus pentosaceus* VJ13 isolated from idly batter. Biomed. Chromatogr. 27: 1497–1502.

Wachsman, M.B., V. Castilla, A.P. de Ruiz Holgado, P.A. de Torres, F. Sesma and C.E. Coto. 2003. Enterocin CRL35 inhibits late stages of HSV-1 and HSV-2 replication *in vitro*. Antiviral. Res. 58: 17–24.

Wedlock, D.N., M.A. Skinner, G.W. De Lisle and B.M. Bubble. 2002. Control of *Mycobacterium bovis* infections and the risk to human populations. Microbes Infect. 4(4): 471–480.

Westhoff, S., G.P. van Wezel and D.E. Rozen. 2017. Distance-dependent danger responses in bacteria. Curr. Opin. Microbiol. 36: 95–101.

Wu, C., H.K. Kim, G.P. Van Wezel and Y.H. Choi. 2015. Metabolomics in the natural products field—a gateway to novel antibiotics. Drug Discov. Today Technol. 13: 11–17.

6

Probiotics, Vitamin D, and Vitamin D Receptor in Health and Disease

Carolina Battistini,[1,2] *Najib Nassani*,[3] *Susana MI Saad*[1,2] and *Jun Sun*[3,*]

Introduction

The gut microbiome, "a newly discovered organ" of the body, plays a critical role in immunity and metabolism in the host's healthy and diseased states. Microbiome products released in the gut, e.g., short chain fatty acids, may reach and influence the function of organs and systems beyond the intestinal tract. Environmental factors, lifestyle, age, and sex influence the profile and function of the microbiome. Its imbalance, termed 'dysbiosis', is directly related to the development of various human diseases (Wang et al. 2016, Costea et al. 2018, Salvucci 2019).

Vitamin D (VitD) and the vitamin D receptor (VDR) are involved in many functions of the body, including but not limited to calcium absorption, immunity, glucose and liver metabolism, in addition to gut microbiota modulation. For example, VitD deficiency and the lack or low expression of the VDR are directly related to inflammatory bowel diseases (IBD), and the treatment with VitD supplements might be effective in some cases (Haussler et al. 2013, Ooi et al. 2013, Wu et al. 2015a,b, National Institute of Health 2016, Celiberto et al. 2018). Therefore, VDR could be

[1] Department of Pharmaceutical and Biochemical Technology, School of Pharmaceutical Sciences, University of São Paulo, São Paulo, SP, Brazil.
Email: carolina.battistini@gmail.com
[2] Food Research Center, University of São Paulo, São Paulo, SP, Brazil.
Email: susaad@usp.br
[3] Department of Medicine, University of Illinois at Chicago, Chicago, IL, USA.
Email: najib.nassani@gmail.com
* Corresponding author: junsun7@uic.edu

used as a predictive biomarker for dysbiosis, and the development of strategies that boost its functions are of utmost importance for the gut microbiome restoration.

Dietary interventions with pre-, pro-, and/or symbiotic foods or supplements have shown to be effective in restoring the gut microbiome to a healthier pre-disease state pattern and are interesting and less harmful alternatives when compared to antibiotic therapies (Costea et al. 2018, Salvucci 2019). The benefits of probiotics are strain dependent and may comprise improvement of intestinal epithelial cells turnover, competition for nutrients and adhesion sites, production of short chain fatty acids, vitamins, bacteriocins, and anti-inflammatory compounds, regulation of the intestinal transit, modulation of the intestinal microbiota, and enhancement of immunity. The most common microorganisms known as having probiotic properties are bacteria from the *Lactobacillus*, *Streptococcus*, *Enterococcus*, and *Bifidobacterium* genera (Holzapfel and Schilinger 2002, Gibson 2004, Williams 2010, Hill et al. 2014).

Probiotics are usually consumed through food products, supplements or nutraceutical capsules. The food vehicle used, the presence of other bioactive substances, such as fibers, for example, and the consumption frequency are factors that influence probiotics efficacy (Ranadheera et al. 2010, Sanders and Marco 2010). Besides, the microorganisms' survival through the gastrointestinal tract passage (resistance to acids, bile, and enzymes) and the antibiotics resistance should also be taken into consideration (Gibson 2004, Ranadheera et al. 2010).

Additionally, another way to increase the population of beneficial bacteria in the intestinal microbiota is through the intake of prebiotic compounds, which are defined as "substrates that are selectively utilized by host microorganisms conferring a health benefit" (Gibson et al. 2017). Some fibers might have prebiotic potential, once they are not digested in the small intestine and reach the colon intact, serving as substrates for the local microbiota (Puupponen-Pimiä et al. 2002, Gibson et al. 2010, Martinez et al. 2015). The combination of prebiotic ingredients and probiotic microorganisms results in synbiotic foods or supplements and may confer a competitive advantage for the probiotics over the intestinal commensal microbiota or pathogenic bacteria (Puupponen-Pimiä et al. 2002, Martinez et al. 2015). Food research should focus on the development of symbiotic products that target the increase of VDR expression, and consequently improve the anti-inflammatory responses and modulate the gut microbiota, which could be applied in a preventive approach for IBD and other diseases.

The clinical trials outcomes of probiotic treatments are still inconsistent and controversial. In this chapter, we will discuss the potential role of VDR in regulating probiotic functions, and whether they can be used as a therapeutic strategy to increase the VDR expression, and consequently, to promote the healthy microbiota.

Vitamin D/Vitamin D Receptor

Chemical structure and functions

VitD (vitamin D) is a fat-soluble vitamin synthesized by the exposure of the skin to sunlight or consumed through nutraceutical supplements, fatty fishes, egg yolks, or fortified foods, such as dairy products. Commonly called pre-VitD, it is found in two chemical arrangements: vitamin D2, or ergocalciferol, and vitamin D3, or cholecalciferol. The main differences among these two forms are a double bond between carbons 22 and 23, and a methyl group bonded to carbon 24 in ergocalciferol (Figure 1). Nonetheless, these compounds are biologically inactive and should run through two hydroxylation before they have biological functions in the body (Ross et al. 2011b, Health Canada 2012, FDA 2016, National Institute of Health 2016).

In blood circulation, pre-VitD is bounded to the VitD binding protein, and then transported to the liver where the first hydroxylation occurs through the action of 25-hydroxylase, an enzyme encoded by the gene CYP2R1, generating the 25-OH-cholecalciferol (25(OH)D). This compound is then released in the bloodstream and is the major circulating form and the main marker of VitD status. Thereafter, the second hydroxylation may occur in the kidneys, brain, lung, prostate, placenta, or in the immune cells, resulting in VitD itself, or 1,25-$(OH)_2$D (Holick 2011a,b, Genetic Home Reference 2017, Bivona et al. 2018).

A great number of biological functions are associated with VitD through the mediation of the VitD receptor (VDR). Macrophages, when in contact with pathogens,

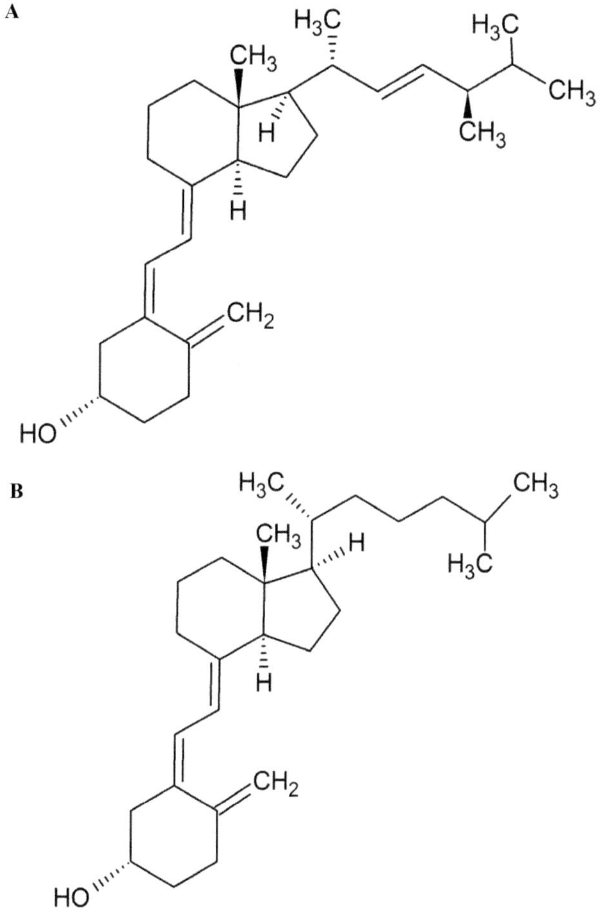

Figure 1: Chemical structure of ergocalciferol (A) and cholecalciferol (B).

induce the enzymatic activation of VitD in the cytoplasm, through the action of Toll-like receptors (TLR). When 1,25-(OH)2D binds to VDR, both migrate to the cell nucleus and antibacterial compounds are produced, such as β-defensin 2 and cathelicidin, and the Th1 immune response is activated with the production of interferon gamma (IFN-γ). This whole process is regulated automatically. When 1,25-$(OH)_2$D is accumulated, the Th1 profile is inhibited and the Th2 immune response is activated. Hence, the production of IFN-γ is reduced, while the production of interleukin 4 (IL-4) increases (Gatti et al. 2016).

The VDR is found in several organs and tissues of the body, especially in the parathyroid gland, small intestines, and colon. In addition, it is also a receptor for the secondary bile acid lithocholic acid, a transcriptional factor associated with immunomodulation, proliferation, intestinal barrier function, and autophagy, and shows a similar sequence with the steroid and thyroid hormone receptor. When bound to the retinoid X receptor (RXR), they form a heterodimer that is involved in physiological effects, not only of VitD but, also, of microbial and dietary metabolites, like secondary bile acids and fatty acids. Furthermore, evidences indicate that VDR is genetically associated with the gut microbiota profile, IBD, liver diseases, cancer, and blood sugar regulation (Wang et al. 2016, Bakke et al. 2018).

In summary, VitD and VDR influence body defenses and inflammatory responses. Thus, it is crucial to understand their mechanisms of action and establish potential therapies targeting the improvement of their status and functions for health maintenance.

Vitamin D requirements

Recommended circulating level and dietary reference intake

VitD status is defined by the serum concentration of 25(OH)D. However, there is no consensus worldwide about the adequate level. The Institute of Medicine (IOM) from the USA considers 20 ng/mL (50 nmol/L) of 25(OH)D sufficient (Ross et al. 2011a,b), while other medical societies adopted a level of 30 ng/mL based on a possible reduction on the risk of falls and fractures in the elderly (Holick et al. 2011, American Geriatrics Society 2014). In fact, there are a lot of controversial results from clinical trials regarding the effectiveness of VitD supplementation in different diseases and, for this reason, it is difficult to establish a suitable acceptance range of 25(OH)D level (Trivedi et al. 2003, Need et al. 2008, Sanders et al. 2010, Bolland et al. 2018).

According to the IOM, assuming minimal sunlight exposure, the daily recommended dietary allowance of VitD is 400 IU (10 mcg) for infants up to 12 months, 800 IU (20 mcg) for elderly (> 70 years of age), and 600 IU (15 mcg) for other age groups (Ross et al. 2011a,b). However, these values can vary depending on the geographic location, season of the year, skin pigmentation, and use of sun blocking agents (Holick 1994).

An effective method to increase the levels of VitD is the exposure of the skin to sunlight. However, this is not often indicated due to the risk of skin cancer development. Alternatively, supplements intake or food fortification might be useful strategies against hypovitaminosis (Pilz et al. 2018). In addition to ergocalciferol and cholecalciferol, VitD metabolites may be administered under specific circumstances. Calcidiol, due to its hydrophilic properties and ability to bypass the hepatic

25-hydroxylation, is useful for individuals with fat malabsorption or liver disease, while calcitriol, which bypasses the one alpha hydroxylation and activation phase, is indicated for patients with chronic kidney disease or type 1 VitD-dependent rickets.

Hypovitaminosis and toxicity

VitD deficiency or resistance can be triggered by several factors, e.g., low availability of the vitamin, resulting from low dietary intake, lack of sunlight exposure, or low absorption; defects on the hydroxylation stage in the liver or activation in the kidneys, which is common in chronic renal disease and VitD-dependent rickets type 1; increased catabolism by the liver; resistance at the end organ level, like in hereditary VitD-resistant rickets type 2.

Of particular concern is the fact that low levels of VitD lead to malabsorption of calcium and phosphorus in the intestines, and when during a long period, this may cause rickets in children and osteomalacia in adults. Even though these diseases are less common in developed countries due to food fortification practices, subclinical VitD deficiency is common and may be associated with osteoporosis and increased risk of falls and fractures. In addition, patients with malabsorptive issues have an increased risk of VitD hypovitaminosis, such as patients suffering celiac disease, inflammatory bowel disease, and a history of gastrectomy.

The prescription of vitamin supplements is a good strategy to increase the circulating VitD, but clinicians should be cautious. High VitD doses, that have been associated with errors in the formulation and excessively fortified dairy products intake, can be toxic and lead to hypercalciuria. In this condition, the excretion of calcium through the urinary tract is increased, inducing kidney stones in some cases, or in worse scenarios, hypercalcemia is developed, when the concentration of calcium in the blood stream is high, and the patients usually report symptoms like fatigue, muscle weakness, weight loss, nausea, vomiting, soft tissue calcification, and tachycardia (Jacobus et al. 1992, Holick 2003, Pilz et al. 2018). In addition, increased levels of 25(OH)D seems to be related to an increased risk of falls and fractures, pancreatic and prostate cancer, and mortality (Wortsman et al. 2000, Sanders et al. 2010).

Hence, medical societies do not have a current consensus about VitD adequacy, and their recommendations are based only on studies about bone health. On the other hand, a lot of other disorders are related to VitD deficiency and might need to be taken into consideration to determine the prescription of supplements. Thus, there is a gap and more studies are needed in the field in order to regulate the adequate levels of VitD, mostly for anti-inflammatory effects.

Probiotics, Vitamin D, and VDR on Health

Probiotics, VDR, and the gut microbiota

The gut microbiota comprises all the microorganisms present in the gut, including bacteria, viruses, archeae, fungi, and yeasts. Its composition reaches maturity at the fourth year of life approximately, and it is influenced by non-genetic factors, such as age, sex, body mass index (BMI), smoking status, and dietary patterns, and at the genetic level by the VDR gene (Cani 2018, Cani et al. 2019, Fouhy et al. 2019). A healthy intestinal environment is associated with the presence of beneficial microorganisms like *Bifidobacterium* spp., *Faecalibacterium prausnitzii*, and *Lactobacillus* spp., higher amounts of butyrate and anti-inflammatory cytokines, a thicker mucus layer, and improved barrier function (Celiberto et al. 2018, Costea et al. 2018).

The VDR is expressed in intestinal epithelial and immune cells, and regulates the transcription of barrier proteins, like claudin 2 (Zhang et al. 2015); the expression of antimicrobial peptides (AMPs), like cathelicidin and β-defensins, promoting mucosal homeostasis and barrier function, protection against inflammation, and preventing epithelial apoptosis. Notably, paneth cells might play a role in preventing dysbiosis as they are responsible for autophagy and for the production of AMPs (Adolph et al. 2013, Wu et al. 2015a). In addition, both VitD and VDR promote a tolerogenic and anti-inflammatory response in Crohn's disease patients, favoring regulator T cells (Tregcells) and increasing the proportions of the Actinobacteria and the Firmicutes phyla (Schäffler et al. 2018).

The VDR status is implicated in inflammation, signal transduction, infections, amino acid and carbohydrate metabolism, and neoplasm (Bakke and Sun 2018, Bakke et al. 2018, Sun 2018). In addition, polymorphisms in VDR gene were associated with susceptibility to IBD. Studies with mice showed that VDR knockout mice presented a significant shift in their microbiota at a phylogenetic level and were more prone to autoimmune diseases (Kongsbak et al. 2013). Moreover, animals incapable of producing the active form of VitD or VDR knockout developed a more severe colitis induced by dextran sulfate sodium (DSS) (Simmons et al. 2000, Azad et al. 2012, Lee and Song 2012, Kongsbak et al. 2013, Ooi et al. 2013, Wu et al. 2015a, Jin et al. 2015, Wang et al. 2016).

In a recent study, *Lactobacillus casei rhamnosus* (Lcr35®) was effective in increasing the populations of *Lactobacillus* and *Bifidobacterium* in the feces of hospitalized children with acute diarrhea. Besides, the probiotic treatment ameliorated the abdominal discomforts and diarrhea, improving the patients' appetite and food intake (Lai et al. 2019). Similarly, *Bifidobacterium animalis* subsp. *lactis* (txid302911) was able to shape the gut microbiota of low birth weight infants to a more diverse and complex profile. The strain increased the counts of the *Bifidobacterium* and *Lactobacillus* genera in their faeces, whereas the control group presented a predominant presence of opportunistic pathogens associated with the risk of necrotizing enterocolitis (Chi et al. 2019).

Interestingly, VitD supplementation was associated with lower counts of *Clostridium difficile* in breast fed infants while increased *Lachnobacterium* and decreased *Lactococcus* in the gut microbiome of infants, age 3–6 months, were linked to cord blood VitD levels (Sordillo et al. 2017).

In conclusion, the VDR status has an important role on the gut microbiota profile and with the development of IBD; dietary interventions with VitD and probiotics could be explored further, aiming at improving the VDR expression and restoration of the gut microbiome.

Probiotics, vitamin D, and VDR in IBD and metabolic disorders

Inflammatory bowel diseases

Inflammatory bowel disease is a chronic inflammation of the gastrointestinal tract (GIT), caused by abnormal immune responses. The main types of IBD are ulcerative colitis (UC), which affects the large intestine, and Crohn's disease (CD), which may affect the whole intestinal tract (Feuerstein et al. 2017), although more recent studies suggest that they may be part of the same spectrum.

VitD deficiency is recurrent in patients with IBD. However, it is unclear whether it is a cause or a consequence. Actually, it has been reported that the consumption of products with lactose may worsen the abdominal discomforts in IBD, which is in completely accordance with the lower concentrations of VitD because dairy products are the main fortified foods with VitD and calcium, and IBD patients stop or decrease consumption of these products to avoid abdominal discomfort. In addition, studies have shown that VitD supplementation is a promising tool to improve the clinical outcome and quality of life of IBD patients, with the potential to inhibit the activity of CD and maintain remission (Miheller et al. 2009, Yang et al. 2013, Kabbani et al. 2016, Williams et al. 2018, Scotti et al. 2019).

The expression of VDR is decreased in IBD. In fact, investigations with mice have shown that VDR knockout animals are more prone to develop severe colitis while transgenic overexpression of VDR may confer a protective effect (Kong et al. 2008, Liu et al. 2013, Ooi et al. 2013, Belizário and Napolitano 2015, Wu et al. 2015a). This anti-inflammatory response seems to be directly related to NF-κB pathway, given that 42 Single Nucleotide Polymorphisms (SNPs) linked with immune disorders are binding sites for both VDR and NF-κB. VDR is also involved in pro-inflammatory response to bacterial endotoxin LPS and with autophagy regulation (Singh et al. 2017).

Interestingly, VDR expression is crucial for probiotic anti-inflammatory effects. In a *Salmonella* infection model, *Lactobacillus plantarum* showed physiological and histological protection only for wild-type mice, whereas no effect was observed in VDR knockout mice (Wu et al. 2015a). Similarly, our research group found that the administration of probiotic fermented milk with *Lactobacillus paracasei* subsp. *paracasei* F19 showed a promising increase in the VDR expression at the mRNA level in wild-type mice, whereas VDR knockout mice presented an exacerbated inflammation induced by DSS when compared to wild-type mice (unpublished data).

Metabolic disorders

Metabolic syndrome, insulin resistance, obesity, and coronary disease have also been related to low levels of VitD and defects in VDR signaling (Oh et al. 2015, Chang and Kim 2017, Moreno-Santos et al. 2017). Nonetheless, there is a lack of studies involving VitD and both type 1 and type 2 diabetes. A few reports have shown controversial, but promising results (Cooper et al. 2011, Dong et al. 2013, Song et al. 2013, Krul-Poel et al. 2017).

Interestingly, experiments with VDR knockout mice showed that lean animals resist weight gain induced by high-fat diet (Peirce et al. 2014), while interventional studies reported that VitD supplementation improved insulin signaling (Vrieze et al. 2012, Jin et al. 2015). Meanwhile, VitD supplementation may improve lipid profile and inflammatory markers in non-alcoholic fatty liver disease (NAFLD). In a study with a Non-Alcoholic Steatohepatitis (NASH) mice model, VitD significantly decreased steatosis and NAFLD activity (Zhu et al. 2013, Gibson et al. 2018).

Probiotics are alternatives to enhance metabolic disease status. The consumption of the probiotic strain *Lactobacillus reuteri* NCIMB 30242 improved VitD status of hypercholesterolemic adults by 22.4% when compared with the placebo group (Jones et al. 2013). In the same trend, obese adults that consumed symbiotic fermented milk produced with *Bifidobacterium lactis* Bb-12 and insulin showed a significant reduction in the serum insulin and blood triglycerides levels, and at the same time, improved their VitD status and insulin sensitivity (Mohammadi-Sartang et al. 2018).

Moreover, individuals with metabolic syndrome who consumed probiotic fermented milk (containing *Bifidobacterium lactis* Bb-12 and *Lactobacillus acidophilus* La5) for eight weeks presented lower levels of blood glucose and vascular cell adhesion molecules, whereas no significant changes were observed for insulin level, homeostasis model of assessment-estimated insulin resistance (HOMA-IR), and for other metabolic and cardiovascular markers (Rezazadeh et al. 2019).

The co-supplementation of VitD and probiotics seems to have synergic effects, and more studies should be conducted in order to investigate these observations. Nevertheless, in a clinical trial, the combination of VitD with a pool of probiotics improved the 25(OH)D levels, whereas it significantly reduced serum insulin levels, HOMA-IR, and high-sensitive C-reactive protein in patients with type 2 diabetes and coronary heart disease (Raygan et al. 2018).

Probiotic anti-inflammatory properties are dependent on VDR expression. Meanwhile, both VitD and probiotics influence the markers for intestinal microbiome and metabolic disease. Their synergic effect is not well explored and more *in vivo* and clinical trials should be carried out to answer this unexplored path.

Conclusions and Future Direction

This chapter discussed how the VDR status influences the gut microbiome and its critical impact on the potential benefit of probiotics against inflammation and infections. Overall, VitD and probiotics have great potential to modulate the gut microbiota, improving the anti-inflammatory response and metabolic markers. Nonetheless, VDR is an important factor for homeostasis and is directly involved in the probiotics' mechanisms of action. It is a double way path, since VDR expression is crucial for probiotics anti-inflammatory potential, while the administration of probiotic may enhance the VDR expression.

Nevertheless, there is promising evidence of synergic effects of co-supplementation with VitD and probiotics. However, more *in vivo* studies and clinical trials combining probiotics and VitD supplementation are of utmost importance for the comprehension of their roles. Therefore, it is possible to establish novel biomarkers and develop alternative treatments for the prevention of inflammation and other diseases.

To the best of our knowledge, only *Lactobacillus* strains were investigated regarding the potential of probiotics to improve VDR expression, and future studies should include other genera like *Bifidobacterium*. Furthermore, the impact of prebiotic compounds and food processing technologies, such as microencapsulation, could be used as strategies to increase the probiotic survival after passage through the gastrointestinal tract, and possibly boost their potential health benefits. IBD patients avoid dairy products due to lactose content (Scotti et al. 2019). Thus, other food matrices should be explored as vehicles for probiotics, such as soy, rice, coconut, almond, among other vegetable beverages. This knowledge can be exploited to develop novel strategies by restoring healthy VDR functions and normal host-microbe interactions.

Acknowledgments

The authors would like to thank São Research Foundation (Project FAPESP #2018/21584-4), FoRC - Food Research Center (Project FAPESP #2013/07914-8), and CAPES Foundation (PDSE #88881.187323/2018-01) for their financial support and fellowships.

References

Adolph, T.E., M.F. Tomczak, L. Niederreiter, H.-J. Ko, J. Bock, E. Martinez-Nave, J.N. Glickman, M. Tschurtschenthaler, J. Hartwig, S. Hosomi, M.B. Flak, J.L. Cusick, K. Kohno, T. Iwawaki, S. Billmann-Born, T. Raine, R. Bharti, R. Lucius, M.-N. Kweon, S.J. Marciniak, A. Choi, S.J. Hagen, S. Schreiber, P. Rosenstiel, A. Kaser and R.S. Blumberg. 2013. Paneth cells as a site of origin for intestinal inflammation. Nature 503: 272–276.

American Geriatrics Society Workgroup on Vitamin D Supplementation for Older Adults. 2014. Recommendations abstracted from the American Geriatrics Society Consensus Statement on vitamin D for Prevention of Falls and Their Consequences. J. Am. Geriatr. Soc. 62: 147–152.

Azad, A.K., W. Sadee and L.S. Schlesinger. 2012. Innate immune gene polymorphisms in tuberculosis. Infect. Immun. 80: 3343–3359.

Bakke, D. and J. Sun. 2018. Ancient nuclear receptor VDR With New functions: Microbiome and inflammation. Inflamm. Bowel Dis. 24: 1149–1154.

Bakke, D., I. Chartterjee, A. Agrawal, Y. Dai and J. Sun. 2018. Regulation of microbiota by vD receptor: A nuclear weapon in metabolic diseases. Nucl. Receptor Res. 5: 2314–5714.

Belizário, J.E. and M. Napolitano. 2015. Human microbiomes and their roles in dysbiosis, common diseases, and novel therapeutic approaches. Front. Microbiol. 6: 1050.

Bivona, G., L. Agnello and M. Ciaccio. 2018. The immunological implication of the new vitamin D metabolism. Cent. Eur. J. Immunol. 43: 331–334.

Bolland, M.J., A. Grey and A. Avenell. 2018. Effects of vitamin D supplementation on musculoskeletal health: a systematic review, meta-analysis, and trial sequential analysis. Lancet Diabetes Endo. 6: 847–858.

Cani, P.D. 2018. Human gut microbiome: hopes, threats and promises. Gut. 67: 1716–1725.

Cani, P.D., M.V. Hul, C. Lefort, C. Depommier, M. Rastelli and A. Everard. 2019. Microbial regulation of organismal energy homeostasis. Nat. Metab. 1: 34–46.

Celiberto, L.S., F.A. Graef, G.R. Healey, E.S. Bosman, K. Jacobson, L.M. Sly and B.A. Vallance. 2018. Inflammatory bowel disease and immunomodulation: novel therapeutic approaches through modulation of diet and the gut microbiome. Immunology 155: 36–52.

Chang, E. and Y. Kim. 2017. Vitamin D insufficiency exacerbates adipose tissue macrophage infiltration and decreases AMPK/SIRT1 activity in obese rats. Nutrients 9: 338.

Chi, C., Y. Xue, R. Liu, Y. Wang, N. Lv, H. Zeng, N. Buys, B. Zhu, J. Sun and C. Yin. 2019. Effects of a formula with a probiotic *Bifidobacterium lactis* supplement on the gut microbiota of low birth weight infants. Eur. J. Nutr. Available at: https://link.springer.com/article/10.1007%2Fs00394-019-02006-4#citeas.

Cooper, J.D., D.J. Smyth, N.M. Walker, H. Stevens, O.S. Burren, C. Wallace, C. Greissl, E. Ramos-Lopez, E. Hyppönen, D.B. Dunger, T.D. Spector, W.H. Ouwehand, T.J. Wang, K. Badenhoop and J.A. Todd. 2011. Inherited variation in vitamin D genes is associated with predisposition to autoimmune disease type 1 diabetes. Diabetes. 60: 1624–1631.

Costea, P.I., F. Hildebrand, M. Arumugan, F. Bäckhed, M.J. Blasé, F.D. Bushman, W.M. de Vos, S.D. Ehrlich, C.M. Fraser, M. Hattori, C. Huttenhower, I.B. Jeffery, D. Knights, J.D. Lewis, R.E. Ley, H. Ochman, P.W. O'Toole, C. Quince, D.A. Relman, F. Shanahan, S. Sunagawa, J. Wang,

G.M. Weinstock, G.D. Wu, G. Zeller, L. Zhao, J. Raes, R. Knight and P. Bork. 2018. Enterotypes in the landscape of gut microbial community composition. Nat. Microbiol. 3: 8–16.

Dong, J.-Y., W. Zhang, J.J. Chen, Z.-L. Zhang, S.-F. Han and L.-Q. Qin. 2013. Vitamin D intake and risk of type 1 diabetes: a meta-analysis of observational studies. Nutrients 5: 3551–3562.

FDA – Food and Drug Administration. 2016. FDA Approves an Increase to the Amount of Vitamin D for Milk and Milk Alternatives. Available at: https://www.fda.gov/Food/NewsEvents/ConstituentUpdates/ucm510556.htm. Accessed Sep 2018.

Feuerstein, J.D., G.C. Nguyen, S.S. Kupfer, Y. Falck-Ytter and S. Singh. 2017. American gastroenterological association institute guideline on therapeutic drug monitoring in inflammatory bowel disease. Gastroenterology 153: 827–834. Available at: https://pdfs.semanticscholar.org/4cad/5440e26b6c35bd70ba45f414bd5c686e2bcf.pdf. Accessed August 2019.

Fouhy, F., C. Watkins, C.J. Hill, C.-A. O'shea, B. Nagle, E.M. Dempsey, P.W. O'toole, R.P. Ross, C.A. Ryan and C. Stanton. 2019. Perinatal factors affect the gut microbiota up to four years after birth. Nat. Commun. 10: 1517.

Gatti, D., L. Idolazzi and A. Fassio. 2016. Vitamin D: not just bone, but also immunity. Minerva Med. 107: 452–460.

Genetic Home Reference. H. CYP2R1 gene. 2017. Available at: https://www.ncbi.nlm.nih.gov/pubmed/. Accessed Jan 2019.

Gibson, G.R. 2004. Fibre and effects on probiotics (the prebiotic concept). Clin. Nutr. Supplements 1: 25–31.

Gibson, G.R., K.P. Scott, R.A. Rastall, K.M. Tuohy, A. Hotchkiss, A. Dubert-Ferrandon, M. Gareau, E.F. Murphy, D. Saulnier, G. Loh, S. Macfarlane, N. Delzenne, Y. Ringel, G. Kozianowski, R. Dickmann, I. Lenoir-Wijnkook, C. Walker and R. Buddington. 2010. Dietary prebiotics: current status and new definition. Food Sci. Tech. Bull. Fuct. Foods 7: 1–19.

Gibson, G.R., R. Hutkins, M.E. Sanders, S.L. Prescott, R.A. Reimer, S.J. Salminen, K. Scott, C. Stanton, K.S. Swanson, P.D. Cani, K. Verbeke and G. Reid. 2017. The International Scientific Association for Probiotics and Prebiotics (ISAPP) consensus statement on the definition and scope of prebiotics. Nat. Rev. Gastroenterol. Hepatol. 14: 491–502.

Gibson, P.S., A. Quaglia, A. Dhawan, H. Wu, S. Lanham-New, K.H. Hart, E. Fitzpatrick and J.B. Moore. 2018. Vitamin D status and associated genetic polymorphisms in a cohort of UK children with non-alcoholic fatty liver disease. Pediatr. Obes. 13: 433–441.

Haussler, M.R., G.K. Whitfield, I. Kaneko, C.A. Haussler, D. Hsieh, J.-C. Hsieh and P.W. Jurutka. 2013. Molecular mechanisms of vitamin D action. Calcif. Tissue Int. 92: 77–98.

Health Canada. Vitamin D. 2012. Available at: https://www.canada.ca/en/health-canada/services/food-nutrition/food-nutrition-surveillance/health-nutrition-surveys/canadian-health-measures-survey/vitamin-nutrition-biomarkers-cycle-1-canadian-health-measures-survey-food-nutrition-surveillance-health-canada-1.html. Accessed August 2018.

Hill, C., F. Guarner, G. Reid, G.R. Gibson, D.J. Merenstein, B. Pot, L. Morelli, R.B. Canani, H.J. Flint, S. Salminen, P.C. Calder and M.E. Sanders. 2014. The international scientific association for probiotics and prebiotics consensus statement on the scope and appropriate use of the term probiotic. Nat. Rev. Gastroenterol. Hepatol. 11: 506–514.

Holick, M.F. 1994. McCollum Award Lecture, 1994: Vitamin D—new horizons for the 21st century. Am. J. Clin. Nutr. 60: 619–630.

Holick, M.F. 2003. Vitamin D: A millenium perspective. J. Cell Biochem. 88: 296–307.

Holick, M.F. 2011a. Vitamin D: A d-lightful solution for health. J. Investig. Med. 59: 872–880.

Holick, M.F. 2011b. Vitamin D: Evolutionary, physiological and health perspectives. Curr. Drug Targets. 12: 4–18.

Holzapfel, W.H. and U. Schilinger. 2002. Introduction to pre- and probiotics. Food Res. Int. 35: 109–116.

Jacobus, C.H., M.F. Holick, Q. Shao, T.C. Chen, I.A. Holm, J.M. Kolodny, G.E. Fuleihan and E.W. Seely. 1992. Hypervitaminosis D associated with drinking milk. N. Engl. J. Med. 326: 1173–1177.

Jin, D., S. Wu, Y.G. Zhang, R. Lu, Y. Xia, H. Dong and J. Sun. 2015. Lack of vitamin D receptor causes dysbiosis and changes the functions of the murine intestinal microbiome. Clin. Ther. 37: 996–1009.

Jones, M.L., C.J. Martoni and S. Prakash. 2013. Oral supplementation with probiotic *L. reuteri* NCIMB 30242 increases mean circulating 25-Hydroxyvitamin D: A post hoc analysis of a randomized controlled trial. J. Clin. Endocrinol. Metab. 98: 2944–2951.

Kabbani, T.A., I.E. Koutroubakis, R.E. Schoen, C. Ramos-Rivers, N. Shah, J. Swoger, M. Regueiro, A. Barrie, M. Schwartz, J.G. Hashash, L. Baidoo, M.A. Dunn and D.G. Binion. 2016. Association of vitamin D level with clinical status in inflammatory bowel disease: A 5-year longitudinal study. Am. J. Gastroenterol. 111: 712–719.

Kong, J., Z. Zhang, M.W. Mush, G. Ning, J. Sun, J. Hart, M. Bissonnette and Y.C. Li. 2008. Novel role of the vitamin D receptor in maintaining the integrity of the intestinal mucosal barrier. Am. J. Physiol. Gastrointest. Liver Physiol. 294: G208–G216.

Kongsbak, M., T.B. Levring, C. Geisler and M.R. von Essen. 2013. The vitamin D receptor and T cell function. Front. Immunol. 4: 148.

Krul-Poel, Y.H., M.M. Ter Wee, P. Lips and S. Simsek. 2017. Management of endocrine disease: The effect of vitamin D supplementation on glycaemic control in patients with type 2 diabetes mellitus: a systematic review and meta-analysis. Eur. J. Endocrinol. 176: R1–R14.

Lai, H.-H., C.-H. Chiu, M.-S. Kong, C.-J. Chang and C.-C. Chen. 2019. Probiotic *Lactobacillus casei*: Effective for managing childhood diarrhea by altering gut microbiota and attenuating fecal inflammatory markers. Nutrients 11: 1150–1164.

Lee, Y.H. and G.G. Song. 2012. Pathway analysis of a genome-wide association study of ileal Crohn's disease. DNA Cell Biol. 31: 1549–1554.

Liu, W., Y. Chen, M.A. Golan, M.L. Annunziata, J. Du, U. Dougherty, J. Kong, M. Mush, Y. Huang, J. Pekow, C. Zheng, M. Bissonnette, S.B. Hanauer and Y.C. Li. 2013. Intestinal epithelial vitamin D receptor signaling inhibits experimental colitis. J. Clin. Invest. 123: 3983–3996.

Martinez, R.C.R., R. Bedani and S.M.I. Saad. 2015. Scientific evidence for health effects attributed to the consumption of probiotics and prebiotics: an update for current perspectives and future challenges. Br. J. Nutr. 114: 1993–2015.

Miheller, P., G. Muzes, I. Hritz, G. Lakatos, I. Pregun, P.L. Lakatos, L. Herszényi and Z. Tulassay. 2009. Comparison of the effects of 1,25 dihydroxyvitamin D and 25 hydroxyvitamin D on bone pathology and disease activity in Crohn's disease patients. Inflamm. Bowel Dis. 15: 1656–1662.

Mohammadi-Sartang, M., N. Bellissimo, J.O.T. Zepetnek, N.R. Brett, S.M. Mazloomi, M. Fararouie, A. Bedeltavana, M. Famouri and Z. Mazloom. 2018. The effect of daily fortified yogurt consumption on weight loss in adults with metabolic syndrome: A 10-week randomized controlled trial. Nutr. Metab. Cardiovas. 28: 565–574.

Moreno-Santos, I., D. Castellano-Castillo, M.F. Lara, J.C. Fernandez-Garcia, F.J. Tinahones and M. Macias-Gonzales. 2017. IGFBP-3 Interacts with the vitamin D receptor in insulin signaling associated with obesity in visceral adipose tissue. Int. J. Mol. Sci. 18: 2349.

National Institute of Health. 2016. Vitamin D Fact Sheet for Consumers. Available at: https://ods.od.nih.gov/factsheets/VitaminD-Consumer/. Acessed August 2019.

Need, A.G., P.D. O'Loughlin, H.A. Morris, P.S. Coates, M. Horowitz and B.E. Nordin. 2008. Vitamin D metabolites and calcium absorption in severe vitamin D deficiency. J. Bone Miner. Res. 23: 1859–1863.

Oh, J., A.E. Riek, I. Darwech, K. Funai, J. Shao, K. Chin, O.L. Sierra, G. Carmeliet, Ostlund, R.E. Carmeliet Jr and C. Bernal-Mizrachi. 2015. Deletion of macrophage vitamin D receptor promotes insulin resistance and monocyte cholesterol transport to accelerate atherosclerosis in mice. Cell Rep. 10: 1872–1886.

Ooi, J.H., Y. Li, C.J. Rogers and M.T. Cantorna. 2013. Vitamin D regulates the gut microbiome and protects mice from dextran sodium sulfate-induced colitis. J. Nutr. 143: 1679–1686.

Peirce, V., S. Carobbio and A. Vidal-Puig. 2014. The different shades of fat. Nature 510: 76–83.

Pilz, S., W. März, K.D. Cashman, M.E. Kiely, S.J. Whiting, M.F. Holick, W.B. Grant, P. Pludowski, M. Hiligsmann, C. Trummer, V. Schwetz, E. Lerchbaum, M. Pandis, A. Tomaschitz, M.R. Grübler, M. Gaksch, N. Verheyen, B.W. Hollis, L. Rejnmark, S.N. Karras, A. Hahn, H.A. Bischoff-Ferrari, J. Reichrath, R. Jorde, I. Elmadfa, R. Vieth, R. Scragg, M.S. Calvo, N.M.

van Schoor, R. Bouillon, P. Lips, S.T. Itkonen, A.R. Martineau, C. Lamberg-Allardt and A. Zittermann. 2018. Rationale and plan for vitamin D food fortification: A review and guidance paper. Front. Endocrinol (Lausanne). 9: 373.

Puupponen-Pimiä, R., A.-M. Aura, K.-M. Oksman-Caldentey, P. Myllärinen, M. Saarela, T. Mattila-Sandholm and K. Poutanen. 2002. Development of functional ingredients for gut health. Trends Food Sci. Tech. 13: 3–11.

Ranadheera, C.S., S.K. Baines and M.C. Adams. 2010. Importance of food in probiotic efficacy. Food Res. Int. 43: 1–7.

Raygan, F., O. Vahidreza, F. Bahmani and Z. Asemi. 2018. The effects of vitamin D and probiotic co-supplementation on mental health parameters and metabolic status in type 2 diabetic patients with coronary heart disease: A randomized, double-blind, placebo-controlled trial. Prog. Neuropsychopharmacol. Biol. Psychiatry 84: 50–55.

Rezazadeh, L., B.P. Gargari, M.A. Jafarabadi and B. Alipour. 2019. Nutrition 62: 162–168.

Ross, A.C., J.E. Manson, S.A. Abrams, J.F. Aloia, P.M. Brannon, S.K. Clinton, R.A. Durazo-Arvizu, J.C. Gallagher, R.L. Gallo, G. Jones, C.S. Kovacs, S.T. Mayne, C.J. Rosen and S.A. Shapses. 2011a. The 2011 report on dietary reference intakes for calcium and vitamin D from the Institute of Medicine: what clinicians need to know. J. Clin. Endocr. Metab. 96: 53–58.

Ross, A.C., C.L. Taylor, A.L. Yaktine and H.B. Del Valle. 2011b. Dietary Reference intakes for Calcium and Vitamin D. National Academies Press,Washington, DC, USA.

Salvucci, E. 2019. The human-microbiome superorganism and its modulation to restore health. Int. J. Food Sci. Nutr. 7: 1–15.

Sanders, K.M., A.L. Stuart, E.J. Williamson, J.A. Simpson, M.A. Kotowicz, D. Young and G.C. Nicholson. 2010. Annual high-dose oral vitamin D and falls and fractures in older women: a randomized controlled trial. JAMA 303: 1815–1822.

Sanders, M.E. and M.L. Marco. 2010. Food formats for effective delivery of probiotics. Ann. Rev. Food Sci. Techonol. 1: 65–85.

Schäffler, H., D.P. Herlemann, P. Klinitzke, P. Berlin, B. Kreikemeyer, R. Jaster and G. Lamprecht. 2018. Vitamin D administration leads to a shift of the intestinal bacterial composition in Crohn's disease patients, but not in healthy controls. J. Dig. Dis. 19: 225–234.

Scotti, G.B., M.T. Afferri, A. De Carolis, V. Vaiarello, V. Fassino, F. Ferrone, S. Minisola, L. Nieddu and P. Vernia. 2019. Factors affecting vitamin D deficiency in active inflammatory bowel diseases. Dig Liver Dis. 51: 657–662.

Simmons, J.D., C. Mullighan, K.I. Welsh and D.P. Jewell. 2000. Vitamin D receptor gene polymorphism: association with Crohn's disease susceptibility. Gut. 47: 211–214.

Singh, P.K., P.R. Van Den Berg, M.D. Long, A. Vreugdenhil, L. Grieshober, H.M. Ochs-Balcom, J. Wang, S. Delcambre, S. Heikkinen, C. Carlberg, M.J. Campbell and L.E. Sucheston-Campbell. 2017. Integration of VDR genome wide binding and GWAS genetic variation data reveals co-occurrence of VDR and NF-κB binding that is linked to immune phenotypes. BMC Genomics 18: 132.

Song, Y., L. Wang, A.G. Pittas, L.C. Del Gobbo, C. Zhang, J.E. Manson and F.B. Hu. 2013. Blood 25-hydroxy vitamin D levels and incident type 2 diabetes: a meta-analysis of prospective studies. Diabetes Care. 36: 1422–1428.

Sordillo, J.E., Y. Zhou, M.J. McGeachie, J. Ziniti, N. Lange, N. Laranjo, J.R. Savage, V. Carey, G. O'Connor, M. Sandel, R. Strunk, L. Bacharier, R. Zeiger, S.T. Weiss, G. Weinstock, D.R. Gold and A.A. Litonjua. 2017. Factors influencing the infant gut microbiome at age 3–6 months: Findings from the ethnically diverse Vitamin D Antenatal Asthma Reduction Trial (VDAART). J. Allergy Clin. Immunol. 139: 482–491.

Sun, J. 2018. Dietary vitamin D, vitamin D receptor, and microbiome. Curr. Opin. Clin. Nutr. Metab. Care. 21: 471–474.

Trivedi, D.P., R. Doll and K.T. Khaw. 2003. Effect of four monthly oral vitamin D3 (cholecalciferol) supplementation on fractures and mortality in men and women living in the community: randomised double blind controlled trial. BMJ 326: 469.

Vrieze, A., E. Van Nood, F. Holleman, J. Salojärvi, R.S. Kootte, J.F. Bartelsman, G.M. Dallinga-Thie, M.T. Ackermans, M.J. Serlie, R. Oozeer, M. Derrien, A. Druesne, J.E. Van Hylckama Vlieg, V.W. Bloks, A.K. Groen, H.G. Heilig, E.G. Zoetendal, E.S. Stroes, W.M. de Vos, J.B. Hoekstra and M. Nieuwdorp. 2012. Transfer of intestinal microbiota from lean donors increases insulin sensitivity in individuals with metabolic syndrome. Gastroenterology 143: 913–916.

Wang, J., L.B. Thingholm, J. Skiecevičienė, P. Rausch, M. Kummen, J.R. Hov, F. Degenhardt, F.-A. Heinsen, M.C. Rühlemann, S. Szymczak, K. Holm, T. Esko, J. Sun, M. Pricop-Jeckstadt, S. Al-Dury, P. Bohov, J. Bethune, F. Sommer, D. Ellinghaus, R.K. Berge, M. Hübenthal, M. Koch, K. Schwarz, G. Rimbach, P. Hübbe, W.-H. Pan, R. Sheibani-Tezerji, R. Häsler, P. Rosenstiel, M. D'Amato, K. Cloppenborg-Schmidt, S. Künzel, M. Laudes, H.-U. Marschall, W. Lieb, U. Nöthlings, T.H. Karlsen, J.F. Baines and A. Franke. 2016. Genome-wide association analysis identifies variation in vitamin D receptor and other host factors influencing the gut microbiota. Nat. Genet. 48: 1396–1406.

Williams, C.E., E.A.Williams and B.M. Corfe. 2018. Vitamin D status in irritable bowel syndrome and the impact of supplementation on symptoms: what do we know and what do we need to know? Eur. J. Clin. Nutr. 72: 1358–1363.

Williams, N.T. 2010. Probiotics. Am. J. Health-Syst Ph. 67: 449–458.

Wortsman, J., L.Y. Matsuoka, T.C. Chen, Z. Lu and M.F. Holick. 2000. Decreased bioavailability of vitamin D in obesity. Am. J. Clin. Nutr. 72: 690–693.

Wu, S., Y.G. Zhang, R. Lu, Y. Xia, D. Zhou, E.O. Petrof, E.C. Claud, D. Chen, E.B. Chang, G. Carmeliet and J. Sun. 2015a. Intestinal epithelial vitamin D receptor deletion leads to defective autophagy in colitis. Gut. 64: 1082–1094.

Wu, S., S. Yoon, Y.G. Zhang, R. Lu, Y. Xia, J. Wan, E.O. Petrof, E.C. Claud, D. Chen and J. Sun. 2015b. Vitamin D receptor pathway is required for probiotic protection in colitis. Am. J. Physiol. Gastrointest. Liver Physiol. 309: G341–G349.

Yang, L., V. Weaver, J.P. Smith, S. Bingaman, T.J. Hartman and M.T. Cantorna. 2013. Therapeutic effect of vitamin d supplementation in a pilot study of Crohn's patients. Clin. Transl. Gastroenterol. 4: e33.

Zhang, Y.G., S. Wu, R. Lu, D. Zhou, J. Wan, J. Zhou, G. Carmeliet, E. Petrof, E.C. Claud and J. Sun. 2015. Tight junction CLDN2 gene is a direct target of the vitamin D receptor. Sci. Rep. 5: 10642.

Zhu, L., S.S. Baker, C. Gill, W. Liu, R. Alkhouri, R.D. Baker and S.R. Gill. 2013. Characterization of gut microbiomes in nonalcoholic steatohepatitis (NASH) patients: a connection between endogenous alcohol and NASH. Hepatology 57(2): 601–609.

7

B-Group Vitamin-Producing Lactic Acid Bacteria

A Tool to Bio-Enrich Foods and Delivery Natural Vitamins to the Host

Marcela Albuquerque Cavalcanti de Albuquerque,[1,2,*] *María del Milagro Teran,*[3] *Luiz Henrique Groto Garutti,*[4] *Ana Clara Candelaria Cucik,*[1,2] *Susana Marta Isay Saad,*[2,4] *Bernadette Dora Gombossy de Melo Franco*[1,2] and *Jean Guy LeBlanc*[3]

Introduction

Vitamins are micronutrients that play an important role as cofactors for several biochemical reactions that occur in all living cells. Although vitamins act as precursors for different coenzymes, humans are not able to produce many of them. Thus, these nutrients must be obtained exogenously, especially from the diet and/or dietary supplements. In this context, B-group vitamins are water-soluble nutrients

[1] Department of Food and Experimental Nutrition, School of Pharmaceutical Sciences, University of São Paulo, São Paulo, Brazil.
[2] Food Research Center, University of São Paulo, São Paulo, Brazil.
Emails: anaclara.candelaria@gmail.com; bfranco@usp.br
[3] Centro de Referencia para Lactobacilos (CERELA-CONICET), San Miguel de Tucumán, Tucumán, Argentina.
Emails: mteran@cerela.org.ar; leblanc@cerela.org.ar
[4] Department of Biochemical-Pharmaceutical Technology, School of Pharmaceutical Sciences, University of São Paulo, São Paulo, Brazil.
Emails: luiz.garutti@gmail.com; susaad@usp.br
* Corresponding author: malbuquerque17@gmail.com

that contribute to maintain the homeostasis of humans including carbohydrate and fat metabolism, which impacts on energy generation, and in the synthesis of DNA and amino acids (Magnúsdóttir et al. 2015, LeBlanc et al. 2017).

In the past few years, people are more concerned about consuming foods that are more natural and that present a health claim. Although B-vitamins are found in many foods, it is known that some of these nutrients are temperature sensitive leading to their destruction during the cooking process (Magnúsdóttir et al. 2015, Albuquerque et al. 2017). Nowadays, it is common to find products fortified with chemically synthesized vitamins. Although this practice is mandatory for some countries such as the USA, Brazil, Argentina, and Canada, it is known that the use of synthetic nutrients to fortify foods may cause some adverse effects to human health (Albuquerque et al. 2016). Consumers who eat balanced diets may ingest a highest amount of the nutrient by consuming fortified foods. In addition, synthetic vitamins require more time to be metabolized by humans than the vitamins in their natural form.

On the other hand, some microorganisms are able to produce some specific bioactive compounds, including B-group vitamins, and studies have shown that different strains of lactic acid bacteria (LAB) and bifidobacteria synthesize *de novo* natural forms of thiamine (B1) (Masuda et al. 2012), riboflavin (B2) (Yépez et al. 2019), folate (B9) (Albuquerque et al. 2016), cobalamin (B12) (Li et al. 2017, Piwowarek et al. 2018), and other vitamins belonging to the B group (LeBlanc et al. 2017). Therefore, the use of vitamin-producing microorganisms may represent a more natural and consumer-friendly alternative to fortification using chemically synthesized vitamins. Additionally, these microorganisms can be used as a tool to increase the vitamin content of foods through the fermentation processes. They can also act as a vitamin bio-supplement by delivering these nutrients throughout the digestion of the host or even by producing B-vitamins directly in the human gut (Rowland et al. 2018).

Considering that the vitamin production by LAB or bifidobacteria is strain-dependent, it is important to point out that depending on the environmental conditions such as pH, temperature, and available nutrients, the microbial vitamin production may be enhanced (Sybesma et al. 2003). Also, the use of genetic engineering techniques may allow these microorganisms to overexpress the production of these nutrients which represents an important technological characteristic for the use of vitamin-producing microorganisms by the food or pharmaceutical industries (LeBlanc et al. 2010).

Therefore, this chapter will discuss the use of LAB as a tool to bio-enrich foods with natural B-vitamins produced during fermentation process and how these beneficial microorganisms can act as bio-supplements to deliver these natural nutrients along the gastrointestinal tract during host digestion. In addition, the use of these vitamin-producing microorganisms to produce vitamins directly in the human gut will be also addressed.

Thiamine

Thiamine is a water-soluble B vitamin also known as vitamin B1 or aneurine. Thiamine is an essential nutrient required by all tissues, and because the human body itself, unlike some microorganisms and plants, cannot synthesize this vitamin, it must be incorporated as part of the diet. This vitamin is mainly found in whole grains, nuts, meats (especially pork), fruits, vegetables and fortified cereals. Nowadays many commercial foods are commonly fortified with thiamine, including breads and cereals. Annual production, by chemical synthesis, is in the order of 3,300 tons (Burdick 1998). Human beings require a minimum of 0.33 milligrams (mg) thiamine for every 1,000 kilocalories of energy they consume (Hoyumpa Jr 1980); therefore a daily intake of about 1.2 mg thiamine is recommended for adults, lower levels are suggested for children, and slightly higher levels for pregnant and breast-feeding women.

Thiamine is needed for the release of energy from carbohydrates and is also involved in the normal functioning of the nervous system and the heart. Thiamine is present in the human body as free thiamine and as different phosphorylated forms; thiamine pyrophosphate (ThDP) is the biologically active form, working as an essential cofactor in all forms of life and has a key role in carbohydrate, branched-chain amino acid and other organic molecules metabolism. ThDP is synthesized *de novo* by certain bacteria, archaea, yeast, fungi, plants, and protozoans (Begley et al. 1999). Other organisms, such as humans, rely upon thiamine transport and salvage to obtain sufficient amounts for their metabolic reactions.

Thiamine deficiency is currently an installed condition in different populations and may result from inadequate thiamine intake, increased requirements, excessive loss of the vitamin from the body, consumption of anti-thiamine factors in food, or a combination of these causes. A reduction in thiamine levels can interfere with numerous cellular functions, which is why this vitamin is implicated in several human diseases including Alzheimer's, diabetes, dementia, depression and Beriberi disease. Chronic alcohol consumption, the main cause of thiamine deficiency in industrialized countries, acts by causing inadequate nutritional thiamine intake, decreased absorption in the gastrointestinal tract, and impairs its utilization in the cells (Martin et al. 2003) that can lead to serious brain disorders, including Wernicke-Korsakoff syndrome, which is found predominantly in alcoholics.

Thiamine content in different foods and products

B-vitamins are found in many food products, but they are water-soluble and many of them are temperature sensitive; thus, these vitamins can easily be removed or destroyed during the cooking process. According to Lebiedzińska and Szefer (2006) the highest contents of thiamine were present in dry sunflower seeds, soybeans and dry sesame seeds (1.05–0.72 mg/100 g). In addition, grain-cereal products were rich in thiamine, especially barley, buckwheat groats and millet (around 0.30 mg/100 g). Rice is a good source of vitamin B1, especially brown rice containing around 0.22 mg/100 g. In other kinds of rice and rice products, the thiamine content decreases significantly and only enriched rice groats are a very good source of thiamine (0.94 mg/100 g). Thiamine is the vitamin most sensitive to canning and its content decreases markedly in processed products, which is in contrary to whole grain-cereal products. Martin-Belloso and Lianos-Barriobero (2001) have reported that vitamin B1 content decreased up to 53% during the canning process.

Consumers are becoming increasingly health conscious and therefore more alert to their food choices. The production of fermented food products with elevated levels of B-vitamins, such as thiamine, increase both their commercial and nutritional value, eliminate the need for subsequent fortification with the essential vitamins and could reduce the incidence of inadequate vitamin intake (Burgess et al. 2009). Moreover, the concept of *in situ* fortification by bacterial fermentation opens a way for development of new food products directed to specific groups in society such as the elderly and adolescents.

In relation to fermented food, a study of fruit milk drinks available on the Polish market (Drywień et al. 2015) showed that a fermented beverage of natural flavour had the highest amount of thiamine (17.4 ± 0.98 μg/100 g), whereas a fermented multi-fruit milk drink had lower concentrations of the vitamin (13.3 ± 1.02 μg/100 g).

Thiamine-producing LAB

Because human cells are not capable of producing B-vitamins, including thiamine, they must obtain such vitamins either from the human diet, including fermented foods, or directly from the gut microbiota, which can help supply its host with several nutrients (Magnúsdóttir et al. 2015).

Lactic acid bacteria (LAB), natural members of the human gastrointestinal microbiota, are an industrially important group of microorganisms used for producing different fermented foods and also, many strains are considered probiotics, which can improve intestinal microbiota and impart beneficial effects on human health (Masuda et al. 2012). Microbial vitamin production could be a suitable strategy to achieve natural enrichment of thiamine in fermented foods, especially from vegetable sources; or for *in situ* vitamin production in the gut with probiotic LAB. Moreover, increasingly LAB are also used, through metabolic engineering, for the production of different compounds, including B-vitamins such as folate and riboflavin, which could be also used to improve microbial production of thiamine (Hugenholtz et al. 2002, Sybesma et al. 2003).

Although it has been reported that fermentation with bifidobacteria can produce an increase in the content of thiamine (Hou et al. 2000), few studies have investigated the thiamine production by LAB (Table 1). Masuda et al. (2012) determined the production of thiamine in cultures of LAB isolated from nukazuke, a traditional Japanese pickle. LAB were screened for their ability to grow in a thiamine-free medium after 24 h of incubation. The results showed that from 283 strains evaluated, 76 were able to grow in absence of thiamine and no isolates produced high quantities of extracellular thiamine, only levels near the detection limit. The 53 reference strains of LAB evaluated did not show high production of this vitamin. In this same study, the intracellular thiamine was examined for some strains showing levels of 0.1–0.2 ng/mg fresh-weight of cells; several times higher than extracellular concentrations. It was show that LAB produced about 10 μg/L of thiamine, 10–20% of this amount was released into the medium and the remaining 80–90% was maintained in the cells.

In another study, the ability to synthesize thiamine was investigated with 24 bifidobacteria strains of five species derived from human faeces, using cultures grown for 48 h in a semi-synthetic medium and determining intracellular thiamine and also extracellular liberation of the vitamin in the supernatant fluid (Deguchi et al. 1985). Most of the studied bifidobacteria strains could synthesize thiamine, a large portion of the vitamin was excreted into the medium, and the concentrations accumulated varied widely among different species or strains. Thiamine was accumulated (0–0.25 μg/mL)

Table 1: Intra- and extra-cellular thiamine production by several strains of LAB.

Microorganisms	Intracellular thiamine	Extracellular thiamine	References
	(μg/L of broth)		
Lactobacillus (*L.*) *curvatus* L-6	7.3	1.2	Masuda et al. 2012
L. curvatus L-20	9.4	1.3	Masuda et al. 2012
L. plantarum L-82	9.8	1.1	Masuda et al. 2012
Lactococcus lactis ssp. *cremoris* L-17	5.9	1.1	Masuda et al. 2012
Leuconostocmesenteroides ssp.*cremoris* L-50	6.8	–	Masuda et al. 2012
Pediococcusparvulus L-12	6.8	0.7	Masuda et al. 2012
Bifidobacterium (*B.*) *longum* 5C-1	130	nd	Deguchi et al. 1985
B. bifidum E-319	250	nd	Deguchi et al. 1985
B. bifidum	230–270	170–210	Deguchi et al. 1985
B. infantis	210–250	180–210	Deguchi et al. 1985
B. breve	100–170	50–100	Deguchi et al. 1985
B. longum	100–150	30–60	Deguchi et al. 1985
B. adolescentis	20–40	10–30	Deguchi et al. 1985

nd: not determined

in all the strains of *Bifidobacterium* (*B.*) *bifidum* and *B. infantis* as well as in many strains of *B. breve* and *B. longum* but the concentrations were significantly higher in the former species (higher-accumulators) than in the latter species (lower-accumulators). On the other hand, thiamine was not detected in most strains of *B. adolescentis* or in some strains of *B. breve* and *B. longum* (non-accumulators), which also required thiamine for maximal growth in a complete synthetic medium. Furthermore, as an approach to determine whether a regulatory mechanism is involved in thiamine synthesis in bifidobacteria, the effect of the addition of exogenous thiamine to the medium on the abilities to synthesize thiamine was investigated. The level of thiamine accumulated during growth was significantly reduced in lower-accumulator strains; whereas the addition of thiamine had almost no effect in high-accumulator strains. These results strongly suggest that the thiamine biosynthetic pathway is controlled by the end-product, thiamine, in lower-accumulator strains but not in higher-accumulator strains.

Genetic focus

Regarding genetic aspects, a few studies about the potential production of thiamine by LAB are available. Thiamine biosynthesis has been well studied in other prokaryotes, including *Escherichia coli*, *Salmonella*, and *Bacillus subtilis* (Begley et al. 1999). The complete thiamine biosynthetic pathway has been described in *B. longum* ssp. *infantis* ATCC15697, isolated from human breast milk (Sela et al. 2008). Additionally, the genome analysis in the strain *L. reuteri* ATCC 55730 predicted the presence of a complete pathway for thiamine biosynthesis and it was the first report of thiamine biosynthesis in lactobacilli (Saulnier et al. 2011). In this latter study, the thiamine synthesis pathway does not seem to be acquired from other bacteria because the genes for the thiamine synthesis pathway are not located in a single operon but rather in multiple locations across the genome. Further studies are necessary to confirm whether thiamine is effectively produced in this strain.

Teusink et al. (2005) tested experimentally the vitamin requirements of *L. plantarum* WCFS1, and compared the results with the known pathway-genome database; in most cases, experimental results agreed with the final reconstruction. The addition of thiamine to the medium was not required for growth, indicating that the pathway related to the biosynthesis of this vitamin should be complete and active. However, the pathway appeared to be incomplete because 3 of 10 required reactions for thiamine biosynthesis were not coupled to a gene, but most gaps could be filled by combining sequence information from different species and other alternative genes might be able to complete thiamine biosynthesis.

Food fermentation and bio-enrichment

One possible advantage of food fermentation is the possibility to increase the content of some nutrients such as thiamine, in the fermented product, improving its nutritional value. Some probiotic bacteria (including *Lactobacillus* and *Bifidobacterium*) are producers of B-vitamins thus, foods fermented by selected strains could be an additional source of these components (Albuquerque et al. 2016). Moreover, prebiotic supplements are capable of intensifying bacterial growth and potentially vitamin synthesis in the gut as well as in foods (LeBlanc et al. 2017).

Bifidobacteria are one of the predominant intestinal microorganisms in human beings and the importance of using these microorganisms as dietary supplements is described for different age groups of humans. These microorganisms were reported to be capable of synthesizing and liberating many kinds of B-vitamins (Rasic and Kurmann 1983). However, Deguchi et al. (1985) found that the concentrations of vitamins accumulated by bifidobacteria in the culture media, especially thiamine, varied widely among different species or strains. Furthermore, feedback repression or inhibition in bacteria may also affect the synthesis. According to Hou et al. (2000), thiamine content (examined through HPLC method) was increased in soymilk fermented for 48 h with two bifidobacteria, *B. longum* B6 and *B. infantis* CCRC 14633, separately.

Thiamine content during food shelf life

Food-grade bacteria hence make possible to fortify some raw food products with B-vitamins by a natural process. On the other hand, the levels of some vitamins can be reduced during fermentation as some LAB absorb or consume them; this rate depend on the type of bacterial strain and on the fermentation conditions (LeBlanc et al. 2011). Drywień et al. (2015) studied the effects of the storage time and the type of prebiotic addition (inulin or oligofructose) on the thiamine concentration in a banana-milk drink fermented with the probiotic strain *L. casei* KNE-1. Thiamine content was determined by thiochrome method and it was found that the storage time after the end of the fermentation process did not increase the content more than the output level. The type of prebiotic affected the synthesis of thiamine by *L. casei* KNE-1, which was higher in the drink when oligofructose was added. The growth of LAB may be stimulated by the addition of certain prebiotics and consequently contribute to increase the vitamin production during fermentation in the cases where thiamine production is growth associated.

Analytical methods for thiamine determination

Nowadays there is a wide range of methods for thiamine detection and quantification, which vary in their sensibility due to different thiamine properties. Nevertheless, highly sensitive and specific detection of thiamine remains an analytical challenge, as very low levels of thiamine need to be detected in environmental and human samples, and many phosphorylated forms need to be differentiated. A few approaches that are available for thiamine analysis include fluorescence detection of the oxidized form of thiamine, thiamine requirement for microbiological growth, sensing based on biorecognition, and high performance liquid chromatography (HPLC) (Edwards et al. 2017).

Analytical methods based on the chemistry of thiamine take advantage of its cationic charge (Herr 1945), detection relying on its activity in the UV range (Gratacós-Cubarsí et al. 2011), ability to be oxidized to thiochrome and detected via fluorescence (Fujiwara and Matsui 1953), as well as polarity differences between thiamine, its phosphate esters, and other species (Zajicek et al. 2005). On the other hand, methods based on the biological function of thiamine, such as microbiological assays and biorecognition by enzymes, take advantage of the role of thiamine as an essential vitamin for the growth of many microorganisms in its diphosphate form, acting as cofactor for numerous enzymes.

In the microbiological assay, the most common test organisms employed are *L. fermentum* ATCC 9338 (Olkowski and Gooneratne 1992) and *Weisellaviridescens* ATCC 12706 (Deibel et al. 1957) because thiamine is an essential micronutrient for growth of these auxotrophic organisms. In this method, a culture medium free from thiamine is used which contain all the other essential nutrients and vitamins for the growth of the test organism; the addition of thiamine in increasing concentrations gives a correlative growth response.

Riboflavin

Vitamin B2, or riboflavin, is a water-soluble vitamin belonging to a group of yellow and fluorescent pigments named flavins, which are organic compounds that have the isoalloxazine tricyclic ring. In addition, this vitamin has a ribitol chain, which is a reduced ribose. Riboflavin is also the precursor of two co-enzymes, namely FMN (Flavin mononucleotide, or Riboflavin-5'-phosphate) and FAD (Flavin-Adenine dinucleotide), both of which are essential to the cell respiratory functions and have roles in cellular growth and development, and in the metabolism of fats and steroids (Ball 2004).

Vertebrates do not commonly synthesize this vitamin; however, it is widely spread in foodstuffs like kidneys, liver, dairy products and eggs, where it is bound to enzymes and proteins. Thus, it is essential to consume this vitamin in sufficient amounts in order to maintain the well-being and functioning of human cells and metabolism (Combs 2008). Compared to other B-group vitamins, riboflavin has a lower solubility in water, and provides a vibrant yellow color when in solution (Ball 2004).

Human riboflavin deficiency

Ariboflavinosis is the deficiency of riboflavin, and is characterized by inflammation of the mucous membranes, alopecia, anemia, corneal vascularization and deficient metabolism. In children, riboflavin is essential for the correct development of the brain and other critical tissue (Thakur et al. 2016).

Although riboflavin is available in several foods, there is a risk in the global population for ariboflavinosis. As dietary customs can vary and certain dieting practices might actually hampers the body ability to absorb riboflavin or even not have available riboflavin for absorption. This condition is primary concern in populations that suffer from substance abuse and/or undernourishment (Wacker et al. 2000, Combs 2008, Thakur et al. 2016).

Production of riboflavin

Riboflavin synthesis is mainly handled by microbial production at an industrial level. Fungi and modified bacteria are used to overproduce the vitamin, which must be purified and separated from other compounds (Ball 2004, Combs 2008). This process, although efficient, brings a certain amount of cost and complexity for the industry, and this cost is then applied to the final product in which this vitamin will be inserted. In contrast, some microorganisms are naturally riboflavin-producers. The riboflavin production is observed in very diverse microorganisms, ranging from yeasts to algae. Recently, the use of lactic acid bacteria as riboflavin-naturally producers has been explored (Thakur et al. 2015, Pacheco da Silva et al. 2016).

Ibrahim et al. (2014) identified producers among lactic acid bacteria isolated from dairy products and found at least four strains that were able to compete with commercially available overproducers of riboflavin. Juarez del Valle et al. (2014) screened different lactic acid bacteria to investigate their ability to produce riboflavin *de novo*. The authors used the riboflavin-producer strains to bio-enrich a soymilk that was administrated for the treatment of colitis in a murine model (Levit et al. 2018).

Food application

To address the problem of ariboflavinosis and other diseases, one strategy is the production of novel foods for the purpose of supplementing the normal diet with riboflavin supplements or even as a means to replenish a deficient diet.

Not all media are ideal for the production of riboflavin, although this might be mitigated by the overproducing nature of the specific strain (Thakur et al. 2017). Juarez del Valle et al. (2016) used a strain of *L. plantarum* to ferment soymilk. When applied to a murine model with ariboflavinosis, the bio-enriched fermented soymilk was capable to revert the animal riboflavin deficiency when a food portion that represents 28% of the daily recommended intake of the vitamin was administered to the animals. Thakur et al. (2016) assessed the technological properties of putative riboflavin producing lactobacilli isolated from dairy and non-dairy products when applied to milk and whey based media. The authors concluded that the strains evaluated can be applied to a fermented dairy product enriched in riboflavin. Pacheco da Silva et al. (2016) isolated lactic acid bacteria from goat cheese and goat milk and their ability to produce riboflavin was assessed. The authors observed that the strains were able to produce both folate and riboflavin and increase the vitamin content when used to ferment goat milk. Rajendran et al. (2017) studied the production of riboflavin in idli batter, a regional cake from India, testing both *Saccharomyces boulardii* and *Lactococcus* (*Lc.*) *lactis* strains for riboflavin and folate production. In this study, they found that *Lc. lactis* can enhance folate concentration, but consumes riboflavin, and *S. boulardii* enhances both folate and riboflavin concentrations.

In this way, the use of LAB, including probiotic strains as riboflavin producers in food can be considered a promising approach, providing a natural fortification of the products.

Folate

As the others B-vitamins, folate (B9) is responsible for many important functions of the human body. It is a micronutrient vital for normal cellular functions, acting as acceptor and donor of carbon units, part of important metabolic pathways such as the DNA cycle. Folate is a term to nominate a family of chemically similar compounds derived from pteroic acid and, among them, the polyglutamates, which are the natural forms present in foods or produced by microorganisms. The most common form is 5-methyltetrahydrofolate (5-MTHF), which is the active form of folate in the body. In contrast, folic acid, a monoglutamate, refers to the fully oxidized chemical compound which does not exist in natural foods and is widely employed to fortify foods (LeBlanc et al. 2011).

Although the natural forms of folates are common in foods, their concentration is low and they are unstable, with up to 30% loss as a result of food processing. The foods with the highest amounts of folates are bovine liver, leafy green vegetables, and some oilseeds. In this way, the daily dietary intake of folates is generally lower than the dosage recommended and the inadequate folate intake lead to deficiency which can cause serious health problems such as neural tube defects, some forms of anemia, and others adverse health conditions such as cardiovascular disease. According to the World Health Organization, the recommended dietary allowance (RDA) of folate for an average adult is 400 µg per day (FAO/WHO 2002).

Therefore, the deficiency of this vitamin is considered a public health problem and to avoid it, food fortifications programs using synthetic form of this vitamin were adopted in many countries such as United States, Canada, Argentina, Brazil, and Chile. However, these fortification programs are controversial and not universally accepted. Countries like Norway and United Kingdom do not adopted mandatory fortification due to concerns about several side effects caused by excessive folic acid intake, such as masking vitamin B12 deficiency as the latter leads to the progression of irreversible neurological symptoms and other effects. In addition, recent studies suggest a relation between the concentration of unmetabolized folic acid and some types of cancer (Delchier et al. 2013, Scaglione and Panzavolta 2014, Laiño et al. 2015, Moll and Davis 2017, Meucci et al. 2018, Naderiand House 2018).

Studies suggest that some lactic acid bacteria are able to produce folate *de novo*. This ability emerges as an alternative to raising the intake of folate by the population, without the adverse effects caused by the consumption of the synthetic folic acid. The use of lactic acid bacteria to improve folate status in food by fermentation process represents an alternative to the development of new functional foods bio-enriched with this natural vitamin bio-synthesized by food grade microorganisms (Albuquerque et al. 2017).

Folate production by lactic acid bacteria

The biosynthesis of folate is a characteristic of some microorganisms such as bacteria. Eukaryotes, such as mammals, have a membrane-associated active transport system for folate uptake and they are unable to produce folate *de novo*. Prokaryotes and plants have *de novo* biosynthesis pathway of folate that involves six key genes for the production: *fol*E (GTP cyclohydrolase I), *fol*Q (dihydroneopterin triphosphate pyrophosphohydrolase),*fol*K (2-amino-4-hydroxy-6-hydroxy-methyldihydropteridine pyrophosphokinase), *fol*P (dihydropteroate synthase), *fol*C, and *fol*A (dihydrofolate reductase).

Many species of microorganisms can produce folate, whereas the LAB are the most studied group due to their functional and technological characteristics, and their GRAS (generally recognized as safe) status. *Streptococcus* (*S.*) *thermophilus* is usually used as starter culture for food production and is one of the most important folate-producing species (Iyer and Tomar 2009). Although all the published genomes demonstrate the presence of the six key genes of folate production in this species, not all members produce folates; once the folate production by LAB is strain-dependent and also dependent of environmental conditions (Albuquerque et al. 2016). Some folate-producing LAB are described in Table 2.

Folate production may be affected by some factors such as temperature, pH conditions, and presence of others strains in the environment (Sybesma et al. 2003). Lãino et al. (2017) evaluated the production of folate by *S. thermophilus* strains in different conditions, and found out that at 42°C, the production of the vitamin was 40% higher than at 37°C. This study also observed the influence of pH and showed that at pH 6, the folate biosynthesis was 2–3 folds higher than at others pH values. The pH effect was also observed by Divya et al. (2015). The authors evaluated the folate production by *Lactococcus lactis* at pH 5 and 6, and observed that the production of the vitamin was 37% higher at pH 6 than at pH 5. In addition, Lãino et al. (2018)

Table 2: Folate-producing lactic acid bacteria.

Microorganism	Folate production (ng/mL)	Reference
Streptococcus thermophilus	44–837	Iyer et al. 2010, Laiño et al. 2013, Pacheco et al. 2016, Albuquerque et al. 2017, Meucci et al. 2017
Lactobacillus spp.	29–748	Lãino et al. 2012, Kodi et al. 2015, Albuquerque et al. 2016, Greppi et al. 2017, Mosso et al. 2018
Lactococcus lactis subsp. *lactis*	162–262	Sybesma et al. 2003, Divya and Nampoothiri 2015, Pacheco et al. 2016
Bifdobacterium spp.	78–1223	Daimmo et al. 2014, Albuquerque et al. 2016, Pompei et al. 2007

demonstrated that the production of folate by *S. gallolyticus* subsp. *macedonicus* under pH 6 was 2-fold higher than in free pH conditions.

It is important to screen the strains and find the best cultivation conditions to improve the production of folate by LAB in order to bio-enrich and add nutritional values to fermented products. Albuquerque et al. (2017) demonstrated that one portion (100 mL) of the soymilk supplemented with passion fruit by-products and the prebiotic fructo oligosaccharides (FOS), and fermented by *St. thermophilus* and *L. rhamnosus* LGG would contribute to approximately 45% riboflavin RDA for adults. Therefore it is possible to increase the intake of the vitamin by the population using a the microbial production of vitamins, which is a sustainable and cheap bioprocess, as a tool to bio-enrich foods (Albuquerque et al. 2017, Dias et al. 2017, Lãino et al. 2017).

Methods of quantification

The analysis of vitamins, as it was explained before, is not an easy process due to the low concentration of this nutrient in food or other samples, complexity of matrix, and low stability of this compound. The quantification of folates is more complex due to the diversity of folate forms. There are several methods for folate quantification including biological, microbiological, radio binding, immuno assay, eletrochemical, spectrophotometric, and chromatographic methods such as gel or high-pressure liquid chromatography (HPLC). Among these methods, the most traditional and the only one that has the official status for the determination of food folates by American Association of Analytical Chemists (AOAC) is the microbiological assay (MA). The microbiological assay consists on the growth of *L. rhamnosus* (NCIMB 10463 or ATCC 7469) which is used as an indicator organism. This method can respond to all forms of folates although the response decreases as the number of glutamyl residues linked to the pteroyl group increases. The HPLC is widely used but has as limitation the complex sample extraction and purification procedures which cause loss of sensitive derivatives of the vitamin that lead to decrease the folate value. The liquid chromatography with mass spectrometry (LQMS/MS) is also used, but it can measure only 5 folate forms: 5-methyltetrahydrofolic acid (5CH3THF), FA, 5-formyltetrahydrofolic acid (5CHOTHF), tetrahydrofolic acid (THF), and 5,10-methenyltetrahydrofolic acid (5,10CHTHF) (Tamura et al. 1997, Shrestha et al. 2000, Iyer et al. 1999).

Bioavailability

The concept of bioavailability is defined as the proportion of a food compound that is absorbed by the organism and achieves the systemic circulation. It depends on two mainly factors: (1) bioaccessibility, that refers to the proportion of a food compound released from the food matrix in a absorbable form in the gastrointestinal tract; (2) bioconversion, that is the fraction of the component converted to the active form. The bioefficacy is the sum of bioavailability and bioconversion (Marze et al. 2017).

The bioavailability of natural folates produced by LAB and folic acid is different. Natural folates can be impaired by the polyglutamate chain present in its structure (Mckillop et al. 2006, LeBlanc et al. 2007). It is estimated that the folate bioavailability is 50% of that presented by folic acid; however, there are many divergences in this regard in the literature and may vary from 10% to 98%, depending on the evaluation methodology. Considering the bioaccessibility, folic acid is more bioaccessible than the natural form due to its smaller polyglutamate chain. In contrast, if the bioconversion is evaluated, the folic acid is less bioavailable so that the liver enzyme converts it into the active form. This reaction is saturable; and as a difference, it is not necessary to convert natural folates, they are already in active form (Scaglione and Panzavolta 2014). These differences can also be attributed to other factors like population heterogeneity, research duration, food matrix composition, in addition to the different forms of folates (Winkels et al. 2007, Ohrvik and Witthoft 2011, Scaglione and Panzavolta 2014).

A growing concern about the bioavailability of folic acid has become the subject of some studies due to the presence of unmetabolized folic acid in the population's blood circulation. It occurs due to polymorphisms of the gene that encodes for the enzyme that converts the folic acid into the active form in the liver, and also due to the excessive intake of folic acid that leads to saturation of the reaction (Saini et al. 2016, Moazen et al. 2017, Rees et al. 2017). Foods or drugs that alter intestinal pH may alter the bioavailability of folate, as well as other factors such as the presence of folate antagonists, and morphological and functional changes in the intestine (LeBlanc et al. 2007).

B Vitamin-Producing LAB as Source of Natural Vitamins for the Host

As stated previously, humans lack the ability to biosynthesize many micronutrients, including B-vitamins. Therefore, to keep the homeostasis of the biochemical reactions, that are vital for the metabolism, humans need to obtain vitamins from exogenous sources such as foods and/or synthetic supplements. Several conditions may increase the human demand for vitamins including malnutrition, poor ingestion of food, low quality of the diet, pathological conditions, and pregnancy, among others (Acevedo-Rocha et al. 2019). In addition, as many B-vitamins are sensitive to different extrinsic conditions such as light, pH, and temperature, large amounts of these nutrients may be lost during cooking and processing contributing to decrease in the recommended daily intake of these micronutrients (Albuquerque et al. 2017).

Processes including chemical synthesis or biotechnology are commonly used to produce vitamins. These processes are important to supply part of nutritional demands of consumers, which is associated with up to annual increase in the global B-vitamin market, as mentioned by Acevedo-Rocha et al. (2019). About half of the adult US population uses some form of dietary supplements, and although there are regional, cultural, and economic differences, a similar prevalence can be verified in many other countries (Maughan et al. 2018). For them, the intake of dietary supplements aims to complement the intake of nutrients from food consumption and the promotion of health and wellness.

Among the dietary supplement categories, vitamin/mineral supplements correspond to 98% of the dietary supplements consumed (CRN Consumer Survey 2018). Dietary supplements are often used by individuals who already have nutrient-rich diets; in particular, older women use multiple supplements, which can increase the potential for oversupplementation with the consequent excessive intake of certain nutrients. Although the chemical synthesis of some B-vitamins is a cheap and fast process, the human body does not process them as efficiently as the natural vitamins, which can cause the accumulation of these synthetic micronutrients in the host's body (Bailey and Ayling 2009). Additionally, many countries stated mandatory programs to fortify foods with micronutrients, aiming to decrease public health problems related to vitamin deficiencies and the use of chemically synthesized forms of vitamins, such as folic acid (B9), are largely employed. In this regard, considering the fortification of many food products with folic acid and that consumers who keeps a nutritional balanced diet may intake an excessive amount of this vitamin, some adverse effects, such as masking the clinical manifestations of B12 vitamin deficiency may occur (Laiño et al. 2015).

Therefore, as previously discussed, an alternative to the consumption of synthetic vitamins as dietary supplements to improve nutrients intake and contribute to the management of nutritional deficiencies is the use of LAB as a cell factory to produce natural forms of different B-vitamins which can be used to formulate bio-enrich foods (Masuda et al. 2012, Albuquerque et al. 2016, Juarez del Valle et al. 2014). Many studies have focused on food bio-enrichment through the fermentation process, where food grade LAB are used to improve B-vitamin content of fermented products (Albuquerque et al. 2019, Yépez et al. 2019) and to improve B-vitamin status in vitamin-deficiency *in vivo* models. Thus, Laiño et al. (2015) used fermented milk bio-enriched with folates produced by LAB to improve the folate status in rodent's depletion-repletion and complete deficiency models. The authors observed that the bio-enriched fermented milk was able to improve the folate status and prevent folate deficiency. Additionally, they proposed that bio-enriched fermented milk could improve the natural folate intake being consumed as part of the normal diet. LeBlanc et al. (2006) used *Propionibacterium freudenreichii* to produce riboflavin (B2) in a bio-enriched fermented milk. This fermented milk was administrated to riboflavin deficient rats and was able to improve their vitamin status eliminating most physiologic manifestations of ariboflavinosis. Juarez del Valle et al. (2016) also observed an improvement in the riboflavin status of deficient rats when these animals were fed with fermented soymilk bio-enriched with riboflavin produced by a strain of *Lactobacillus plantarum*.

Beyond the use of food grade microorganisms, such as LAB, to improve B-vitamin content of foods and contribute to improve the B-vitamin status of consumers, the use of these beneficial microorganisms as bio-supplements to deliver natural B-vitamins to the host is still underexplored. The use of LAB as source of natural B-vitamins could add economic and functional value to new bio-products. Considering that there is a movement among consumers for the consumption of more natural products, the use of microorganisms as source of natural nutrients could contribute to improve-vitamin levels in deficient people. These vitamins are normally stored inside the cells and released by direct diffusion, using specific transporters in the cell membrane or via cellular lysis either in their growth media or inside the gastrointestinal tract of the host, making these strains ideal candidates for the *in situ* delivery of B-group vitamins (LeBlanc et al. 2017).

LeBlanc et al. (2006) used a depletion-repletion rat bioassay to evaluate the impact of *Lactococcus lactis* strains intake on the bioavailability of riboflavin. The use of the *Lactococcus lactis* as source of natural riboflavin, especially the engineered strains that overproduced B2, improved the vitamin status and the growth rates of depleted rats, which represents a possibility to use food grade microorganisms containing nutrients produced in their intracellular compartment as bio-supplements to improve B-vitamin status in deficient individuals. Additionally, LeBlanc et al. (2010) also observed that the production of folate could be overstimulated in engineered *Lactococcus lactis*. These microorganisms could be used as bioavailable sources of folates, improving the status of this vitamin in rats deficient in this vitamin. The use of LAB as source of intracellular B-vitamins could be extended to humans as bio-supplements to deliver natural B-vitamins to them.

It is important to point out that studies regarding the use of LAB as a natural source of thiamine to deliver this nutrient to the host are lacking. The use of genetic engineering to overexpress the thiamine production by LAB could contribute to produce bio-enriched foods and to use the microorganisms as bio-supplement to deliver natural thiamine to the host.

The use of probiotic LAB and/or bifidobacteria also represents a strategy to improve B-vitamin content in the gut. These functional microorganisms are able to adhere to the intestinal mucosal modulating the host immunity and improving the colonic microbiota homeostasis. Considering that vitamin deficiency may increase the risk of developing diseases, such as allergy and inflammatory diseases, human gut microbiota may supply its host with several nutrients, including B-vitamins. Although the effects of vitamins produced by human gut microbiota need to be further studied, especially the amounts of vitamins produced in the gastrointestinal tract, Magnúsdóttir et al. (2015) suggests that a symbiosis exists amongst gut microorganisms for B-vitamin biosynthesis.

Conclusion

This chapter discussed the use of LAB as cell factories to produce natural B-vitamins (thiamine, riboflavin, and folates). The use of these vitamin-producing microorganisms can be applied as tools to bio-enrich foods through fermentation process. This technology represents an alternative to the chemical fortification of foods with synthetic vitamins. We also discussed that bio-enriched foods are promising alternatives for the low intake of vitamins, especially folates and riboflavin, improving the vitamins status in *in vivo* vitamin deficiency models. This is a promising field and should be further explored by the food industry. Additionally, the use of LAB to delivery natural B-vitamins to the host was also addressed. Although information regarding the use of these microorganisms as bio-supplements is scarce, they could be explored by the food supplement industry as an alternative to the use of synthetic multivitamin products. Also, many of these microorganisms may be potential probiotics candidates which could contribute to modulate the human intestinal microbiota, improving vitamin levels in the gut as well as providing the host beneficial effects.

Acknowledgments

The authors would like to thank CONICET and ANPCyT (Argentina), and FAPESP (project #2018/10540-6) for their financial support and fellowships. This chapter was developed with the support of the Food Research Center (FoRC/FAPESP/CEPID 2013/07914-8).

References

Acevedo-Rocha, C.G., L.S. Gronenberg, M. Mack, F.M. Commichau and H.J. Genee. 2019. Microbial cell factories for the sustainable manufacturing of B vitamins. Curr. Opin. Biotech. 56: 18–29.

Albuquerque, M.A.C., R. Bedani, A.D.S. Vieira, J.G. Leblanc and S.M.I. Saad. 2016. Supplementation with fruit and okara soybean by-products and amaranth flour increases the folate production by starter and probiotic cultures. Int. J. Food Microbiol. 236: 26–32.

Albuquerque, M.A.C., R. Bedani, J.G. LeBlanc and S.M.I. Saad. 2017. Passion fruit by-product and fructooligosaccharides stimulate the growth and folate production by starter and probiotic cultures in fermented soymilk. Int. J. Food Microbiol. 261: 35–41.

Albuquerque, M.A.C., D.S. Yamacita, R. Bedani, J.G. LeBlanc and S.M.I. Saad. 2019. Influence of passion fruit by-product and fructooligosaccharides on the viability of *Streptococcus thermophilus* TH-4 and *Lactobacillus rhamnosus* LGG in folate bio-enriched fermented soy products and their effect on probiotic survival and folate bio-accessibility under *in vitro* simulated gastrointestinal conditions. Int. J. Food Microbiol. 292: 126–136.

Bailey, S.W. and J.E. Ayling. 2009. The extremely slow and variable activity of dihydrofolate reductase in human liver and its implications for high folic acid intake. PNAS 106: 1524–1529.

Ball, G.F.M. 2004. Vitamins: Their Role in the Human Body. [s.l.] Blackwell Publishing Ltd., Oxford, UK.

Begley, T.P., D.M. Downs, S.E. Ealick, F.W. McLafferty, A.P.G.M. Van Loon, S. Taylor, N. Campobasso, H.-J. Chiu, C. Kinsland and J.J. Reddick. 1999. Thiamin biosynthesis in prokaryotes. Arch. Microbiol. 171(5): 293–300.

Burdick, D. 1998. Thiamine. Kirk-Othmer Encyclopedia of Chemical Technology 25: 152–171.

Burgess, C.M., E.J. Smid and D. van Sinderen. 2009. Bacterial vitamin B2, B11 and B12 overproduction: An overview. Int. J. Food Microbiol. 133(1-2): 1–7.

Capozzi, V., P. Russo, M. Dueñas, P. López and G. Spano. 2012. Lactic acid bacteria producing B-group vitamins: a great potential for functional cereals products. Appl. Microbiol. Biot. 96: 1383–1394.

Combs, G. 2007. The Vitamins: Fundamental Aspects in Nutrition and Health, 3rd edn. Academic Press.

CRN Consumer Survey. 2018. Available at https://www.crnusa.org/CRNConsumerSurvey.

Deguchi, Y., T. Morishita and M. Mutai. 1985. Comparative studies on synthesis of water-soluble vitamins among human species of bifidobacteria. Agric. Biol. Chem. 49: 13–19.

Deibel, R.H., J.B. Evans and C.F. Niven. 1957. Microbiological assay for thiamin using *Lactobacillus viridescens*. J. Bacteriol. 74: 818–821.

Delchier, N., C. Ringling, J. Le Grandois, D. Aoudé-werner, R. Galland, S. Georgé, M. Rychlik and C.M. Renard. 2013. Effects of industrial processing on folate content in green vegetables. Food Chem. 139: 815–824.

Divya, J.B. and K.M. Nampoothiri. 2015. Encapsulated *Lactococcus lactis* with enhanced gastrointestinal survival for the development of folate enriched functional foods. Bioresour. Technol. 188: 226–230.

Drywień, M., J. Frackiewicz, M. Górnicka, J. Gadek and M. Jałosińska. 2015. Effect of probiotic and storage time of thiamine and riboflavin content in the milk drinks fermented by *Lactobacillus casei* KNE-1. Rocz Panstw Zakl Hig. 66: 373–377.

Edwards, K.A., N. Tu-Maung, K. Cheng, B. Wang, A.J. Baeumner and C.E. Kraft. 2017. Thiamine assays-advances, challenges, and caveats. ChemistryOpen 6: 178–191.

Fujiwara, M. and K. Matsui. 1953. Determination of thiamine by thiochrome reaction. Anal. Chem. 25(5): 810–812.

Gratacós-Cubarsí, M., C. Sárraga, M. Clariana, J.A.G. Regueiro and M. Castellari. 2011. Analysis of vitamin B1 in dry-cured sausages by hydrophilic interaction liquid chromatography (HILIC) and diode array detection. Meat Sci. 87(3): 234–238.

Greppi, A., Y. Hemery, I. Berrazaga, Z. Almaksour and C. Humblot. 2017. Ability of lactobacilli isolated from traditional cereal-based fermented food to produce folate in culture media under different growth conditions. Food Sci. Technol. 86: 277–284.

Herr, D.S. 1945. Synthetic ion exchange resins in the separation, recovery, and concentration of thiamine. J. Ind. Eng. Chem. 3(7): 631–634.

Hou, J.W., R.C. Yu and C.C. Chou. 2000. Changes in some components of soymilk during fermentation with bifidobacteria. Food Res. Int. 33(5): 393–397.

Hoyumpa, Jr, A.M. 1980. Mechanisms of thiamin deficiency in chronic alcoholism. The Am. J. Clin. Nutr. 33(12): 2750–2761.

Hugenholtz, J., W. Sybesma, M.N. Groot, W. Wisselink, V. Ladero, K. Burgess, D. van Sinderen, J.C. Piard, G. Eggink, E.J. Smid, G. Savoy, F. Sesma, T. Jansen, P. Hols and M. Kleerebezem. 2002. Metabolic engineering of lactic acid bacteria for the production of nutraceuticals. Antonie Van Leeuwenhoek. 82(1-4): 217–235.

Hugenschmidt, S., M. Schwenninger, N. Gnehm and C. Lacroix. 2010. Screening of a natural biodiversity of lactic and propionic acid bacteria for folate and vitamin B12 production in supplemented whey permeate. Int. Dairy J. 20: 852–857.

Ibrahim, G.A., K. El-Shafei, H.S. El-Sayed and O.M. Sharaf. 2014. Isolation, identification and selection of lactic acid bacterial cultures for production of riboflavin and folate. Middle East J. Appl. Sci. 4: 924–930.

Iyer, R., S.K. Tomar, S. Kapila, J. Mani and R. Singh. 2010. Probiotic properties of folate producing *Streptococcus thermophilus* strains. Food Res. Int. 43: 103–110.

Iyer, R. and S.D. Tomar. 2013. Determination of folate/folic acid level in milk by microbiological assay, immuno assay and high performance liquid chromatography. J. Dairy Res. 80: 233–239.

Juarez del Valle, M., J.E. Laiño, G.S. de Giori and J.G. LeBlanc. 2014. Riboflavin producing lactic acid bacteria as a biotechnological strategy to obtain bio-enriched soymilk. Food Res. Int. 62: 1015–1019.

Juarez del Valle, M., J.E. Laiño, G. Savoy de Giori and J.G. LeBlanc. 2016a. Factors stimulating riboflavin produced by *Lactobacillus plantarum* CRL 725 grow in a semi-defined medium. J. Basic Microbiol. 57(3): 245–252.

Juarez del Valle, M., J.E. Laiño, A. de Moreno de LeBlanc, G. Savoy de Giori and J.G. LeBlanc. 2016b. Soyamilk fermented with riboflavin-producing *Lactobacillus Plantarum* CRL 2130 reverts and prevents ariboflavinosis in murine models. Br. J. Nutr. 116(07): 1229–1235.

Laiño, J.E., J.G. Leblanc and G. Savoy De Giori. 2012. Production of natural folates by lactic acid bacteria starter cultures isolated from artisanal Argentinean yogurts. Can. J. Microbiol. 58: 581–588.

Laiño, J.E., H. Zelaya, M. Juarez del Valle, G. Savoy de Giori and J.G. LeBlanc. 2015. Milk fermented with selected strains of lactic acid bacteria is able to improve folate status of deficient rodents and also prevent folate deficiency. J. Funct. Foods. 17: 22–32.

Laiño, J.E., M. Juárez Del Valle, E.M. Hébert, G. Savoy De Giori and J.G. Leblanc. 2017. Folate production and fol genes expression by the dairy starter culture *Streptococcus thermophilus* CRL803 in free and controlled pH batch fermentations. Food Sci. Technol. 85: 146–150.

Lebiedzińska, A. and P. Szefer. 2006. Vitamins B in grain and cereal-grain food, soy-products and seeds. Food Chem. 95(1): 116–122.

LeBlanc, J.G., G. Rutten, P. Bruinenberg, F. Sesma, G. Savoy de Giori and E.J. Smid. 2006. A novel dairy product fermented with *Propionibacterium freudenreichii* improves the riboflavin status of deficient rats. Nutrition. 22: 645–651.

Leblanc, J.G., G. Savoy de Giori, E.J. Smid and J. Hugenholtz. 2007. Folate production by lactic acid bacteria and other food-grade microorganisms. Comunicating Current Research and Educational Topics and Trends in Applied Microbiology 2: 329–339.

LeBlanc, J.G., W. Sybesma, M. Starrenburg, F. Sesma, W.M. de Vos, G. Savoy de Giori and J. Hugenholtz. 2010. Nutrition. 26: 835–841.

LeBlanc, J.G., J.E. Laiño, M. Juarez del Valle, V.V. Vannini, D. van Sinderen, M.P. Taranto, G.F. de Valdez, G. Savoy de Giori and F. Sesma. 2011. B-Group vitamin production by lactic acid bacteria-current knowledge and potential applications. J. Appl. Microbiol. 111(6): 1297–1309.

LeBlanc, J.G., F. Chain, R. Martín, L.G. Bermúdez-Humarán, S. Courau and P. Langella. 2017. Beneficial effects on host energy metabolism of short-chain fatty acids and vitamins produced by commensal and probiotic bacteria. Microb. Cell Fact. 16: 79.

Levit, R., G. Savoyde Giori, A. de Moreno de LeBlanc and J.G. LeBlanc. 2018. Effect of riboflavin-producing bacteria against chemically induced colitis in mice. J. Appl. Microbiol. 124: 232–240.

Maughan, R.J., L.M. Burke, J. Dvorak, D.E. Larson-Meyer, P. Peeling, S.M. Philips, E.S. Rawson, N.P. Walsh, I. Garthe, H. Geyer, R. Meeusen, L.J.C. van Loon, S.M. Shirreffs, L.L. Spriet, M. Stuart, A. Vernec, K. Currel, V.M. Ali, R.G.M. Budgett, A. Ljungqvist, M. Mountjoy, Y.P. Pitsiladis, T. Soligard, U. Erdener and L. Engebretsen. 2018. IOC consensus statement: dietary supplements and the high-performance athlete. Sport Med. 52: 439–455.

Magnúsdóttir, S., D. Ravcheev, V. de Crécy-Lagard and I. Thiele. 2015. Systematic genome assessment of B-vitamin biosynthesis suggests co-operation among gut microbes. Front. Genet 6: 148.

Martin-Belloso, O. and E. Llanos-Barriobero. 2001. Proximate composition, minerals and vitamins in selected canned vegetables. Euro Food Res. Technol. 212(2): 182–187.

Martin, P.R., C.K. Singleton and S. Hiller-Sturmhöfel. 2003. The role of thiamine deficiency in alcoholic brain disease. Alcohol Res. Health. 27(2): 134–142.

Marze, S. 2017. Bioavailability of nutrients and micronutrients: advances in modeling and *in vitro* approaches. Annu. Rev. Food Sci. Technol. 8: 35–55.

Masuda, M., M. Ide, H. Utsumi, T. Niiro, Y. Shimamura and M. Murata. 2012. Production potency of folate, vitamin B12, and thiamine by lactic acid bacteria isolated from Japanese pickles. Biosci. Biotechnol. Biochem. 76: 2061–2067.

Mckillop, D.J., H. Mcnulty, J.M. Scott, J.M. Mcpartlin, J. Strain, I. Bradbury, J. Girvan, L. Hoey, R. Mccreedy, J. Alexander, B.K. Patterson, M. Hannon-Fletcher and K. Pentieva. 2006. The rate of intestinal absorption of natural food folates is not related to the extent of folate conjugation. Am. J. Clin. Nutr. 84: 167–17.

Meucci, A., L. Rossetti, M. Zago, L. Monti, G. Giraffa, D. Carminati and F. Tidona. 2018. Folates biosynthesis by *Streptococcus thermophilus* during growth in milk. Food Microbiol. 69: 116–122.

Moazzen, S., R. Dolatkhah, J. Sadegh, J. Shaarba, B.Z. Alizadeh, G.H. Bock and S. Dastgiri. 2018. Folic acid intake and folate status and colorectal cancer risk: A systematic review and meta-analysis. Clin. Nutr. 37: 1926–1934.

Moll, R. and B. Davis. 2017. Iron, vitamin B 12 and folate. Medicine 45: 198–203.

Mosso, A.L., M.E. Jimenez, G. Vignolo, J.G. LeBlanc and N.C. Samman. 2018. Increasing the folate content of tuber based foods using potentially probiotic lactic acid bacteria. Food Res. Int. 108: 168–174.

Naderi, Nassim and J.D. House. 2018. Recent developments in folate nutrition. Adv. Food Nutr. Res. 83: 195–213.

Ohrvik, V.E. and C.M. Witthoft. 2011. Human folate bioavailability. Nutrients. 3: 475–490.

Orsomando, G., G.G. Bozzo, R.D. de La Garza, G.J. Basset, E.P. Quinlivan, V. Naponelli, F. Rébeillé, S. Ravanel, J.F. Gregory and A.D. Hanson. 2006. Evidence for folate-salvage reactions in plants. Plant J. 46: 426–435.

Olkowski, A.A. and S.R. Gooneratne. 1992. Microbiological methods of thiamine measurement in biological material. Int. J. Vitam. Nutr. Res. 62(1): 34–42.

Pacheco da Silva, F.F., V. Biscola, J.G. Leblanc and B.D.G.M. Franco. 2016. Effect of indigenous lactic acid bacteria isolated from goat milk and cheeses on folate and riboflavin content of fermented goat milk. Food Sci. Technol. 71: 155–161.

Rajendran, S.C., B. Chamlagain, S. Kariluoto, V. Piironen and P.E.J. Saris. 2017. Biofortification of riboflavin and folate in Idli batter, based on fermented cereal and pulse, by *Lactococcus lactis* N8 and *Saccharomyces boulardii* SAA655. J. Appl. Microbiol. 122(6): 1663–1671.

Rasic, J.L. and J.A. Kurmann. 1983. *Bifidobacteria* and their role. Microbiological, nutritional-physiological, medical and technological aspects and bibliography. Experientia Suppl. 39: 1–295.

Rees, J.R., C.B. Morris, J.L. Peacock, P.M. Ueland, E.L. Barry, G.E. Mckeown-Eyssen, J.C. Figueiredo, D.C. Snover and J. Baron. 2017. Unmetabolized folic acid, tetrahydrofolate, and colorectal adenoma risk. Cancer Prev. Res. 10: 451–458.

Saini, R.K., S.H. Nile and Y. Keum. 2016. Folates: Chemistry, analysis, occurrence, biofortification and bioavailability. FRIN 89: 1–13.

Saulnier, D.M., F. Santos, S. Roos, T.A. Mistretta, J.K. Spinler, D. Molenaar, B. Teusink and J. Versalovic. 2011. Exploring metabolic pathway reconstruction and genome-wide expression profiling in *Lactobacillus reuteri* to define functional probiotic features. PLoS One 6(4): e18783.

Scaglione, F. and G. Panzavolta. 2014. Folate, folic acid and 5-methyltetrahydrofolate are not the same thing. Xenobiotica 44: 480–488.

Sela, D.A., J. Chapman, A. Adeuya, J.H. Kim, F. Chen, T.R. Whitehead, A. Lapidus, D.S. Rokhsar, C.B. Lebrilla and J.B. German. 2008. The genome sequence of *Bifidobacterium longum* subsp. *infantis* reveals adaptations for milk utilization within the infant microbiome. Proc. Natl. Acad. Sci. USA 105(48): 18964–18969.

Shrestha, A.K., J. Arcot and J. Paterson. 2000. Folate assay of foods by traditional and tri-enzyme treatments using cryoprotected *Lactobacillus casei.* Anal. Nutr. Clin. Methods Section 71: 545–552.

Sybesma, W., M. Starrenburg, L. Tijsseling, M.H.N. Hoefnage and J. Hugenholtz. 2003. Effects of cultivation conditions on folate production by lactic acid bacteria. Metab. Eng. 69: 4542–4548.

Sybesma, W., M. Starrenburg, M. Kleerebezem, I. Mierau, W.M. de Vos and J. Hugenholtz. 2003. Increased production of folate by metabolic engineering of *Lactococcus lactis*. Appl. Environ. Microbiol. 69(6): 3069–3076.

Tamura, T. 1997. Determination of food folate. Methods Nutr. Biochem. 9: 285–293.

Teusink, B., F.H. van Enckevort, C. Francke, A. Wiersma, A. Wegkamp, E.J. Smid and R.J. Siezen. 2005. *In silico* reconstruction of the metabolic pathways of *Lactobacillus plantarum*: comparing predictions of nutrient requirements with those from growth experiments. Appl. Environ. Microbiol. 71(11): 7253–7262.

Thakur, K., S.K. Tomar and S. De. 2016. Lactic acid bacteria as a cell factory for riboflavin production. Microb. Biotechnol. 9(4): 441–451.

Thakur, K. and S.K. Tomar. 2016. *In vitro* study of riboflavin producing lactobacilli as potential probiotic. LWT - Food Sci. Technol. 68: 570–578.

Thakur, K., S.K. Tomar, A.K. Singh, S. Mandal and S. Arora. 2017. Riboflavin and health: A review of recent human Research. Crit. Rev. Food Sci. Nutr. 57(17): 3650–3660.

Wacker, J., J. Frühauf, M. Schulz, F.M. Chiwora, J. Volz and K. Becker. 2000. Riboflavin deficiency and preeclampsia. Obstet. Gynecol. 96(1): 38–44.

Yépez, A., P. Russo, G. Spano, I. Khomenko, F. Biasioli, V. Capozzi and R. Aznar. 2019. *In situ* riboflavin fortification of different kefir-like cereal-based beverages using selected Andean LAB strains. Food Microbiol. 77: 61–68.

Zajicek, J.L., D.E. Tillitt, S.B. Brown, L.R. Brown, D.C. Honeyfield and J.D. Fitzsimons. 2005. A rapid solid-phase extraction fluorometric method for thiamine and riboflavin in salmonid eggs. J. Aquat. Anim. Health. 17(1): 95–105.

8

Effect of Short-Chain Fatty Acids Produced by Probiotics

Functional Role Toward the Improvement of Human Health

Milena Fernandes da Silva,[1] *Meire dos Santos Falcão de Lima*[1] and *Attilio Converti*[2,*]

Introduction

The intestine is the largest immune system in the human body, whose microbiota is composed of trillions of microbial cells belonging to thousands of different bacterial species. Due to their synergistic interactions, such a microbiota as a whole behaves qualitatively and quantitatively as an organ; in this sense, the idea that the intestinal microbiota constitutes an additional organ has re-emerged (Sivaprakasam et al. 2016, 2018, Pekmez et al. 2019). Since the overall wet weight of intestinal microbes of an individual is estimated to be 0.175–1.5 kg, their genome as a whole contains a number of genes about 150-fold that of the body; therefore, it has also been referred to as our second genome (Pluznick 2016, Sivaprakasam et al. 2016, 2018).

Since colonization of gut begins soon after birth and continues throughout the entire life of the individual, its microbial population is dynamic and changes in composition and activity either over time (Pluznick 2016, Sivaprakasam et al.

[1] Federal University of Pernambuco, Av. Prof. Moraes Rego 1235, Cidade Universitária, 50670-901 Recife, PE, Brazil.
Emails: milen_sil@hotmail.com; meirefalcao.ufrpe@gmail.com

[2] Department of Civil, Chemical and Environmental Engineering, University of Genoa, Pole of Chemical Engineering, via Opera Pia 15, 16145 Genoa, Italy.

* Corresponding author: converti@unige.it

2016, Yousuf and Mishra 2019), or due to diseases (dysbiosis) or even in response to variations in host habits (Daliri and Lee 2015, Pluznick 2016, Sivaprakasam et al. 2016, Pekmez et al. 2019, Yousuf and Mishra 2019).

Members of Bacteroidetes and Firmicutes phyla predominate in gut microbiota, while those of Actinobacteria, Proteobacteria, Fusobacteria, Cyanobacteria and Verrucomicrobia are less abundant (Sivaprakasam et al. 2016). Bacteria of gut microbiota produce a large number of different metabolites acting as messengers between them and the host (Pluznick 2016, Sivaprakasam et al. 2018).

Among them, the so-called probiotics are defined, according to the International Scientific Association for Probiotics and Prebiotics, as living non-pathogenic microorganisms that confer a health benefit on the host when administered in adequate amounts (10^6 to 10^9 CFU/mL) (Hill et al. 2014, Daliri et al. 2019). The term "probiotic", a word of Greek origin that means 'for life,' was first introduced in 1953 (Falcinelli et al. 2018). It has been estimated that the global market of probiotics will exceed $64 billion by 2022 (Quigley 2019). Although different and inter-related mechanisms are involved in probiotic activity, their positive effects on humans seem to depend especially on the immune system control, the balance between saccharolytic and proteolytic species in the intestinal microbiota, and the production of antimicrobials and other bioactive compounds (Pessione et al. 2015, Kerry et al. 2018).

On the other hand, prebiotics are defined as 'nondigestible compounds that modulate the composition and/or activity of the gut microbiota, thus conferring a beneficial physiological effect on the host' (Gibson et al. 2017, Rivero-Gutiérrez et al. 2017, Pekmez et al. 2019). Carbohydrates such as inulin, frutooligosaccharides, galactooligosaccharides, lactulose, resistant starches, and other nondigestible oligosaccharides along with complex dietary fibres are the main examples of prebiotics, whose capability to escape absorption in the small bowel may be due either to their non-digestibility or because of bypassing the digestive system (Roy et al. 2006). The main health benefits of prebiotics are then related to their ability to stimulate the growth of specific bacteria, particularly lactobacilli and bifidobacteria (Pekmez et al. 2019).

Nutritional supplements simultaneously containing probiotics and prebiotics, which are called synbiotics, impart health benefits in a synergistic way (Rivero-Gutiérrez et al. 2017). Such a synergism during anaerobic bacterial fermentation (Pluznick 2017) leads to the formation of key metabolites such as short-chain fatty acids (SCFAs) (Pessione et al. 2015, Pekmez et al. 2019), which are readily absorbed in the large bowel and are generally thought to promote health (Roy et al. 2006). SCFAs are carboxylic acids with a low number of carbons (from 1 to 6) in their hydrocarbon chain, mainly acetate (C2:0), propionate (C3:0), and butyrate (C4:0) (Rodrigues et al. 2016).

Given the growing body of evidence that SCFAs play a crucial role in orchestrating the trigger-response relationship among host diet, microbiota, and homeostasis, there is a great deal of interest in exploiting such an exciting and rapidly emerging area in order to shape a healthier microbiota able to promote general health and well-being.

Under this premise, this chapter provides an overview of the current state-of-the-art in the impact of SCFAs produced by probiotic lactic acid bacteria (LAB) on human health. Herein, we discuss SCFA production by these microorganisms, including types and either indirect or direct mechanisms through which they contribute to their accumulation in the colon, as well as recent advances in the development of rapid and selective analytical methods capable of identifying and quantifying SCFAs. Finally, the functional role of SCFAs in the improvement of human health with special focus on obesity and cancer is highlighted.

Production, Transport and Metabolic Function of SCFAs

Production of SCFAs

Since most dietary fibres are not degraded in the upper gastrointestinal tract by human enzymes, it arrives unbroken to the colon and cecum, where it is consumed by the gut microbiota (Pluznick 2016, Richards et al. 2016) and may be partially converted to SCFAs by probiotic LAB. SCFA concentration ranges from 70 to 140 mM in the proximal colon, whereas lower concentrations (20–70 mM) have been reported in the distal colon (Rodrigues et al. 2016).

SCFA production is influenced by various factors, among which are the type of microorganisms, their concentration and substrates, and the transit rate in the intestine (Pekmez et al. 2019). Probiotic LAB may also indirectly stimulate SCFA production by endogenous colonic microbiota (Pessione et al. 2015). It has been reported that SCFA concentration in the serum depends on dietary fiber level, and the one in the cecum is increased by more than 100-fold by the presence of gut microbiota (Pluznick 2016), which in this way modulates the circulating SCFA level. The plasmatic concentration of acetate, which is the most abundant SCFA, ranges from 0.1 mM on a standard diet to 10 mM (Pluznick 2016), but the acetate/butyrate/propionate ratio is mainly dependent on diet (Pluznick 2017).

Transport of SCFAs

Over the past 10 years, SCFAs have been the subject of numerous physiological and clinical studies, which demonstrated that they can be a significant source of calories and that their trophic effects on both the small and large bowel may be useful for the prevention and treatment of several acute and chronic diseases (Roy et al. 2006). An in-depth understanding of SCFA physiological actions requires knowledge of cellular mechanisms underlying their signalling. As described by Pluznick (2016, 2017), SCFA mediate effects on the host through different mechanisms, involving cell-surface G-protein-coupled receptors (GPCRs), intracellular receptors and enzymes (Sivaprakasam et al. 2018). Until now, 4 cell-surface GPCRs-mediated mechanisms have been described: GPR41, GPR43, GPR109A, and Olfr78 (Figure 1).

SCFA level in the ileum is considered enough to activate signalling either by GPR41 or by GPR43 (Sivaprakasam et al. 2016), which are the best-known SCFA GPCRs (Pluznick 2016, Sivaprakasam et al. 2016) able to bind to the main three SCFAs, namely acetate, propionate and butyrate. However, GPR43 has preference for acetate and propionate, while GPR41 for propionate and butyrate (Sivaprakasam et al. 2016). In addition, β-hydroxybutyrate has been reported to be either a GPR41 agonist or an antagonist (Pluznick 2016).

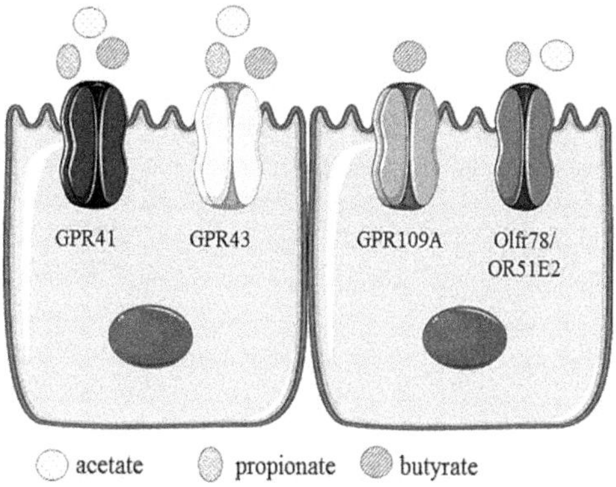

Figure 1: Four receptors for short chain fatty acids have been reported, all being 7-transmembrane G-protein-coupled receptors. GPR41 and GPR43 respond to acetate, propionate, and butyrate, GPR109A respond only to butyrate, while Olfr78 responds to acetate and propionate.

Unlike GPR41 and GPR43, butyrate is the only SCFA able to activate GPR109A, a receptor for niacin and vitamin B3, whereas β-hydroxybutyrate is produced as an alternative GPR109A ligand, result of ketogenesis under starvation (Pluznick 2016, Sivaprakasam et al. 2016).

According to Pluznick (2016), Olfr78 (mouse)/OR51E2 (human) respond to acetate, propionate, and lactate, even though the last acid appears to be a partial agonist compared with the others.

Among the non-GPCR mechanisms of SCFA transport across the apical membrane, we can mention: (a) the electroneutral cotransport with H^+ via monocarboxylate transporter 1 (MCT1, also referred to as SLC16A1) that may occur for a given SCFA, depending on the transmembrane concentration gradient, in either influx or efflux direction; (b) the electroneutral H^+/monocarboxylate symport via monocarboxylate transporter 4 (MCT4, also referred to as SLC16A3); and (c) the electrogenic transport via sodium-coupled monocarboxylate transporters 1 (SMCT1 or SLC5A8) and 2 (SMCT2 or SLC5A12), with Na^+ influx and monocarboxylate substrates leading to depolarization (Sivaprakasam et al. 2018) (Figure 2).

SCFAs released in the colonic lumen cross the epithelium thanks to different transport mechanisms such as diffusion, low-affinity transport via bicarbonate/SCFA exchange, medium-affinity transport via MCT1 and high-affinity transport via SMCT1. Metabolization of most of SCFAs by colonic epithelium results in a significant reduction of the fraction entering the portal circulation, and those few that reach the liver are metabolized into lipids. Even though only negligible concentrations of acetate (~ 100 μM), propionate (~ 4–10 μM), and butyrate (4–10 μM) are detected in peripheral blood (Sivaprakasam et al. 2016), they can exert different cellular effects in many ways (Pluznick 2016).

Metabolic functions of SCFAs

Bacteria explicit their beneficial effects on the host cells mainly through chemical communication, and in this sense SCFAs are among the major components of bacterial metabolites, either quantitatively or qualitatively (Sivaprakasam et al. 2018).

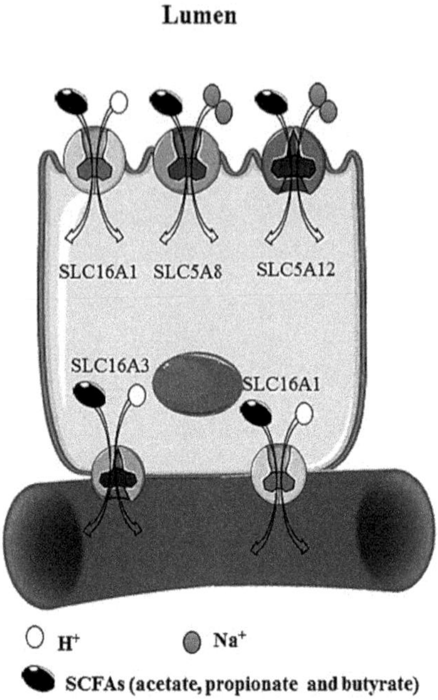

Figure 2: Expression of short-chain fatty acid transporters in colonic epithelium. SCFAs, short-chain fatty acids; SLC16A1, monocarboxylate transporter 1 (MCT1); SLC16A3, monocarboxylate transporter 4 (MCT4); SLC5A8, sodium-coupled monocarboxylate transporter 1 (SMCT1); SLC5A12, sodium-coupled monocarboxylate transporter 2 (SMCT2).

Clostridium propionicum and members of the genera *Propionibacterium*, *Selenomonas*, *Veillonella*, and *Desulfovibrio* are able to metabolize lactic acid produced by LABs to acetate and propionate, while *Megasphaera elsdenii*, *Eubacterium hallii*, *Anaerostipes caccae,* and some other clostridia do so to butyrate (Pessione et al. 2015).

However, probiotic LABs are also able to directly produce SCFAs in the colon. In fact, in addition to the well-known production of acetic acid by either hetero-fermenters or facultative homo-fermenters, it is known the capability of some lactobacilli to directly produce propionic and butyric acids (Pessione et al. 2015).

Acetate

Nearly all anaerobic heterotrophic intestinal eubacteria produce acetate (Hoyles and Wallace 2010) mainly through two different pathways, one heterotrophic and the other autotrophic. In the former pathway, acetate comes from pyruvate decarboxylation to acetyl-CoA and subsequent hydrolysis of this intermediate. In the latter, referred to as Wood-Ljungdahl pathway, carbon dioxide is reduced to carbon monoxide, which then reacts with both a methyl group and a coenzyme A molecule to form acetyl-CoA (Richards et al. 2016, Tungland 2018). Acetate is also the main SCFA produced in the mammalian colon (Roy et al. 2006).

Uptake and use of acetate by many tissues constitute the main route whereby the body gets energy from carbohydrates not digested and not absorbed in the small bowel

(Hoyles and Wallace 2010). Evidence suggests that it provides the body with about 1.5–2 kcal/g (Roy et al. 2006). Acetate is the most abundant SCFA in plasma, where it acts as an energetic substrate (Pessione et al. 2015); its formation is associated with substrate-level phosphorylation of ADP to ATP (Hoyles and Wallace 2010).

After absorption in the colon, acetate is metabolized by the liver, muscles and other peripheral tissues (Roy et al. 2006). Approximately 60% of this SCFA produced by intestinal bacteria is retained in the liver. Hoyles and Wallace (2010) have reported the secretion of acetate by the liver when portal blood concentrations are below a critical level. Finally, acetate is also involved in the control of inflammation and pathogen invasion (Pessione et al. 2015).

Propionate

Propionate can be formed directly from sugars by single species of the intestinal microbiota, or indirectly by cross-feeding by producers of succinate and lactate; however it has not yet been established which the main route is. There are two metabolic pathways for propionate formation, the "randomizing" pathway via succinate and the acrylate pathway (Hoyles and Wallace 2010). The former pathway, called methylmalonyl-CoA pathway, uses a primitive anaerobic electron transport chain. At low carbon dioxide partial pressures, succinate is converted into methylmalonate, which is then split to produce propionate and carbon dioxide, while the released carbon dioxide is recycled to carboxylate phosphoenolpyruvate. In the latter, which is called acryloyl-CoA pathway, pyruvate is reversibly reduced by lactate dehydrogenase to L-lactate, and then subsequently converted to lactyl-CoA, acryloyl-CoA, propionyl-CoA and finally propionate (Richards et al. 2016, Tungland 2018).

In bacteria, propionate formation meets two needs, namely the disposal of excess reducing power mainly in the form of NADH, and the synthesis of ATP (Hoyles and Wallace 2010). Propionate, which is metabolized in the human colon in less extent than butyrate, seems to impact favourably on cholesterol level, unlike acetate. Moreover, it has been reported that propionate may favour colonocyte proliferation and differentiation, albeit less efficiently than butyrate (Roy et al. 2006). Both propionate and butyrate seem to suppress appetite, prevent diet-induced obesity and enhance insulin sensitivity (Pessione et al. 2015).

Butyrate

Butyrate production begins with the condensation of two acetyl-CoA molecules to form of acetacetyl-CoA, which is then progressively reduced to butyryl-CoA. Butyrate can be synthesized from this intermediate in two different ways. Whereas the former mechanism implies direct enzymatic conversion of butyryl-CoA to butyrate, the latter involves the reaction of butyryl-CoA and exogenously produced acetate, with formation of butyrate and acetyl-CoA along with energy recovery in the form of ATP (Hoyles and Wallace 2010, Richards et al. 2016, Tungland 2018).

Butyrate, which is considered the SCFA for excellence (Roy et al. 2006), is likely to be involved in immune homeostasis by either suppression of lipopolysaccharide-induced metabolic reprogramming of human dendritic cells or inhibition of NF-kB signalling pathway in macrophages (Pekmez et al. 2019). It is the preferred energy source for colon epithelial cells, while acetate and propionate are mainly used as carbon sources for gluconeogenesis and lipogenesis by peripheral tissues and liver (Pekmez et al. 2019).

Besides being implicated as a factor for both differentiation and trophic activity, butyrate stimulates the growth of ileal and jejunal cells delivered in the colon (Roy et al. 2006). In addition, by promoting cell differentiation, cell-cycle arrest and apoptosis of colonocytes mainly altering gene transcription by inhibition of histone deacetylase activity, it plays a key role in the treatment and prevention of colon cancer (Pessione et al. 2015).

Since SCFAs interact with multiple signalling molecules expressed by colon immune and epithelial cells, it is difficult to understand the complex molecular mechanisms involved in their protective effects on colonic health. Histone deacetylases (HDACs) are the most widely studied targets of propionate and butyrate (Pluznick 2017). By inhibiting HDACs activity an increase in the acetylation of histones occurs, hence decreasing their positive charge and preventing them from binding to negatively-charged DNA. Without bound, histones remain open and favour gene transcription. Thus, histone acetylation should result in overall changes in chromatin structure, which could affect the expression pattern of various genes, and this change in gene expression could induce apoptosis and/or cell cycle arrest in colon cancer cells (Sun and O'Riordan 2013, Sivaprakasam et al. 2016).

Functional Role of SCFAs in Human Health

An overview of the literature published over the last five years up to the end of December 2018 has shown an increase in the contributions related to "short chain fatty acids and probiotics and human health", with no less than 2955 scientific articles (based on Science Direct search: http://www.sciencedirect.com). A substantial portion of them deals with "short chain fatty acids and probiotic lactic acid bacteria and human health" (1269 scientific articles); thus, this topic is an active and expanding field of research. Owing to the biological significance of SCFAs, recent advances have been made in the development of rapid and selective analytical methods capable of identifying and quantifying them especially in human faecal samples (Primec et al. 2019), which are summarized in Table 1. Readers are directed elsewhere for a comprehensive review of the SCFAs analysis in human faeces (Primec et al. 2017). On the other hand, Table 2 summarizes the more recent examples of current scientific reports on the functional role of SCFAs produced by probiotic bacteria in the improvement of human health.

Several bacteria inhabiting human intestine are able to ferment dietary fibre and not otherwise digestible starches into SCFAs. In particular, those belonging to the Bacteroidetes phylum release propionate and acetate in large amounts, while those belonging to the Firmicutes, including probiotic species, mainly butyrate (Haase et al. 2018).

Lee et al. (2017) reported that probiotics have beneficial effects on cancer, diarrhoea and irritable bowel disease. In addition, in another study conducted by Meng et al. (2017) probiotics were shown to exert an anti-inflammatory effect by reducing the secretion of pro-inflammatory cytokine tumour necrosis factor-α. These activities of specific probiotics, for example *Bifidobacterium animalis* subsp. *lactis*, were attributed to different possible mechanisms including production of SCFAs.

Table 1: Overview of current scientific reports on the development of SCFAs analytical methods.

SCFAs analytical methods	References
GC/MS SCFA profiling method combining BSTFA derivatization and sodium sulphate dehydration pretreatment using faecal and serum samples from mice.	Zhang et al. (2019)
Measuring SCFAs by GC in SIM mode using 3.5–12 h long lyophilization; extraction temperature from 4 to 25°C; extraction time from 15 min to 45 min; extraction solvents acetic, propionic and butyric acid. Biological samples used: faeces and colonic contents of mice. This method may also be employed in liquid biological samples like urine, fermentation broth or blood by appropriate confirmation.	Han et al. (2018)
O-Benzylhydroxylamine derivatization of SCFAs and ketone bodies for fast LC-MS/MS using faeces and plasma as biological matrices.	Zeng and Cao (2018)
GC-MS with isobutyl chloroformate derivatization in aqueous solution without drying the samples. This method has high potential for clinical purposes (e.g., diagnosis of chronic intestinal disease).	Furuhashi et al. (2018)
Ultrasound digestion for 40 min at 35°C; time of shaking 4 min with 3 subsequent extractions, followed by isotachophoretic determination. This method may be useful for investigating SCFA influence on the energy balance, anticancer processes and antibacterial effects as well as other biological processes related to SCFA clinical implications like bowel disease prevention.	Dobrowolska-Iwanek et al. (2016)

SCFAs: short-chain fatty acids; GC-MS: gas chromatography-mass spectrometry; BSTFA: N,O-Bis(trimethyl-silyl)-trifluoroacetamide; LC-MS/MS: Liquid chromatography-tandem mass spectrometry.

Total SCFA composition and concentration have implications for human metabolism and health besides influencing intestinal motility and pH. Constituting a relevant energy source not only for intestinal microbiota but also for the gut epithelial cells (Haase et al. 2018), SCFAs do in fact contribute to better absorption of nutrients, making the enterocytes healthier (Hoyles and Wallace 2010), as well as normal functioning of the large intestine, by stimulating blood flow and consequently improving absorption of water and salts. SCFAs also reduce appetite, through an effect on leptin that acts as a satiety signal for human body (Remely et al. 2014).

Patients affected by inflammatory bowel and autoimmune diseases, type 2 diabetes or obesity often exhibit abnormal reduction of levels of SCFAs in gut or SCFA-producing gut bacteria (Rivière et al. 2016, Sun et al. 2017, Hu et al. 2018). Nagpal et al. (2018) recently suggested that such perturbations may be the result of a reduction in the population of gut bacteria that grow in syntrophy with SCFA producers or of an increase in the release of harmful compounds by either the gastrointestinal tract or other microbes. Thus, the following sessions will provide some examples of SCFA regulatory functions on host physiology, highlighting the functional role of these acids in improving human health, with special emphasis on obesity and cancer.

SCFAs and obesity

It is recognized that obesity is as a condition that has a negative impact on the overall human state, life quality and work ability, plays a very serious role in the development

Table 2: Examples of current scientific reports on expected functional role of SCFAs produced by probiotic bacteria in the improvement of human health.

Probiotic bacteria	Bioactive compounds	Main findings about functional role of SCFAs in the improvement of health	References
5 Lactobacilli (*L. paracasei* D3-5, *L. rhamnosus* D4-4, *L. plantarum* D6-2, D7-5 and D13-4) and 5 enterococci (*E. raffinosus* D24-1, *E.* INBio D24-2, *E. avium* D25-1, D25-2 and D26-1) - Novel human-origin probiotic cocktail.	Propionate and butyrate	Probiotic cocktail (oral gavage) in mice (a) showed potential to inhibit the growth of uropathogenic Enterobacteriaceae (*Escherichia coli* CFT07332 and *Klebsiella pneumonia* KPPR133), (b) modulated gut microbiome, (c) increased SCFA production, (d) reduced gut microbiome dysbiosis, and (e) may be a potential therapeutic for human diseases involving decreased SCFA production in the gut.	Nagpal et al. (2018)
L. acidophilus NCIMB 30175, *L. plantarum* NCIMB 30173, *L. rhamnosus* NCIMB 30174 and *E. faecium* NCIMB 30176 (Symprove™, 4-strain aqueous probiotic supplement).	Butyrate	Probiotic cocktail (a) exerted positive immunomodulatory effects *in vitro*, (b) increased and reduced production of anti-inflammatory cytokines (IL-6 and IL-10) and inflammatory chemokines (MCP-1, CXCL 10 and IL-8), respectively, (c) soon integrated with the existing microbiota, (d) increased proximal and distal colonic lactate concentrations, (e) stimulated growth of lactate-consuming bacteria, (f) altered microbiota bacterial diversity, (g) increased SCFA production, mainly butyrate, and (h) is expected, contrary to the individual probiotics used alone, to promote health benefits in humans with improvement in clinical symptoms of gut infections.	Moens et al. (2019)

Vigiis 101 – *L. paracasei* subsp. *paracasei* NTU 101 (commercial product).	Acetate, propionate and butyrate	Vigiis 101 treatment on gastric mucosa and intestinal bacterial flora in rats (a) increased the levels of beneficial *Bifidobacterium* spp. and *Lactobacillus* spp., (b) reduced the ones of detrimental *Clostridium perfringens* and Enterobacteriaceae, (c) reduced gastric lesion areas, (d) increased the levels of SCFAs (acetic, propionic and butyric acids) in cecal samples, (e) up-regulated SCFAs as well as prostaglandin E2 levels in the gastric mucosa, and (f) is expected to stimulate mucus secretion, increase transmucosal resistance and contribute to protection against gastric mucosal lesions induced by pylorus.	Kao et al. (2018)
Probiotic mixture containing 6 lactobacilli and 3 bifidobacteria (not specified).	Butyrate	Probiotic cocktail alleviated systemic adiposity and inflammation in non-alcoholic fatty liver disease rats through SCFA receptor (Gpr109a) and commensal metabolite butyrate in response to the insult of high-fat diet.	Liang et al. (2018)
Probiotic formulation containing *Bifidobacterium breve* BR03 (DSM 16604) B632.	Acetate, propionate and butyrate	Clinical intervention after administration of probiotic formulation containing *Bifidobacterium breve* BR03 and B632 in celiac disease children revealed novel microbial modulators of serum tumour necrosis factor alpha (TNF-α) and SCFAs.	Primec et al. (2019)

of human diseases and is responsible for a reduction of human reproductive potential (Kobyliak et al. 2018). Recent data have also shown a correlation between obesity and intestinal microbiota (Remely et al. 2014, Dugas et al. 2018, Kobyliak et al. 2018, Gholizadeh et al. 2019). The current status of research on microbiome as it relates to understanding obesity as a whole, with particular emphasis on the development of microbiome-targeted therapies to prevent and treat it, has been discussed extensively by Maruvada et al. (2017); therefore, readers are invited to refer to this review for more details. Since obese people have intestinal dysbiosis, to lose weight they need to improve the endogenous intestinal microbiome, which seems to be one of the mechanisms underlying the effects of probiotics. Nonetheless, whether and how probiotics modulate intestinal microbiome and SCFA profile in healthy hosts is still a matter of debate (Nagpal et al. 2018, Quigley 2019).

Obesity induced by diet is also related to modifications on intestinal microbiota and microbiome. Indeed, recent studies have shown that human microbiota has no less than 150 times more genes than human genome, influences human physiology and provides additional metabolic capabilities (Ley et al. 2005, Remely et al. 2014, Mulders et al. 2018).

The intestinal microbiota plays a critical role in establishing and maintaining human health (Remely et al. 2014). Most bacteria of human gut microbiota belong to the phylum Firmicutes (60 to 80%), which includes the main probiotics (*Lactobacillus* and *Bifidobacterium*), and that of Bacteroidetes (15 to 25%), whereas others, present in smaller amounts, belong to the phyla Actinobacteria, Proteobacteria, and Verrucomicrobia (Mulders et al. 2018).

Metabolic diseases such as obesity are associated with imbalance of diversity, abundance, and metabolic function in the intestinal microbiota (Remely et al. 2014). In particular, imbalance in the composition of the major phyla has shown interference in metabolism (Mulders et al. 2018). An increased Firmicutes/Bacteroidetes ratio detected in obese people (Koblyliak et al. 2018) has been ascribed in many studies to the increased SCFA production and recognized to give a significant contribution to obesity pathophysiology (Riva et al. 2017). Contrariwise, Schwiertz et al. (2010) reported a lower proportion of Firmicutes in overweight and obese adult humans compared to lean controls.

The production of SCFAs by the intestinal microbiota fermenting non-digestible prebiotic fibres is even altered under obesity conditions (Mulders et al. 2018). As many Firmicutes produce butyrate, a higher Firmicutes/Bacteroidetes ratio is expected to increase the levels of butyrate and acetate as final fermentation products (Remely et al. 2014). Up to 70% of the energy spent by the normal colonic epithelium comes from SCFAs, particularly butyrate; much of the propionate is absorbed by the liver acting as a precursor for gluconeogenesis, liponeogenesis, and protein synthesis; while acetate, upon entering the peripheral circulation, can either be metabolized by tissues or serve as a substrate for cholesterol synthesis (Schwiertz et al. 2010).

There are several ways in which SCFAs protect the individual against obesity, via the activation of nutrient-specific receptors (Mulders et al. 2018). SCFAs' affinity for the GPR41 receptor (Ffar3, free fatty acid receptor) depends on the organic acid chain length. Since the receptor GPR40, which is located near the GPR41 gene, binds to long chain fatty acids instead of SCFAs (Remely et al. 2014), it has been hypothesized that GPR41 might have no promoter region and be transcribed by the same GPR40 promoter (Bahar Halpern et al. 2012). These receptors are closely linked to the biological functions of lipid metabolism modulation, regulate the production of leptin and other hormones of satiety, and control the energy metabolism of the host (Mulders et al. 2018).

SCFAs are also involved in the epigenetic control of gene expression, which occurs via histone acetylation, and can thus prevent obesity and exert anti-inflammatory effects (Remely et al. 2014). Folate produced by the microbiota is among the best-known DNA methyl donors, even though other compounds such as choline, methionine, and butyrate are also capable of influencig the degree of DNA methylation (Canani et al. 2011). To reduce the transcriptional activity of the downstream gene, highly methylated promoter regions are proposed. In this sense, different types of intestinal microbiota lead to different metabolic syntheses of SCFAs. Butyrate has proven to influence the epigenetic methylation of SCFA receptors, particularly the Ffar3 promoter region, hence regulating gene expression and function. These changes in epigenetic regulation may cause changes in the expression and signalling of Ffars and influence cycles of hunger and satiety (Remely et al. 2014).

In the process of dysbiosis, the pathogenic microbiota present in the intestines of obese individuals, extracts more energy from the diet, increases intestinal permeability, causes inflammation and decreases SCFAs' production. The inflammation caused from this disequilibrium of microbiota can affect the nervous system by increasing the activation of microglia, altering energy homeostasis of hypothalamic neurons, and increasing hypothalamic inflammation through humoral, cellular and neuronal pathways. Thus, SCFAs have an important protective role in this context because they can inhibit or modulate different pathways that contribute to the obesity pathogenesis (Mulders et al. 2018).

Previous studies with axenic cultures of probiotics belonging mainly to *Lactobacillus* and *Bifidobacterium* genera have demonstrated the ability of probiotic therapy to decrease body weight and to mitigate insulin resistance, chronic systemic inflammation and hepatic steatosis in experimental obesity models (Nova et al. 2016). Fifteen *Lactobacillus* and 3 *Bifidobacterium* strains have been tested by Cani and Van Hul (2015) in order to check their efficacy as obesity treatments; only 10 reduced total body and/or visceral adipose tissue weight, while 12 mitigated liver and/or fat tissue inflammation, with strain-specific variations attributable to different action mechanisms. The same authors mentioned the following mechanisms underlying the therapeutic action of probiotic strains: (a) regulation of absorption and excretion of fat; (b) increase in primary cholic acids due to the action of bile salt hydrolase (Renga et al. 2010); (c) increase in the level of glucagon-like peptides, resulting in alleviation of hunger, decreased energy consumption, and improved sensitivity to insulin and β-cell function; (d) regulation of gene expression to decrease *de novo* lipogenesis and accelerate β-oxidation; (e) regulation of the expression of proteins (ZO-1 and ZO-2) responsible for restoration/maintenance of gut barrier function as well as reduction of lipopolysaccharide absorption and metabolic endotoxemia; (f) increase in the intestinal levels of SCFAs, mostly butyrate, with consequent mitigation of chronic systemic inflammation, release of anti-inflammatory cytokines by fat tissue, reduction of insulin resistance, enhancement of β-cell differentiation, proliferation and development (Koblyliak et al. 2018).

SCFAs and cancer

Immunity is shown by highly integrated cells susceptible to surrounding factors. This system is organized in various regions of the body, but approximately 70% of cells of the immune system and more than 90% of the immunoglobulin (Ig)-producing cells in the human body are found in the intestine. These cells, in turn, are sensitive to metabolites derived from nutrients and products released by the microbiota such as SCFAs (Galloway-Peña et al. 2017) and modulate cell activation and function (Çehreli 2018).

The proper functioning of the immune system is closely related to functions of the gastrointestinal system (Çehreli 2018). It has been shown that when a small amount of SCFAs is present in the intestine, there is a low production of regulatory T (Treg) cells, whose number increases after treatment with SCFAs (Eckburg et al. 2005). Therefore, more transforming growth factor β (TGF-β) and less pro-inflammatory cytokines (interleukin-6 [IL-6], IL-17, interferon-γ) are produced, thereby improving the immune responsiveness (Maslowski and Mackay 2011). Butyrate regulation takes place through the stimulation of dendritic cells and macrophages, which, via the activation of Gp43 and Gpr109, produce IL-10 and retinoic acid, and stimulate the development of IL-10-producing Treg cells in the colon. Butyrate also stimulates histone acetylation, increasing the expression of $FoxP3^{+}$, which drives the initial differentiation of T cells in Treg cells, crucial for the maintenance of immune homeostasis (Galloway-Peña et al. 2017).

The immune system is closely linked to cancer control; defects in immunity may be associated not only with carcinogenesis and cancer progression, but also to poor responses to cancer therapy (Gopalakrishnan et al. 2018). Cancer can also be influenced in terms of incidence and progression by components of the diet. Changes in the microbiota may affect oncogenesis and tumour progression at multiple levels. Some specific probiotic bacteria and their metabolites as cancer therapeutic agents have shown an entirely new approach to the understanding and treatment of this disease (Sharma 2019). The release of SCFAs may be one of the mechanisms implicated in these effects, so it is important to know how they can improve functioning of the immune system (with reference to intestinal diseases) as well as cancer prevention or therapy (Sharma 2019).

A condition of chronic inflammation linked to nutritional factors with pro-inflammatory properties is associated with at least 25% of cases of colorectal carcinoma (CRC) (Niederreiter et al. 2018). The correlation between SCFAs and CRC was demonstrated in a study conducted among native Africans, where higher levels of SCFAs in faeces indicated lower risk of CRC as compared to African and Caucasians Americans (O'Keefe et al. 2009).

High concentrations of SCFAs in faeces have been considered a mechanism by which probiotics exert their anticarcinogenic activity (Reis et al. 2017). To check this hypothesis, the same researchers (Reis et al. 2019) tested kefir as a food to reduce the incidence of pre-neoplastic lesions in an animal model for colorectal cancer. They observed that the consumption of this complex fermented dairy product increased the total concentration of SCFAs compared to the control group.

An increase in total concentration of SCFAs indicates that these organic acids act synergistically when exerting their beneficial effects (Kilner et al. 2012). There are reports in the literature on the ability of propionic acid to stimulate apoptosis and inhibit the proliferation of neoplastic cells. Moreover, propionic and acetic acids equal butyric acid in their anti-inflammatory activity, which is of great importance to hinder the progression of tumor development. Butyric acid has been more widely studied in experiments involving CRC, thanks to its ability to regulate colonocyte proliferation, differentiation, and apoptosis (Encarnação et al. 2018, Reis et al. 2019).

SCFAs are produced naturally by the intestinal microbiota thanks to the fibres coming from the diet; however, the amount of these fatty acids may not be enough to inhibit the development of CRC. The literature indicates that the daily consumption of probiotics and a balanced diet may contribute to the increase of SCFAs' level (LeBlanc et al. 2017, Reis et al. 2017, Falcinelli et al. 2018). Symbiotic supplements may also increase SCFAs' production (Gu and Roberts 2019), hence contributing in a variety of ways to the overall health and well-being of the consumer.

Conclusions and Future Perspectives

In this review, we have focused on an overview of the current state-of-the-art in the positive impact of SCFAs produced by probiotic lactic acid bacteria on the improvement of human health. Probiotics have important functional attributes and can exert positive effects to improve human health and well-being through various mechanisms, including the production of SCFAs. According to the scientific consensus, such key metabolites, mainly acetate, propionate, and butyrate, play a crucial role in the trigger-response relationship among host diet, microbiota, and homeostasis. They act, for example, on the regulation of energetic metabolism, immune inflammation, and blood pressure, recognizing its receptors and inhibiting HDACs. In this sense, there is a great interest in exploring such an exciting and emerging area in order to shape a healthier microbiota.

Evidences have shown that manipulation of the intestinal microbiota through diet and treatment with probiotics, as well as prebiotics and synbiotics, whose beneficial effects are far more pronounced when used together than alone, may provide a novel approach to the treatment of obesity, cancer, and other diseases. In addition, studies aiming to develop new mixtures of probiotics and prebiotics are vital to exploring other possibilities for improving health benefits. It is worth mentioning that advances in SCFA analytical methods presented in this chapter provide the tools to explore the composition and metabolic capacity of the microbiota with previously unattended precision.

Further developments in this field are expected such as (i) evaluation of new probiotic strains and their applicability in clinical research, (ii) design of engineered probiotics with specific properties and their use for the prevention and treatment of diseases, and (iii) design of improved approaches combining *in vitro*, *in vivo* and "omic" technologies, such as the study of signaling molecules in metagenomic, metabolomic, and epigenetic pathways that may contribute to current mechanistic understandings. These challenging approaches should be considered as future research efforts, with emphasis on human intervention tests, in order to explore SCFAs production and action mechanisms, thus opening the possibility of finding highly promising strategies in this area.

References

Bahar Halpern, K., A. Veprik, N. Rubins, O. Naaman and M.D. Walker. 2012. GPR41 gene expression is mediated by internal ribosome entry site (IRES)-dependent translation of bicistronic mRNA encoding GPR40 and GPR41 proteins. J. Biol. Chem. 287: 20154–20163.

Canani, R.B., M.D. Costanzo, L. Leone, G. Bedogni, P. Brambilla, S. Cianfarani, V. Nobili, A. Pietrobelli and C. Agostoni. 2011. Epigenetic mechanisms elicited by nutrition in early life. Nutr. Res. Rev. 24: 198–205.

Cani, P.D. and M. Van Hul. 2015. Novel opportunities for next-generation probiotics targeting metabolic syndrome. Curr. Opin. Biotechnol. 32: 21–27.

Çehreli, R. 2018. Moleculer nutritional immunology and cancer. J. Oncol. Sci. 4: 40–46.

Daliri, E.B.-M. and B.H. Lee. 2015. New perspectives on probiotics in health and disease. Food Sci. Human Wellness 4(2): 56–65.

Daliri, E.B.-M., B.H. Lee and D.H. Oh. 2019. Safety of probiotics in health and disease. pp. 603–622. *In*: Singh, R.B., R.R. Watson and T. Takahashi (eds.). The Role of Functional Food Security in Global Health. Academic Press, London, UK.

Dobrowolska-Iwanek, J., P. Zagrodzki, M. Woźniakiewicz, A. Woźniakiewicz, M. Zwolińska-Wcisło, D. Winnicka and P. Paśko. 2016. Procedure optimization for extracting short-chain fatty acids from human faeces. J. Pharm. Biomed. Anal. 124: 337–340.

Dugas, L.R., L. Lie, J. Plange-Rhule, K. Bedu-Addo, P. Bovet, E.V. Lambert, T.E. Forrester, A. Luke, J.A. Gilbert and B.T. Layden. 2018. Gut microbiota, short chain fatty acids, and obesity across the epidemiologic transition: the METS-Microbiome study protocol. BMC Public Health 18(1): 978.

Eckburg, P.B., E.M. Bik, C.N. Bernstein, E. Purdom, L. Dethlefsen, M. Sargent, S.R. Gill, K.E. Nelson and D.A. Relman. 2005. Diversity of the human intestinal microbial flora. Science 308: 1635–1638.

Encarnação, J.C., A.S. Pires, R.A. Amaral, T.J. Gonçalves, M. Laranjo, J.E. Casalta-Lopes, A.C. Gonçalves, A.B. Sarmento-Ribeiro, A.M. Abrantes and M.F. Botelho. 2018. Butyrate, a dietary fiber derivative that improves irinotecan effect in colon cancer cells. J. Nutr. Biochem. 56: 183–192.

Falcinelli, S., A. Rodiles, A. Hatef, S. Picchietti, L. Cossignani, D.L. Merrifield, S. Unniappan and O. Carnevali. 2018. Influence of probiotics administration on gut microbiota core: a review on the effects on appetite control, glucose, and lipid metabolism. J. Clin. Gastroenterol. 52 (suppl.1): S50–S56.

Furuhashi, T., K. Sugitate, T. Nakai, Y. Jikumaru and G. Ishihara. 2018. Rapid profiling method for mammalian feces short chain fatty acids by GC-MS. Anal. Biochem. 543: 51–54.

Galloway-Peña, J., C. Brumlow and S. Shelburne. 2017. Impact of the microbiota on bacterial infections during cancer treatment. Trends Microbiol. 25: 992–1004.

Gholizadeh, P., M. Mahallei, A. Pormohammad, M. Varshochi, K. Ganbarov, E. Zeinalzadeh, B. Yousefi, M. Bastami, A. Tanomand, S.S. Mahmood, M. Yousefi, M. Asgharzadeh and H.S. Kafil. 2019. Microbial balance in the intestinal microbiota and its association with diabetes, obesity and allergic disease. Microb. Pathog. 127: 48–55.

Gibson, G.R., R. Hutkins, M.L. Sanders, S.L. Prescott, R.A. Reimer, S.J. Salminen, K. Scott, C. Stanton, K.S. Swanson, P.D. Cani, K. Verbeke and G. Reid. 2017. Expert consensus document: The International Scientific Association for Probiotics and Prebiotics (ISAPP) consensus statement on the definition and scope of prebiotics. Nat. Rev. Gastroenterol. Hepatol. 14: 491–502.

Gopalakrishnan, V., B.A. Helmink, C.N. Spencer, A. Reuben and J.A. Wargo. 2018. The influence of the gut microbiome on cancer, immunity, and cancer immunotherapy. Cancer Cell. 33: 570–580.

Gu, J. and K. Roberts. 2019. Probiotics and prebiotics. pp. 67–80. *In*: Corrigan, M., K. Roberts and E. Steiger (eds.). Adult Short Bowel Syndrome. Academic Press, London, UK.

Haase, S., A. Haghikia, N. Wilck, D.N. Muller and R.A. Linker. 2018. Impacts of microbiome metabolites on immune regulation and autoimmunity. Immunology 154: 230–238.

Han, X., J. Guo, Y. You, M. Yin, C. Ren, J. Zhan and W. Huang. 2018. A fast and accurate way to determine short chain fatty acids in mouse feces based on GC–MS. J. Chromatog. B Analyt. Technol. Biomed. Life Sci. 1099: 73–82.

Hill, C., F. Guarner, G. Reid, G.R. Gibson, D.J. Merenstein, B. Pot, L. Morelli, R.B. Canani, H.J. Flint, S. Salminen, P.C. Calder and M.E. Sanders. 2014. Expert consensus document: the international scientific association for probiotics and prebiotics consensus statement on the scope and appropriate use of the term probiotic. Nat. Rev. Gastroenterol. Hepatol. 11(8): 506–514.

Hoyles, L. and R.J. Wallace. 2010. Gastrointestinal tract: intestinal fatty acid metabolism and implications for health. pp. 3119–3132. *In:* Timmis, K.N. (ed.). Handbook of Hydrocarbon and Lipid Microbiology. Springer, Berlin, Heidelberg, Germany.

Hu, J., S. Lin, B. Zheng and P.C.K. Cheung. 2018. Short-chain fatty acids in control of energy metabolism. Crit. Rev. Food Sci. Nutr. 58: 1243–1249.

Kao, L., T.H. Liu, T.Y. Tsai and T.M. Pan. 2018. Beneficial effects of the commercial lactic acid bacteria product, Vigiis 101, on gastric mucosa and intestinal bacterial flora in rats. J. Microbiol. Immunol. Infect. (in press).

Kerry, R.G., J.K. Patra, S. Gouda, Y. Park, H.-S. Shin and G. Das. 2018. Benefaction of probiotics for human health: a review. J. Food Drug Anal. 26: 927–939.

Kilner, J., J.S. Waby, J. Chowdry, A.Q. Khan, J. Noirel, P.C. Wright, B.M. Corfe and C.A. Evans. 2012. A proteomic analysis of differential cellular responses to the short-chain fatty acids butyrate, valerate and propionate in colon epithelial cancer cells. Mol. Biosyst. 8: 1146–1156.

Kobyliak, N., T. Falalyeyeva, N. Boykoc, O. Tsyryuk, T. Beregova and L. Ostapchenko. 2018. Probiotics and nutraceuticals as a new frontier in obesity prevention and management. Diabetes Res. Clin. Pract. 141: 190–199.

LeBlanc, J.G., F. Chain, R. Martín, L.G. Bermúdez-Humarán, S. Courau and P. Langella. 2017. Beneficial effects on host energy metabolism of short-chain fatty acids and vitamins produced by commensal and probiotic bacteria. Micro. Cell Fact. 16(1): 79.

Lee, Y., Z. Ba, R.F. Roberts, C.J. Rogers, J.A. Fleming, H. Meng, E.J. Furumoto and P.M. Kris-Etherton. 2017. Effects of *Bifidobacterium animalis* subsp. *lactis* BB-12® on the lipid/lipoprotein profile and short chain fatty acids in healthy young adults: a randomized controlled trial. Nutr. J. 16: 39.

Ley, R.E., F. Bäckhed, P. Turnbaugh, C.A. Lozupone, R.D. Knight and J.I. Gordon. 2005. Obesity alters gut microbial ecology. Proc. Natl. Acad. Sci. USA 102: 11070–11075.

Liang, Y., C. Lin, Y. Zhang, Y. Deng, C. Liu and Q. Yang. 2018. Probiotic mixture of *Lactobacillus* and *Bifidobacterium* alleviates systemic adiposity and inflammation in non-alcoholic fatty liver disease rats through Gpr109a and the commensal metabolite butyrate. Inflammopharmacology 26(4): 1051–1055.

Maruvada, P., V. Leone, L.M. Kaplan and E.B. Chang. 2017. The human microbiome and obesity: moving beyond associations. Cell Host Microbe 22(5): 589–599.

Maslowski, K.M. and C.R. Mackay. 2011. Diet, gut microbiota and immune responses. Nat. Immunol. 12: 5–9.

Meng, H., Z. Ba, Y. Lee, J. Peng, J. Lin, J.A. Fleming, E.J. Furumoto, R.F. Roberts, P.M. Kris-Etherton and C.J. Rogers. 2017. Consumption of *Bifidobacterium animalis* subsp. *lactis* BB-12 in yogurt reduced expression of TLR-2 on peripheral blood-derived monocytes and pro-inflammatory cytokine secretion in young adults. Eur. J. Nutr. 56: 649–661.

Moens, F., P. Van den Abbeele, A.W. Basit, C. Dodoo, R. Chatterjee, B. Smith and S. Gaisford. 2019. A four-strain probiotic exerts positive immunomodulatory effects by enhancing colonic butyrate production *in vitro*. Int. J. Pharm. 555: 1–10.

Mulders, R.J., K.C.G. de Git, E. Schéle, S.L. Dickson, Y. Sanz and R.A.H. Adan. 2018. Microbiota in obesity: interactions with enteroendocrine, immune and central nervous systems. Obes. Rev. 19: 435–451.

Nagpal, R., S. Wang, S. Ahmadi, J. Hayes, J. Gagliano, S. Subashchandrabose, D.W. Kitzman, T. Becton, R. Read and H. Yadav. 2018. Human-origin probiotic cocktail increases short-chain fatty acid production via modulation of mice and human gut microbiome. Sci. Rep. 8: 12649.

Niederreiter, L., T.E. Adolph and H. Tilg. 2018. Food, microbiome and colorectal cancer. Dig. Liver Dis. 50: 647–652.

Nova, E., F. Pérez de Heredia, S. Gómez-Martínez and A. Marcos. 2016. The role of probiotics on the microbiota: effect on obesity. Nutr. Clin. Pract. 31: 387–400.

O'Keefe, S.J., J. Ou, S. Aufreiter, D. O'Connor, S. Sharma, J. Sepulveda, T. Fukuwatari, K. Shibata and T. Mawhinney. 2009. Products of the colonic microbiota mediate the effects of diet on colon cancer risk. J. Nutr. 139: 2044–2048.

Pekmez, C.T., L.O. Dragsted and L.K. Brahe. 2019. Gut microbiota alterations and dietary modulation in childhood malnutrition—The role of short chain fatty acids. Clin. Nutr. 38(2): 615–630.

Pessione, A., G. Lo Bianco, E. Mangiapane, S. Cirrincione and E. Pessione. 2015. Characterization of potentially probiotic lactic acid bacteria isolated from olives: Evaluation of short chain fatty acids production and analysis of the extracellular proteome. Food Res. Int. 67: 247–254.

Pluznick, J.L. 2016. Gut microbiota in renal physiology: focus on short-chain fatty acids and their receptors. Kidney Int. 90: 1191–1198.

Pluznick, J.L. 2017. Microbial short-chain fatty acids and blood pressure regulation. Curr. Hypertens. Rep. 19(4): 25.

Primec, M., D. Mičetić-Turk and T. Langerholc. 2017. Analysis of short chain fatty acids in human feces: A scoping review. Anal. Biochem. 526: 9–21.

Primec, M., M. Klemenak, D. Di Gioia, I. Aloisio, N. Bozzi Cionci, A. Quagliariello, M. Gorenjak, D. Mičetić-Turk and T. Langerholc. 2019. Clinical intervention using *Bifidobacterium* strains in celiac disease children reveals novel microbial modulators of TNF-α and short-chain fatty acids. Clin. Nutr. 38: 1373–1381.

Quigley, E.M.M. 2019. Prebiotics and probiotics in digestive health. Clin. Gastroenterol. Hepatol. 17(2): 333–334.

Reis, S.A., L.L. Conceição, N.P. Siqueira, D.D. Rosa, L.L. Silva and M.D. Peluzio. 2017. Review of the mechanisms of probiotic actions in the prevention of colorectal cancer. Nutr. Res. 37: 1–19.

Reis, S.A., L.L. Conceição, M.M. Dias, N.P. Siqueira, D.D. Rosa, L.L. Oliveira, S.L.P. Matta and M.C.G. Peluzio. 2019. Kefir reduces the incidence of pre-neoplastic lesions in an animal model for colorectal cancer. J. Funct. Foods. 53: 1–6.

Remely, M., E. Aumueller, C. Merold, S. Dworzak, B. Hippe, J. Zanner, A. Pointner, H. Brath and A.G. Haslberger. 2014. Effects of SCFA producing bacteria on epigenetic diabetes and obesity. Gene 537: 85–92.

Renga, B., A. Mencarelli, P. Vavassori, V. Brancaleone and S. Fiorucci. 2010. The bile acid sensor FXR regulates insulin transcription and secretion. Biochim. Biophys. Acta 1802: 363–372.

Richards, L.B., M. Li, B.C.A.M. van Esch, J. Garssen and G. Folkerts. 2016. The effects of short-chain fatty acids on the cardiovascular system. PharmaNutrition 4: 68–111.

Riva, A., F. Borgo, C. Lassandro, E. Verduci, G. Morace, E. Borghi and D. Berry. 2017. Pediatric obesity is associated with an altered gut microbiota and discordant shifts in Firmicutes populations. Environ. Microbiol. 19: 95–105.

Rivero-Gutiérrez, B., R. Gámez-Belmonte, M.D. Suárez, J.L. Lavín, A.M. Aransay, M. Olivares, O. Martínez-Augustin, F. Sánchez de Medina and A. Zarzuelo. 2017. A synbiotic composed of *Lactobacillus fermentum* CECT5716 and FOS prevents the development of fatty acid liver and glycemic alterations in rats fed a high fructose diet associated with changes in the microbiota. Mol. Nutr. Food Res. 61: 1600622.

Rivière, A., M. Selak, D. Lantin, F. Leroy and L. De Vuyst. 2016. *Bifidobacteria* and butyrate-producing colon bacteria: importance and strategies for their stimulation in the human gut. Front. Microbiol. 7: 979.

Rodrigues, H.G., F. Takeo Sato, R. Curi and M.A.R. Vinolo. 2016. Fatty acids as modulators of neutrophil recruitment, function and survival. Eur. J. Pharmacol. 785: 50–58.

Roy, C.C., C.L. Kien, L. Bouthillier and E. Levy. 2006. Short-chain fatty acids: ready for prime time? Nutr. Clin. Pract. 21: 351–366.

Schwiertz, A., D. Taras, K. Schäfer, S. Beijer, N.A. Bos, C. Donus and P.D. Hardt. 2010. Microbiota and SCFA in lean and overweight healthy subjects. Obesity 18: 190–195.

Sharma, A. 2019. Importance of probiotics in cancer prevention and treatment. pp. 33–45. *In*: Viswanath, B., J. Fedor, L. Versteeg-Buschman and T. Bennett (eds.). Recent Developments in Applied Microbiology and Biochemistry. Elsevier, San Diego, CA, USA.

Sivaprakasam, S., P.D. Prasad and N. Singh. 2016. Benefits of short-chain fatty acids and their receptors in inflammation and carcinogenesis. Pharmacol. Ther. 164: 144–151.

Sivaprakasam, S., Y.D. Bhutia, S. Yang and V. Ganapathy. 2018. Short-chain fatty acid transporters: role in colonic homeostasis. Compr. Physiol. 8: 299–314.

Sun, M., W. Wu, Z. Liu and Y. Cong. 2017. Microbiota metabolite short chain fatty acids, GPCR, and inflammatory bowel diseases. J. Gastroenterol. 52: 1–8.

Sun, Y. and M.X.D. O'Riordan. 2013. Regulation of bacterial pathogenesis by intestinal short-chain fatty acids. pp. 93–118. *In*: Seriaslani, S. and G.M. Gadd (eds.). Advances in Applied Microbiology. Academic Press, San Diego, CA, USA.

Tungland, B. 2018. Short-chain fatty acid production and functional aspects on host metabolism. pp. 37–106. *In*: Haley, M., S. Masucci and F. Coulthurst (eds.). Human Microbiota in Health and Disease. From Pathogenesis to Therapy. Academic Press, Cambridge, Massachusetts, USA.

Yousuf, B. and A. Mishra. 2019. Exploring human bacterial diversity toward prevention of infectious disease and health promotion. pp. 519–533. *In*: Das, S. and H.R. Dash (eds.). Microbial Diversity in the Genomic Era. Academic Press, Cambridge, Massachusetts, USA.

Zeng, M. and H. Cao. 2018. Fast quantification of short chain fatty acids and ketone bodies by liquid chromatography-tandem mass spectrometry after facile derivatization coupled with liquid-liquid extraction. J. Chromatogr. B Analyt. Technol. Biomed. Life Sci. 1083: 137–145.

Zhang, S., H. Wang and M. Zhu. 2019. A sensitive GC/MS detection method for analyzing microbial metabolites short chain fatty acids in fecal and serum samples. Talanta 196: 249–254.

9

Impact of Probiotics on Human Gut Microbiota and the Relationship with Obesity

Fernanda Bianchi and *Katia Sivieri**

Introduction

The prevalence of obesity has grown to such proportions that it has become one of the greatest public health concerns worldwide. This pathology can cause serious metabolic, social, and psychological consequences, reaching populations of different ages (Apovian 2016, Chu et al. 2018). The World Health Organization (WHO) defines obesity as "abnormal or excessive fat accumulation that may impair health" and sets a body mass index (BMI) of > 30 kg/m^2 for this population (WHO 2018).

Several co-morbidities, such as type 2 diabetes mellitus, cancer, reproductive disorders and cardiovascular diseases, are associated with obesity (WHO 2018). Other consequences, including social discrimination, lower quality of life, and economic loss, are also connected to obesity, emphasizing the severity of this pathology (Puhl and Heuer 2010, Tremmel et al. 2017), which kills at least 2.8 million people per year (WHO 2018).

The main factors leading to obesity are poor lifestyle and diet, genetics, use of medicines, lack of physical activity, family influence, and cultural aspects (Dahiya et al. 2017, WHO 2018). Recent studies have also pointed to the gut microbiota composition as an important contributor to the onset and development of obesity (Bomhof and Reimer 2015, Kasai et al. 2015, Peters et al. 2018), providing a new vision and focus in obesity treatment.

The human gut microbiota is a complex and dynamic community, composed of trillions of microorganisms exerting significant influence over the host's health

Department of Food and Nutrition, School of Pharmaceutical Sciences, State University of São Paulo (UNESP), Araraquara, SP, Brazil.
Email: febianchi@hotmail.com
* Corresponding author: katia.sivieri@unesp.br

(Thursby and Juge 2017). Different factors, such as lifestyle and diet, can affect the composition and activity of the gut microbiota, leading to the improvement of health or to the development of various diseases, such as obesity (Rajoka et al. 2017). In this sense, several probiotic strains have been studied as a potential therapeutic approach in obesity prevention (Dahiya et al. 2017). The high majority of probiotics used for this purpose belong to *Lactobacillus* and *Bifidobacterium* genera and, although the complete action mechanism is still not fully elucidated, various hypotheses have been proposed and several bacterial strains have had their beneficial action proved under the obesity-related microbiota (Aoki et al. 2017, Chen et al. 2018a).

This chapter begins with a review of the gut microbiota structure, composition, and its main functions, and then covers the relationship between gut microbiota, diet, and obesity. This chapter also includes a discussion of the factors that play a role in linking obesity to gut microbiota. Finally, the effects of probiotics on intestinal microbiota as well as the possible mechanisms involved in probiotic's anti-obesity effects are presented.

The Human Gut Microbiota

Gastrointestinal tract structure and microbial composition

The human gastrointestinal tract hosts at least 10^{14} microorganisms, which is approximately 10 times the total number of human body cells (Riaz 2015, Bäumler and Sperandio 2016). This population encodes 3 to 4 million genes, exceeding the numbers in the human genome by approximately 150 times (Dave et al. 2012, Lynch and Pedersen 2016).

The colonization of the digestive tract starts at birth, with reduced diversity and high instability. Different factors such as birth type (caesarean section or vaginal delivery), mother's intrauterine microbiota, method of postnatal feeding (formula or breast milk), and genetics, as well as intestinal immunity, biliary secretion, mucosal barrier composition, digestive enzymes, and intestinal motility can affect the composition of bacteria in the child's gut (Doré and Corthier 2010, Gosalbes et al. 2013, Goodrich et al. 2014). Although the microbiota of adult individuals is relatively stable, some changes can occur due to the lifestyle, type of diet, environmental influences, place of residence, and antibiotic use (Lozupone et al. 2012).

The gastrointestinal microbiota distribution varies according to its location in the digestive tube (Dave et al. 2012, Robles-Alonso and Guarner 2013). As shown in Figure 1, a low bacterial density is found in the stomach and duodenum regions. This fact occurs because of the presence of acidic gastric juice and pancreatic enzymes (O'Hara and Shanahan 2006, Dave et al. 2012). On the other hand, a gradual increase of bacterial density is reported in the distal small intestine, reaching its highest concentration in the colon (O'Hara and Shanahan 2006, Dave et al. 2012), where the intestinal transit time, availability of nutrients, and pH are favourable (Gibson and Roberfroid 1995), totalling 1–2 kg of our body weight (Payne et al. 2012).

In the literature, the gut microbiota is referred to as a "separate organ" (Possemiers et al. 2011) or "forgotten organ" (O'Hara and Shanahan 2006, Clemente et al. 2012) because of the exorbitant number of microorganisms inhabiting this location. The gut microbiota is composed of both anaerobic and aerobic microbial communities although

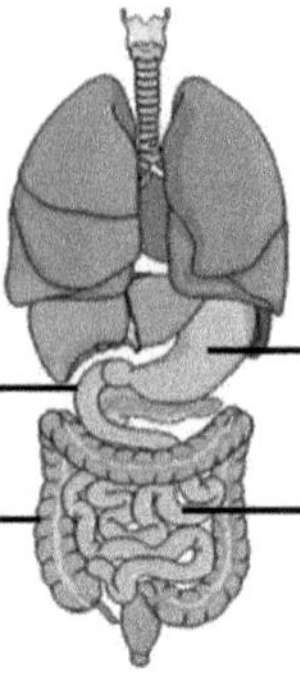

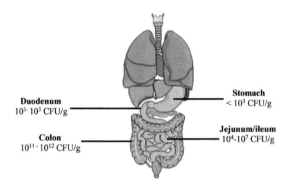

Figure 1: Total number (CFU/g) of microorganisms in the gastrointestinal tract.

the vast majority are strictly anaerobic and difficult to be cultured *in vitro* (Dave et al. 2012). The gut microbiota consists of more than 1,500 species distributed in at least seven different phyla (Dave et al. 2012, Alonso and Guarner 2013). The most dominant bacterial phyla found in the human gut are *Firmicutes* (including for example the genera *Lactobacillus, Ruminococcus, Clostridium,* and *Eubacteria*) and *Bacteroidetes* (including genera *Bacteroides, Porphyromonas, Prevotella* among others), followed by *Proteobacteria* (including genera *Klebsiella, Enterobacter, Succinivibrio*) and *Actinobacteria* (including, genera *Bifidobacterium* and *Coreobacteria*). Other phyla can be found in minor proportions such as *Fusobacteria, Verrucomicrobia, Tenericutes, Cyanobacteria, Spirochaetes,* and *Synergistetes* with less than 2% (Gill et al. 2006, Robles-Alonso and Guarner 2013, Caesar et al. 2015). All these microorganisms interact with each other and with their host, exerting influence on its physiology and health (Dave et al. 2012).

Despite the inter-individual variability of microbiota composition, determined by place of residence, lifestyle, disease conditions, hygiene, sex and age, it has been proposed that the human gut microbiome can be classified into two main specific microbial enterotypes. The identification of each enterotype is based on a relatively high abundance of a single microbial genus: *Bacteroides* (enterotype 1) or *Prevotella* (enterotype 2) (Vieira-Silva et al. 2016, Christensen et al. 2018, Hjorth et al. 2018). Each enterotype is linked with long-term diets. *Bacteroides* enterotype is associated with high fat or protein diets while *Prevotella* is linked with a high carbohydrate and dietary fibre diet (Christensen et al. 2018).

Gut microbiota function and the relationship with diet

Several studies have shown the influence of diet on the microbiota composition and, consequently, on health (Pendyala et al. 2012, Haro et al. 2016, Zhang and Yang 2016, Guo et al. 2017, Willson and Situ 2017). According to the cited authors, a diet high in fat, especially in trans and saturated fat, is related with gut dysbiosis, leading to obesity and metabolic syndrome. Pendyala et al. (2012), for example, observed that a high-fat Western diet increases the concentration of endotoxin in the plasma (endotoxemia), probably due to failure in intestinal barrier function, leading to an inflammatory process. According to Willson and Situ (2017), diets with low fat and high in fruits, complex fibres, vegetables, and supplemented with probiotics are key to a healthy gut microbiota.

The gut microbiota uses ingested dietary components not absorbed in the small intestine, such as carbohydrates, lipids, and proteins, by means of fermentation leading to the production of several metabolites, which can directly or indirectly affect the human metabolism and health (Ramakrishna 2013). Short-chain fatty acids (SCFAs), especially propionic, acetic and butyric acids, are the main metabolites produced by the microbiota in the presence of non-digestible carbohydrates (Simpson and Campbell 2015, Kumari and Kozyrskyj 2017). Besides the SCFAs, the gut microbiota can produce other metabolites like choline, bile acid metabolites (Yatsunenko et al. 2012), indole, and phenolic derivatives (Ramakrishna 2013, Simpson and Campbell 2015, Kumari and Kozyrskyj 2017).

Short-chain fatty acids are the main source of energy for the colon epithelium. Moreover, they stimulate the proliferation of epithelium cells, the visceral blood flow, and enhance the absorption of sodium and water (Almeida et al. 2009). Studies have shown that besides nourishing microorganisms and enterocytes, SCFAs, including acetic, butyric, and propionic acids, have other refined functions (David et al. 2014, Kaji et al. 2014, Ji et al. 2016). They are considered important regulators of immunity, energy metabolism and adipose tissue expansion and can prevent the growth of bacterial pathogens due to the acidification of the colonic lumen (David et al. 2014, Kaji et al. 2014). According to Ji et al. (2016), butyric acid is able to decrease the expression of pro-inflammatory cytokines such as tumour necrosis factor (TNF-α and TNF-β), interleukin-6 (IL-6) and IL-1β, alleviating thus the inflammatory process. It is also able to inhibit the genotoxic capacity of nitrosamines, reducing the risk of colon cancer (Schwabe and Jobin 2013). Moreover, it has been proposed that butyric acid has the ability to reduce food intake, moderating weight gain. Propionate and acetate were also reported to have an action on appetite regulation and, consequently, on body weight. The mechanism connecting the production of intestinal SFCAs to food intake regulation includes the stimulation of anorectic gut hormones' release, such as the peptide YY (PYY) and the glucagon-like peptide-1 (GLP-1), which can inhibit appetite, partly by regulating soluble leptin receptors (David et al. 2014, Kaji et al. 2014, Iepsen et al. 2015, Psichas et al. 2015, Christiansen et al. 2018).

Although the carbohydrates represent the main substrates leading to the production of SCFAs, fatty acids with an aliphatic chain (C1-C5) have also been able to produce acetate, propionate, butyrate, as well as isobutyrate, valerate, and isovalerate (Yatsunenko et al. 2012).

Besides SCFAs' production, the gut microbiota performs many other functions like supplying vitamin K and several B group vitamins (Ramakrishna 2013, Simpson and Campbell 2015), the biotransformation of polyphenols into metabolites, the prevention of pathogenic bacteria colonization by means of competition for essential resources and ecological niche, and exerts a significant influence over immunological processes (Gibson and Roberfroid 1995, Ozdal et al. 2016, Yarandi et al. 2016). Moreover, the gut microbiota attends to the function of the gut-brain axis (GBA). GBA is a bidirectional communication between the central nervous system and the enteric nervous system. This communication links peripheral intestinal functions with the emotional and cognitive centres of the brain (Carabotti et al. 2015).

Some of the functions of the gut microbiota are represented in Figure 2.

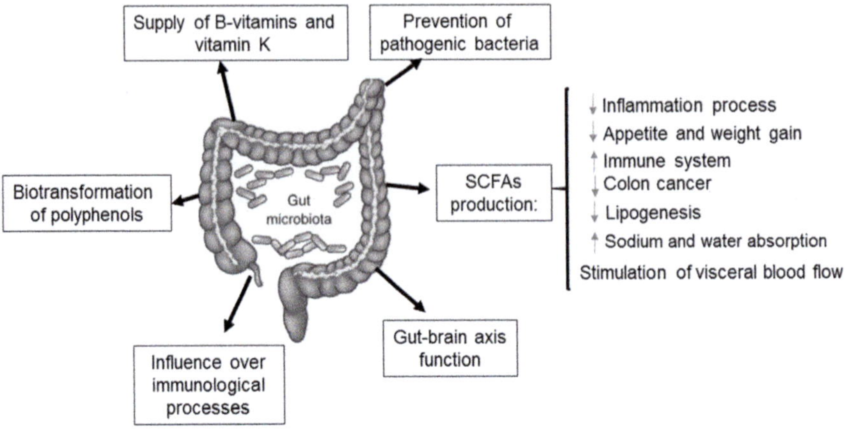

Figure 2: Different functions of gut microbiota.

It is also important to highlight that the anaerobic metabolism of the gut bacteria can include proteolytic fermentation, which will produce nitrogenous derivatives like amines and ammonium, and some of them have carcinogenic effects (Macfarlane and Macfarlane 2012).

Several factors such as lifestyle and diet can affect the composition and activity of gut microbiota and, consequently, interfere in the various already-cited functions of the microbiota, leading to the improvement of health or to the development of various diseases such as obesity, as illustrated in Figure 3.

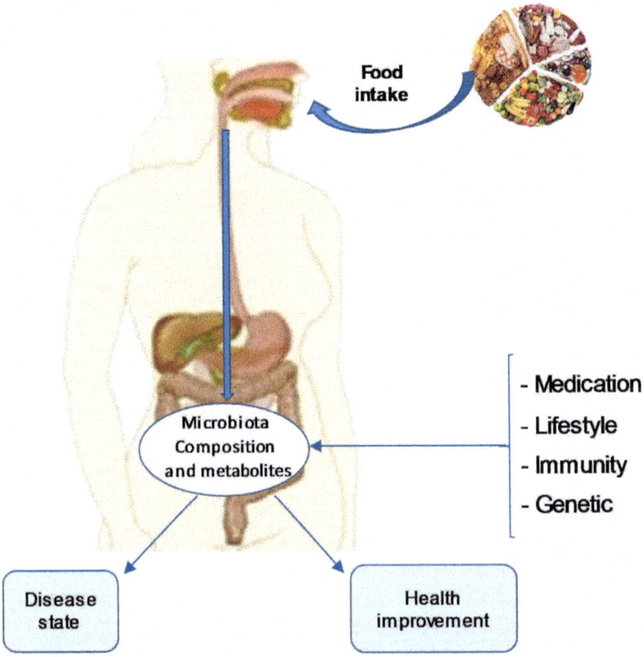

Figure 3: Interaction between diet, gut microbiota and host.

Relationship between Gut Microbiota and Obesity

Obesity has been one of the main public health concerns of the 21st century. According to Swinburn et al. (2011), more than 500 million people across the world are considered obese, which highlights the severity of the disease. Many factors such as lifestyle, host genetics, diet and metabolism as well as the side-effects of drugs are known to contribute to the progression or development of obesity (Ness-Abramof and Apovian 2005, Dahiya et al. 2017). Beyond these factors, some evidence suggests that the gut microbiota composition also has a high influence on obesity onset (Ley et al. 2006, Fleissner et al. 2010, Bomhof and Reimer 2015, Kasai et al. 2015, Rosenbaum et al. 2015, Sommer and Bäckhed 2016, Kootte et al. 2017).

The first evidence linking the gut microbiota to obesity came from Bäckhed et al. (2004) study. These authors transplanted the faecal microbiota from conventionally-raised mice to germ-free (GF) mice. They observed that GF mice gained more fat pad mass and body weight despite the reduction of food consumption, which led to insulin resistance along with higher glucose and leptin levels in the blood. Other studies have also shown that transplantation of caecal or faecal microbiota from obese to lean mice significantly increased body fat mass, insulin resistance, adiposity, and obesity-associated metabolic phenotypes compared to controls (Turnbaugh et al. 2006, Bäckhed et al. 2007, Ridaura et al. 2013, Zhang et al. 2015). On the other hand, clinical trials showed that obese volunteers receiving faecal microbiota from lean donors had an improvement of serum insulin sensitivity over a 6-week period (Kootte et al. 2017, Vrieze et al. 2012). These findings suggest that the obese phenotype is transmissible from one microbiota to another, and that the microbiota composition can interfere with the host's health (Araújo et al. 2017).

Several hypotheses exist linking the gut microbiota composition/metabolites to obesity. Among these hypotheses, we can highlight the role of lipopolysaccharides (LPS), bile acid production, and the SCFAs produced by microbiota (Swann et al. 2011, Go et al. 2013, Chakraborti 2015).

Role of lipopolysaccharides (LPS), bile acid production, and short-chain fatty acids (SCFAs) in linking obesity to gut microbiota

Aiming to investigate the relationship between gut microbiota and obesity, several studies have demonstrated that LPS and short-chain fatty acids SCFAs play a role in linking obesity to gut microbiota. The studies have also shown that the gut microbiota has an influence on bile acid production and on lipid metabolism (Swann et al. 2011, Go et al. 2013, Chakraborti 2015).

Lipopolysaccharide

LPS is an endotoxin found in the cell wall of Gram-negative bacteria (Zhao 2013). It is normally found at low and high concentrations in the blood of healthy and obese individuals, respectively (DiBaise et al. 2012). High concentrations of plasma LPS have been associated with increased levels of cluster of differentiation 14 (CD14, receptor for the LPS found in the surface of monocytes and macrophage) and IL-6, which are markers of inflammation (Graham et al. 2015).

Studies suggest that the ingestion of a high fat diet (HFD) alters the gut microbiota composition, and consequently increases the luminal levels of bacterial LPS. The LPS in contact with Toll-like receptor 4 (TLR4, receptor for the LPS found in the surface of monocytes and macrophage) triggers inflammatory signalling pathways and pro-inflammatory cytokines secretion in the intestine. This inflammatory state can subsequently induce deficiencies in the secretion, production and thickness of the mucus layer and can act by reducing the expression of genes that code the intestinal tight junction proteins such as ZO-1 and occludin. This will consequently favour the increase of intestinal epithelium permeability, allowing the passage of bacterial components, such as LPS from the intestinal lumen to the adipose tissue and circulation, where it can promote systemic inflammation, hyperglycaemia, insulin resistance and adipogenesis (Hotamisligil 2006, Moreira et al. 2012), as shown in Figure 4.

According to Graham et al. (2015), the main source of LPS is the *Prevotellaceae* family (Gram-negative bacterial subgroup belonging to *Bacteroidetes* phylum), which is found in great abundance in the gut microbiota of obese people (Graham et al. 2015).

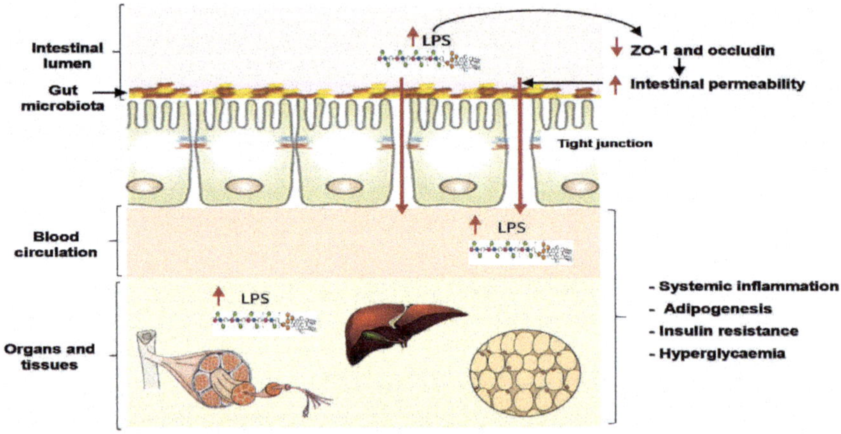

Figure 4: Role of lipopolysaccharides (LPS) in microbiota profile and consequent metabolic complications.

Bile acid production

The gut microbiota composition has also an influence on bile acid production and on the lipogenesis process, which can have an impact on obesity. Swann et al. (2011) demonstrated that mice with a distinct gut microbial structure have a divergent energy metabolism due to different bile acid metabolites in their organs.

Bile acids are usually classified into primary and secondary. The gut microbiota plays a role in the secondary production of bile acids. In the colon, taurocholate (TCA) and taurodeoxycholate (TDCA), primary conjugated bile acids produced in the liver, are de-conjugated by bacterial bile salt hydrolases (BSH), and dehydroxylated by bacterial 7α-dehydroxylases to form the secondary bile acids, deoxycholic acid (DCA) and lithocholic acid (LCA), respectively (Chiang et al. 2017).

The gut microbiota modulates bile acid metabolism by influencing the nuclear bile acid receptor farnesoid X (FXR)/G protein-coupled receptor (TGR5) signalling that indirectly contribute to obesity development (Dahiya et al. 2017). FXR is a receptor that negatively regulates the expression of the gene cholesterol 7α-hydroxylase (CYP7A1), which encodes the enzyme cholesterol 7α-hydroxylase. The enzyme cholesterol 7α-hydroxylase catalyses the initial step of cholesterol catabolism and bile acid synthesis in liver (Chiang 2009). In the intestine, FXR induces fibroblast growth factor 15/19 (FGF15/FGF19), which will indirectly signal to inhibit CYP7A1 gene transcription and, consequently, the initial step of cholesterol catabolism and bile acid synthesis in liver. Decreasing BSH activity in gut microbiota increases tauro-β-muricholic acid (TβMCA), which antagonizes FXR activity, reducing FGF15/19 and stimulating CYP7A1, which will consequently stimulate bile acid synthesis in hepatocytes (Figure 5) (Chiang et al. 2017). The G protein-coupled receptor (TGR5) is a membrane receptor sensitive to the presence of bile acids expressed in the ileum and colon. This receptor promotes an intracellular elevation of cyclic Adenosine MonoPhosphate (cAMP). When cAMP is elevated in the adipose and muscle tissue cells, it triggers mechanisms of energy expenditure and stimulates insulin secretion in pancreatic β cells to improve insulin sensitivity (Figure 5) (de Fabiani et al. 2003, Watanabe et al. 2006, Chiang et al. 2017). Because of these facts, it is possible to state that the gut-liver axis has an important role in metabolic homeostasis regulation and bile acid synthesis (Chiang et al. 2017).

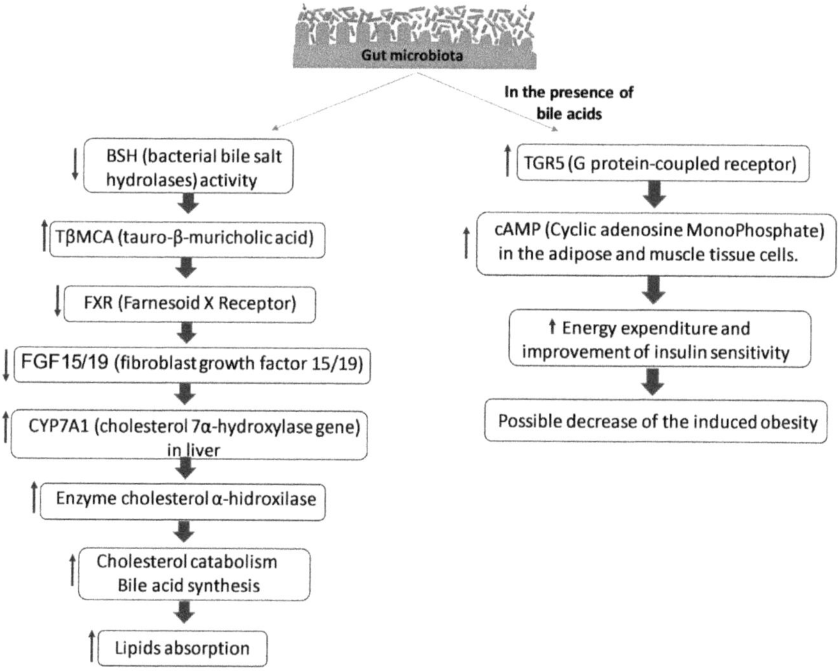

Figure 5: Influence of gut microbiota on bile acid production and lipid metabolism.

Short-chain fatty acids

Regarding SCFAs, there has been growing evidence that specific metabolites, such as acetic, propionic and butyric acids, have important physiological functions (Chakraborti 2015). Although some studies have linked SCFAs with obesity—hypothesizing that the SCFAs can contribute to additional calories through fermentation, causing an imbalance in energy regulation and thus contributing to obesity (Turnbaugh et al. 2006, Schwiertz et al. 2010, Murugesan et al. 2018)—according to Chakraborti (2015) and Morrison and Preston (2016), this hypothesis is not supported by the evidence that high fibre diets, which normally cause an increase in SCFA production, are able to protect against weight gain. Moreover, according to Chakraborti (2015), Canfora et al. (2015), and Morrison and Preston (2016), SCFAs, especially butyrate and propionate, are considered predominantly anti-obesogenic metabolites. In agreement, Brahe et al. (2013) suggested that reduced levels of butyrate produced by the gut bacteria might be associated with metabolic risk in humans.

Butyrate has been found to improve insulin sensitivity (Hartstra et al. 2015), to increase leptin gene expression by regulating food intake and body weight (Harris et al. 2012), and to possess obesity-related anti-inflammatory action by regulation of tight junction gene expression, increasing, consequently, intestinal barrier function (Brahe et al. 2013, Hartstra et al. 2015). Propionic acid also has a favourable effect on leptin gene expression reducing the intake of food and regulating weight gain (Xiong et al. 2004, Lin et al. 2012). Additionally, propionate is able to reduce cholesterol synthesis (Harris et al. 2012). Although acetate serves as a substrate for synthesis of cholesterol and has an influence on the synthesis of lipids in liver (Sanz et al. 2010), it has been demonstrated that acetic acid also has the ability to inhibit weight gain (Lin et al. 2012, Christiansen et al. 2018).

In the face of all these cited factors, it is clear that the composition of and metabolites produced by gut microbiota play a role in obesity and metabolic diseases, although some mechanisms have not been fully elucidated. In this way, the key for positive microbial changes is certainly the consumption of an adequate diet. Evidence that different probiotic strains can modulate the human gut microbiota in a positive way has been growing and is, therefore, considered a good strategy for the management of obesity (Reyes et al. 2016, Dahiya et al. 2017).

Probiotics

Probiotics are live microorganisms, mainly from the bacterial genera *Bifidobacterium* and *Lactobacillus* as well as yeasts that, "*when administered in regularly and adequate amounts, can confer beneficial effects to the host health*" (Hill et al. 2014, Kootte et al. 2012).

The available probiotic foods on the market include milk-based desserts, fermented milks, powdered milks, ice creams, yogurts and various types of cheeses. There are also probiotics in capsule form or powdered products to be dissolved in cold drinks, fortified juices, vegetable drinks and mayonnaises (Saad 2006).

A number of beneficial effects have been related to probiotic consumption; however, to confer the benefits to the host's health, there should be an adequate probiotic consumption in terms of both quantity and duration (WHO/FAO 2006). The minimum amount of viable probiotic strain required to exert beneficial effects on the host should be in the range of 10^8 to 10^9 Colony Forming Units (CFU) of the product's daily consumption recommendation. If the food product's daily recommendation is, for example, 200 mL, there should be a count of 10^6 to 10^7 CFU/mL in the product. Smaller values can be accepted, but the company must prove its effectiveness (ANVISA 2012). In addition, to be considered a probiotic strain, the microorganism must resist the manufacturing and storage process, as well as the physical-chemical conditions of the gastrointestinal tract, such as gastric acid and biliary secretions (Fooks and Gibson 2002). Moreover, the consumption of probiotic strains must be primarily safe (de Simone 2018).

Probiotic effects are strain-specific, which means that the effects described for specific strains cannot always be assumed for other microorganisms from the same specie (Athalye-Jape et al. 2018). The different beneficial probiotic effects include: immunomodulation, improvement of gut barrier function, prevention or treatment of several diseases and disorders such as diarrhoea associated with rotavirus, food allergies, irritable bowel syndrome, diabetes, cancer, and obesity, amongst others (Kerry et al. 2018).

Probiotic, intestinal microbiota, and management of obesity

Several recent pre-clinical studies have shown the beneficial impact of different probiotic strains, especially belonging to *Lactobacillus* spp. and *Bifidobacterium* spp., on obesity (Rather et al. 2014, Alard et al. 2016, Aoki et al. 2017, Bagarolli et al. 2017, Park et al. 2017, Chen et al. 2018b, Roselli et al. 2018, Wanchai et al. 2018).

As shown in Table 1, a reduction in weight gain, plasma lipids, and pro-inflammatory genes as well as an improvement in insulin resistance can be observed in most of the listed studies during treatment with different probiotic strains compared with the controls.

Although the results were very similar between the studies, the beneficial effects on obesity will depend on the bacterial strain and, therefore, we can state that the probiotic effects cannot be generalized to all probiotic strains. Alard et al. (2016), for example, found a reduction in body weight, epididymal adipose tissue, and on blood leptin levels by *Bifidobacterium* (*B.*) *animalis* subsp. *lactis* LMG P-28149, but not by the probiotic *Lactobacillus* (*B.*) *rhamnosus* LMG S-28148 (Table 1). Aoki et al. (2017) also observed different results for different *Bifidobacterium* strains. Although both *B. animalis* ssp. *lactis* GCL2505 (BlaG) and *B. longum* ssp. *longum* CM1217T (BloJ) have shown no effect on weight gain and energy intake, only the strain BlaG was found to have a positive impact on visceral fat accumulation, levels of propionate, acetate, glucagon-like peptide-1, and on glucose tolerance as well as on intestinal *Bifidobacterium* and *Lactobacillus* genera growth (Table 1).

Interestingly, Alard et al. (2016) found an increased abundance of intestinal *Akkermansia* (*A.*) *muciniphila* during treatment with *B. animalis* LMG S-28148 and a combination of *L. rhamnosus* LMG P-28149 and *B. animalis* LMG S-28148. *Akkermansia muciniphila* was reported to have a negative correlation with obesity (Louis et al. 2010, Schneeberger et al. 2015, Dao et al. 2016, Hippe et al. 2016, Remely et al. 2016). Dao et al. (2016), as well as Remely et al. (2016), for example, showed weight lowering effects under *A. muciniphila* administration.

Table 1: Impact of different probiotic strains on obesity parameters using animal studies.

Probiotic	Subjects	Concentration/ Duration	Observed effects	Authors
Lactobacillus (*L.*) *casei* NCDC 19	Male C57BL/6 high-fat diet obesity mice. Control consuming high-fat diet (HFD) without probiotic	8 $\log^{10}$ CFU/mL for 8 wk	Compared to Control: - Reduction of body weight gain and epididymal fat weights - Reduction of blood glucose, plasma lipids and expression level of leptin - Increase of ceacal *Bifidobacterium* spp. and on adiponectin expression levels	Rather et al. (2014)
- *L. rhamnosus* LMG S-28148 - *Bifidobacterium* (*B.*) *animalis* subsp. *lactis* LMG P-28149 - Mix of both strains	Male C57BL/6J mice fed with HFD. Control consuming HFD and Low fat diet (LFD) without probiotics	~ 10^9 CFU of each strain for 7 wk	Compared to control (HFD-fed mice): - *B. animalis* subsp. *lactis* LMG P-28149 reduced body weight, epididymal adipose tissue (EWAT) and blood leptin levels as efficiently as the probiotic mix. *L. rhamnosus* LMG S-28148 did not yield such effects - Both strains decreased the expression levels of Cd11c, Cd11b, F4/80 and Cd68 in the EWAT - Only the mix and *B. animalis* increased the *A. muciniphila* ↑ Butyrate and propionate production by the probiotic mix	Alard et al. (2016)
- *L. plantarum* HAC01 - *L. rhamnosus* GG (strain reference)	Male C57BL/6J mice fed with HFD. Control consuming high-fat diet without probiotic	1×10^8 CFU of each strain suspended in 20 µL of PBS for 8 wk	Compared to control: ↓ Mesenteric adipose depot (by HACO1) ↓ Body weight gain (by HACO1 and GG) ↓ Ruminococcaceae (by HACO1) and increase of this family by GG ↓ *Bacteroides* (by HACO1)	Park et al. (2017)
Pool of probiotics: *L. rhamnosus*, *L. acidophilus* and *B. bifidum*	Male Swiss mice fed with HFD. Control consuming HFD and LFD without probiotic pool	6×10^8 CFU/mL of each strain during 5 wk	Compared to control (HFD-fed mice): - Improvement in hypothalamic insulin and leptin resistance - Decrease of intestinal permeability, LPS translocation and systemic low-grade inflammation and food ingestion - Increase of glucose tolerance	Bagarolli et al. (2017)

L. mali APS1	Male C57BL/6 J diet induced obesity mice. Control consuming normal diet without probiotic	5 × 10^8 CFU in saline by gavage daily for 3 wk	Compared to control: - Body weight loss and reduction of caloric intake and fat accumulation - Return of intestinal microbiota toward a pre-obese state	Chen et al. (2018b)
L. paracasei HII01	Male Wistar high-fat diet obesity rats. Control: normal standard diet with and without probiotic and HFD without probiotic	1 × 10^8 CFU/mL by oral gavage for 12 wk	Compared to control (HFD): - Attenuation of hyperlipidemia, systemic inflammation (as for example reduction of serum LPS levels) and insulin resistance	Wanchai et al. (2018)
- *B. animalis* ssp. *lactis* GCL2505 (BlaG) - *B. longum* ssp. *longum* CM1217T (BloJ)	Male C57BL/6 J mice fed with HFD. Control: HFD without probiotic and normal standard diet	1 × 10^9 FCU/mL of both strains daily for 7 wk	Compared to control (HFD): - No effect on weight gain and energy intake ↓ Visceral fat accumulation and glucose tolerance (by BlaG) ↑ *Bifidobacterium* and *Lactobacillus* genera (by BlaG) ↓ *Clostridium* genus (by BlaG and BloJ) ↑ Levels of propionate, acetate and glucagon-like peptide-1 (by BlaG) - Caecal acetate correlated negatively with the amount of visceral fat and positively with plasma GLP-1 levels - Positive correlation between *Bifidobacterium* and acetate levels and colonic GLP-1 levels and negative correlation between *Bifidobacterium* and visceral fat	Aoki et al. (2017)
Mix of *Bifidobacterium*: *B. lactis* Bi1 *B. breve* Bbr8 *B. breve* BL10	Male C57BL/6J mice fed a high fat diet. Control: HFD without probiotic	1 × 10^9 CFU for 12 wk	Reduction of weight gain, adipose tissue fat accumulation, adipocyte size, and macrophage and CD4+ T cell infiltration, improving lipid profile and regulating leptin and cytokine secretion compared to HFD control	Roselli et al. (2018)

Different clinical studies have also demonstrated the positive effects of one or more probiotic strains on obesity symptoms (Jung et al. 2013, Mazloom 2013, Jung et al. 2015, Chung et al. 2016, Stenman et al. 2016, Szulinska et al. 2018, Razmpoosh et al. 2019) as shown in Table 2.

Different from the animal studies (Table 1), four of the seven studies in human trials presented in Table 2 showed no significant effects of probiotics on body weight or/and on lipid and glycaemic profile. Only Jung et al. (2015), Stenman et al. (2016), and Szulinska et al. (2018) showed positive effects in the mentioned aspects. This fact reinforces the idea that results obtained in animal models cannot always be extrapolated to humans because of the wide physiological differences (Parvova et al. 2011). On the other hand, the simple use of different bacterial species and strains applied in the experiments can be one possible explanation for this finding; different probiotic strains can provide distinct effects on humans' health (Athalye-Jape et al. 2018). The probiotic dose also seems to be important within species. Szulinska et al. (2018), for example, showed that the studied probiotic mix had more pronounced effects on obesity when tested at a concentration of 1×10^{10} CFU per day than a concentration of 2.5×10^{9} CFU per day.

Moreover, none of the studies presented in Table 2 (human studies) analysed changes in the microbiota composition. Changes in the obesity-related microbiota composition due to probiotic intervention are more easily found in pre-clinical studies and are, unfortunately, still controversial. Some studies have shown a decrease of *Firmicutes* (Brahe et al. 2013, Yadav et al. 2013), while others observed no changes in *Firmicutes* (Kim et al. 2017, Murphy et al. 2013, Park et al. 2013) during a regular probiotic consumption. Some authors demonstrated a significant increase of *Lactobacillus* and/or *Bifidobacterium* spp. (Cani and Delzenne 2009, Clarke et al. 2013, Aoki et al. 2017, Singh et al. 2017), while others observed no changes or reduction in *Lactobacillus* spp. and/or in *Bifidobacterium* spp. (Yadav et al. 2013, Alard et al. 2016, Park et al. 2017) during a regular probiotic consumption. Rouxinol-Dias et al. (2016) performed a systematic review aiming to investigate if a probiotic diet can provide a significant difference in weight change of obese and non-obese individuals using experimental models. They concluded that the probiotic effect is specie- and strain-specific, which can probably be one explanation to the cited controversies.

Different food products supplemented with probiotics, such as yogurts (Nabavi et al. 2015), cheese (Sperry et al. 2018), non-dairy beverage (Marchesin et al. 2018), and low fat ice cream (Chaikham and Rattanasena 2017), have also been shown to have beneficial effects on obesity parameters. Nabavi et al. (2015) demonstrated that yogurt supplemented with *B. lactis* Bb12 and *L. acidophilus* La5 (~ 10^{6} CFU/g) was able to reduce body mass index, body weight, and serum levels of fasting insulin in both men and women compared to conventional yogurt after 8 weeks of consumption. Sperry et al. (2018) showed that the consumption of Minas Frescal cheese supplemented with *L. casei* 01 (8 log CFU/g) during 28 days was able to improve the high-density lipoprotein-cholesterol, total cholesterol, triacylglycerides, low-density lipoprotein-cholesterol, haematocrit count, diastolic and systolic pressure and haemoglobin in hypertensive overweight women when compared to controls (Regular Minas Frescal cheese). Marchesin et al. (2018) evaluated the effects of soy-based probiotic drink fermented with *Enterococcus faecium* CRL 183 and added with *B. longum* ATCC 15707 (8 log CFU/g) on body weight, inflammatory parameters, and on faecal

Table 2: Impact of different probiotic strains on obesity parameters using clinical studies.

Probiotic	**Subjects**	**Concentration/Duration**	**Observed effects**	**Authors**
Probiotic capsules containing: *L. acidophilus, L. bulgaricus, L. bifidum*, and *L. casei*	Types 2 diabetic patients aged between 25 to 65 years. Control: Placebo containing 1000 mg magnesium stearate	1500 mg probiotic capsules twice daily for 6 wk	Comparing to placebo: - No differences between anthropometric data including body mass index and waist to hip ratio - No significant difference in serum TG concentration total cholesterol and LDL-C levels - HDL-C Levels were slightly elevated (but no significant) - Insulin, MDA and IL-6 levels were reduced (but no significant)	Mazloon et al. (2013)
L. gasseri BNR17	Obese volunteers aged 19 to 60. Control: placebo capsules without probiotic	Capsules with 10^{10} CFU (6 capsules per day) for 12 wk	Comparing to placebo: - Decrease of waist and hip circumferences - Reduction in body weight (but no significant)	Jung et al. (2013)
L. curvatus HY7601 and *L. plantarum* KY1032	Nondiabetic and overweight subjects. Control: Placebo without probiotic containing crystalline cellulose, lactose, and blueberry-flavoring agent	Each probiotic at 2.5×10^9 CFU, twice a day for 12 wk	Comparing to placebo: - Reductions in body weight, body fat mass and in body fat percentage - Reductions in L1 subcutaneous fat area - Reductions in oxidized low density lipoprotein (ox-LDL) - Increase in low density lipoprotein (LDL) particle size	Jung et al. (2015)
L. reuteri JBD301	Adults with $25 \leq BMI < 35$. Control: Placebo capsule containing vegetable cream	1×10^9 CFU per day for 12 wk	Comparing to placebo: - Reduction of body weight and on BMI - No differences in abdominal fat, blood lipid profile, fasting and blood glucose, blood insulin	Chung et al. (2016)

Table 2 contd. ...

... *Table 2 contd.*

Probiotic	**Subjects**	**Concentration/Duration**	**Observed effects**	**Authors**
B. animalis ssp. *lactis* 420 (B420)	Adults with BMI between 28–34.9. Control: Placebo containing microcrystalline Cellulose	10^{10} CFU per day for 6 mon	Comparing to placebo: - Reduction in abdominal fat mass and in waist circumference - Reduction in energy intake - Significant change in serum zonulin	Stenman et al. (2016)
Probiotic mix: *B. bifidum* W23, *B. lactis* W51, *B. lactis* W52, *L. acidophilus* W37, *L. brevis* W63, *L. casei* W56, *L. salivarius* W24, *Lactococcus lactis* W19, and *Lactococcus lactis* W58	Obese Caucasian postmenopausal women. Control: Placebo sachets containing only the excipients	- Low dose (LD) (2.5×10^9 CFU per day) of each probiotic or - high dose (HD) (1×10^{10} CFU per day) of each probiotic for 12 wk	Comparing to placebo: - Improvement of lipopolysaccharide, waist, fat mass, subcutaneous fat, uric acid, total cholesterol, triglycerides, low-density lipoprotein cholesterol, glucose, insulin, and insulin-resistant index (HOMA-IR) with HD treatment - Improvement of waist, fat mass, subcutaneous fat, total cholesterol, low-density lipoprotein cholesterol, insulin, and insulin-resistant index (HOMA-IR) with LD treatment - Improvement of fat percentage and visceral fat with both LD and HD treatment	Szulinska et al. (2018)
Probiotic mix: *Lactobacillus acidophilus* *Lactobacillus casei* *Lactobacillus rhamnosus* *Lactobacillus bulgaricus* *Bifidobacterium breve* *Bifidobacterium longum*	Patients with type 2 diabetes mellitus. Control: placebo capsules consisted of fructo-oligosaccharide and magnesium stearate	10^9 to 10^{11} CFU/g of each microorganism twice a day for 6 wk	Comparing to placebo: - Increase in the levels of fasting plasma glucose and HDL-C - No significant alterations in the levels of insulin, triglycerides, total cholesterol, insulin resistance and anthropometric measurements, including weight, waist circumference and body mass index	Razmpoosh et al. (2019)

microbiota composition of diet-induced obese mice. They found that the probiotic drink was able to reduce body weight gain and the size of adipocytes. Moreover, soy-based probiotic drink modulated cytokine expression related to obesity control and increased the proportion of *Bifidobacterium* spp. Aiming to investigate the effects of a low fat ice cream supplemented with *L. casei* 01 and *L. acidophilus* LA5 on the colon microbiome using an *in vitro* gut model, Chaikham and Rattanasena (2017) showed that this product increased the production of SCFAs, such as butyric and proprionic acids, and reduced ammonia in the colon region.

It has been suggested that the modulation of the expression of genes involved in inflammation and fatty acid oxidation, such as tumour necrosis factor alpha (TNFα), Interleukin 6 (IL6), PPARγ coactivator-1 alpha (PGC1α), IL1B, chemokine ligand 2 (CCL2), Carnitine palmitoyltransferase I and II (CPT1 and CPT2), and Acyl-CoA Oxidase 1 (ACOX1), are the most frequently-involved mechanisms in the probiotic's anti-obesity effects. The exact genes modulated are dependent on the strain (Park et al. 2013, Miyoshi et al. 2014).

Besides the effects on gut intestinal barrier function and stimulating the expression of several genes involved in the obesity process, probiotics can directly or indirectly increase the production of SCFAs, which can have many positive effects on health. As already mentioned in this chapter, certain SCFAs, especially butyrate and propionate, can decrease the expression of pro-inflammatory markers alleviating the inflammatory process. Moreover, SFCAs can be involved in the release of anorectic gut hormones, such as the peptide YY (PYY) and glucagon-like peptide-1 (GLP-1) (David et al. 2014, Kaji et al. 2014, Iepsen et al. 2015, Psichas et al. 2015, Christiansen et al. 2018). Both hormones are also directly related to hunger and satiety, insulin sensitivity, lipolysis inhibition and adipocyte differentiation (David et al. 2014, Kaji et al. 2014, O'Connor et al. 2017) as represented in Figure 6.

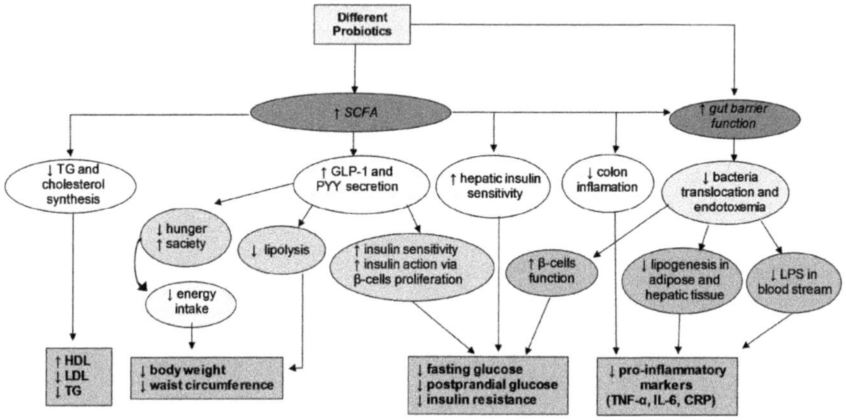

Figure 6: Possible effects of probiotics on obesity parameters. Different probiotic strains can have an effect on SCFAs' production and on gut barrier function, which by a cascade of events will culminate in improvement of obesity.

Interplay between the gut-brain axis, obesity, and probiotic consumption

The gut-brain axis (GBA) consists of complex bidirectional communication between the gut and the brain. This communication is mediated by neural, hormonal, and immunological signals between the central nervous system and the enteric innervation, which includes intrinsic neurons of the enteric nervous system, and extrinsic fibres of the autonomous nervous system (Rhee et al. 2009, Browning and Travagli 2014, Foster et al. 2017).

The gut microbiota plays an important function in the GBA communication, possible by modulating vagal nerve signalling and immune mediators, such as cytokines and chemokines, and by regulating the synthesis and endocrine secretions of neuroactive metabolites, such as glucocorticoids and neuropeptides, or their receptors (Moloney et al. 2014). According to Torres-Fuentes et al. (2017), as a result of dietary nutrient fermentation, the gut microorganisms produce metabolites such as SCFAs, serotonin (5-HT), γ-aminobutyric acid (GABA), and other neurotransmitters, which will have a direct or indirect impact on host metabolism, appetite, and satiety regulation via hypothalamic neuroendocrine pathways.

Metabolic diseases, such as obesity, can be associated with the dysregulation of GBA communication (Daly et al. 2011, Grasset et al. 2017). Besides the production of metabolites implicated in appetite and food ingestion, it was suggested that the human gut microbiota is involved with inflammation and depressive symptoms as well as with eating behaviour and stress, which are, many times, correlated with obesity (Abraham et al. 2013, Champaneri et al. 2013, Guillemot-Legris and Muccioli 2017). Oliveira et al. (2015), for example, showed that long-term obesity can be caused by overeating induced by chronic stress, which increases the consumption of sugar and fat as a compensatory way to reduce the stress related to anxiety. Another recent study showed that the combination of chronic unpredictable mild stress and obesity in rats can induce anxiety-like behaviours and depression as well as the down-regulation of leptin/LepRb signalling (Yang et al. 2016). In addition, levels of systemic inflammatory markers are known to be increased in obese individuals (Schachter et al. 2018).

Several studies have demonstrated the potential of probiotics in reducing inflammation and improving depressive and anxiety symptoms in both human and animal studies (Vaghef-Mehrabany et al. 2014, Pinto-Sanchez et al. 2017, Schachter et al. 2018). Strains of *Bifidobacterium* and *Lactobacillus* are the microorganisms most used to study beneficial effects on neurological and metabolic diseases (Kim et al. 2018). Probiotic *Bacteroides fragilis* has also been associated with improvement of gut permeability and microbiota composition as well as with sensorimotor and anxiety-like behaviours (Hsiao et al. 2013). Besides bacteria, some yeasts, such as *Saccharomyces boulardii* have also been shown to have beneficial effects on obesity, reducing fat mass, body weight, and serum concentrations of different pro-inflammatory cytokines in mice (Everard et al. 2014). According to Sonnenburg et al. (2016), the different beneficial bacteria interact with each other and with different dietary components leading to the release of bioactive metabolites, which signal to the host via the gut-brain axis.

Table 3 shows the effects of different probiotics on inflammation, anxiety and depression symptoms in different disorders.

In summary, studies show that different probiotics have an important role in the bidirectional communication between the brain and gut. Although many studies

Table 3: Effects of different probiotics on inflammation, anxiety and depression.

Probiotic	**Type of study**	**Concentration/ Duration**	**Observed effects**	**Authors**
L. rhamnosus GG	Animal study	1×10^8 CFU per mouse for 4 wk	- Reduction of endoplasmic reticulum stress - Suppression of inflammatory macrophage activation - Enhanced insulin sensitivity	Park et al. (2015)
Multispecies probiotic containing: *B. bifidum* W23, *B. lactis* W52, *L. acidophilus* W37, *L. brevis* W63, *L. casei* W56, *L. salivarius* W24, and *Lactococcus lactis* (W19 and W58)	Human intervention	2.5×10^9 CFU/g of each strain for 4 wk	Reduction of overall cognitive reactivity to sad mood, largely accounted for by reduced rumination and aggressive thoughts	Steenbergen et al. (2015)
L. rhamnosus JB-1	Animal study	1×10^9 CFU of JB-1 per day for 4 wk	Consistent changes in GABA-A and -B receptor sub-types in specific brain regions, accompanied by reductions in anxiety and depression-related behaviors	Janik et al. (2016)
Probiotic capsule daily containing: *L. acidophilus, L. casei, B. bifidum*	Human intervention	2×10^9 CFU/g of each bacteria for 8 wk	- Reduced Beck Depression index scores - Reduction in insulin, HOMA-IR, hs-CRP levels, and glutathione concentrations	Akkasheh et al. (2015)
Multi-species probiotic formulation *B. bifidum* W23, *B. lactis* W52, *L. acidophilus* W37, *L. brevis* W63, *L. casei* W56, *L. lactis* W19, *L. Lactis* W58	Animal study	4.5 g (2.5×10^9 CFU/g) of freeze-dried powder dissolved in 30 mL of tap water for 5 wk	- Reduction on depression-like behavior as measured by the forced swim test (FST) - Reduction of pro-inflammatory cytokines (IL-6, TNF-α)	Abildgaard et al. (2017)
L. johnsonii CJLJ103	Caco-2 cells and mice	1×10^9 CFU per mouse for 5 d	- Inhibition of TNBS-induced TNF and IL-1β expression in the colon - Increase of TNBS-suppressed expression of IL-10 - Increase of tight-junction proteins - Inhibition of fecal LPS levels and *Proteobacterial/ Bacteroides* ratio	Lim et al. (2017)

Table 3 contd. ...

... *Table 3 contd.*

Probiotic	**Type of study**	**Concentration/ Duration**	**Observed effects**	**Authors**
B. longum NCC3001	Human intervention	1×10^{10} CFU/g in 100–200 ml of lactose-free milk, soy milk or rice milk for 6 wk	- Most patients had reduction in depression scores - No significant effect on anxiety - Reduced responses to negative emotional stimuli in multiple brain areas, including amygdala and fronto-limbic regions	Pinto-sanchez et al. (2017)
Multi-species probiotic: *L. plantarum* subsp. *plantarum* MTCC 5422 and CFR MFT1, *L. paraplantarum* MTCC 9483 and *L. rhamnosus* GG or ATCC 53,103	*In vitro* study using Caco-2 cells	10^{8} CFU/mL. Probiotic exposition to Caco-2 cells for 4 hr	- Down-regulation of the gene expression of pro-inflammatory cytokines (TNF-α, IL-1α, IL-1β, IL-6 and IL-8) - Up-regulation of the anti-inflammatory CKs like TGFβ1, IL-4, and IL-10	Devi et al. (2018)
Lactobacillus plantarum P8	Human intervention	2×10^{10} CFU per sachet per day for 12 wk	- Reduction in scores of stress and anxiety - Reduction in pro-inflammatory cytokines IFN-γ and TNF-α - Enhanced memory and cognitive traits such as social emotional cognition and verbal learning	Lew et al. (2018)
Multi-species probiotic formulation: *L. bulgaricus* I-1632 and I-1519 *L. lactis* subsp. *lactis* I-1631 *L. acidophilus* *L. plantarum* *B. lactis* CNCM I-2494 *L. reuteri* DSM 17938	Animal study	1.5×10^{10} of each bacteria for 4 wk	- Reduction of stress-like behaviors - Decrease of hypothalamic expression levels of the pro-neuroinflammatory factors like IL-1β, NLRP3, Caspase-1 and N–kB - Decrease in plasmatic levels of IL-1β, NLRP3 and caspase-1	Avolio et al. (2019)

have tried to show possible mechanisms of probiotics in the modulation of the gut-brain axis in normal and obesity conditions, the exact mechanisms have still not been fully elucidated (Kim et al. 2018). Moreover, more studies evaluating the impact of probiotics in the relief of stress and depression symptoms specifically on obese individuals are necessary, once the majority of studies focusing on the link between probiotics, gut microbiota, stress and depression symptoms are performed in healthy-weight individuals with neurological and psychiatric disorders.

Future Perspective

Knowledge of gut microbiota composition has become key for the management of several diseases, including obesity. This chapter showed that the use of probiotic microorganisms has been considered a promising treatment category for obesity, as far as several researches have demonstrated the positive effects of different strains on obesity parameters. However, for the most effective use of probiotics in favour of obesity management, a more detailed elucidation of the probiotic action mechanisms, especially in terms of the interactions between gut and brain, is necessary. Finally, the selection of specific strains, with proven action in obese humans, is also a challenge for obesity prevention and treatment.

References

Abildgaard, A., B. Elfving, M. Hokland, G. Wegener and S. Lund. 2017. Probiotic treatment reduces depressive-like behaviour in rats independently of diet. Psychoneuroendocrinology 79: 40–40.

Abraham, S., D. Rubino, N. Sinaii, S. Ramsey and L. Nieman. 2013. Cortisol, obesity and the metabolic syndrome: A cross-sectional study of obese subjects and review of the literature. Obes. (Silver Spring) 21: 1–26.

Akkasheh, G., Z. Kashani-Poor, M. Tajadadi-Ebrahimi, P. Jafari, H. Akbari, M. Taghizadeh, M.R. Memarzadeh, Z. Asemi and A. Esmaillzadeh. 2015. Clinical and metabolic response to probiotic administration in patients with major depressive disorder: a randomized, double-blind, placebo-controlled trial. Nutrition 32: 315–320.

Alard, J., V. Lehrter, M. Rhimi, I. Mangin, V. Peucelle, A. Abraham, M. Mariadassou, E. Maguin, B. Pot, I. Wolowczuk and C. Grangette. 2016. Beneficial metabolic effects of selected probiotics on diet-induced obesity and insulin resistance in mice are associated with improvement of dysbiotic gut microbiota. Environ. Microbiol. 18: 1484–1497.

Almeida, L.B., C.B Marinho, C. da Silva Souza and V.B.P. Cheib. 2009. Disbiose Intestinal. Rev. Bras. Nutr. Clín. 24: 58–65.

Alonso, V.R. and F. Guarner. 2013. Linking the gut microbiota to human health. Br. J. Nutr. 109: 21–26.

[ANVISA] Agência Nacional de Vigilância Sanitária. 2012. Resolução da Diretoria colegiada – RDC nº 54, de 12 de novembro de 2012. Diário Of. da União, Brazil: 16 pp.

Aoki, R., K. Kamikado, W. Suda, H. Takii and Y. Mikami. 2017. A proliferative probiotic *Bifidobacterium* strain in the gut ameliorates progression of metabolic disorders via microbiota modulation and acetate elevation. Nat. Publ. Gr. 1–10.

Apovian, C.M. 2016. Obesity: Definition, comorbidities, causes, and burden. Am. J. Manag. Care 22: 176–185.

Araújo, J.R., J. Tomas, C. Brenner and P.J. Sansonetti. 2017. Impact of high-fat diet on the intestinal microbiota and small intestinal physiology before and after the onset of obesity. Biochimie 141: 97–106.

Athalye-Jape, G., S. Rao, K. Simmer and S. Patole. 2018. *Bifidobacterium breve* M-16V as a probiotic for preterm infants: A strain-specific systematic review. J. Parenter. Enter. Nutr. 42: 677–688.

Avolio, E., G. Fazzari, M. Zizza, A. De Lorenzo, L. Di, R. Alò, R. Maria and M. Canonaco. 2019. Probiotics modify body weight together with anxiety states via pro-inflammatory factors in HFD-treated Syrian golden hamster. Behav. Brain Res. 356: 390–399.

Bäckhed, F., H. Ding, T. Wang, L.V. Hooper, G.Y. Koh, A. Nagy, C.F. Semenkovich and J.I. Gordon. 2004. The gut microbiota as an environmental factor that regulates fat storage. Proc. Natl. Acad. Sci. USA 101: 15718–15723.

Bäckhed, F., J.K. Manchester, C.F. Semenkovich and J.I. Gordon. 2007. Mechanisms underlying the resistance to diet-induced obesity in germ-free mice. Proc. Natl. Acad. Sci. 104: 979–984.

Bagarolli, R.A., A.G. Oliveira, G. Tiago, B.M. Carvalho, G.Z. Rocha, J.F. Vecina, D. Guadagnini, O. Prada, A. Santos, S.T.O. Saad and M.J.A. Saad. 2017. Probiotics modulate gut microbiota and improve insulin sensitivity in DIO mice. J. Nutr. Biochem. 50: 16–25.

Bäumler, A.J. and V. Sperandio. 2016. Interactions between the microbiota and pathogenic bacteria in the gut. Nature 535: 85–93.

Bomhof, M.R. and R.A. Reimer. 2015. Pro- and prebiotics: The role of gut microbiota in obesity. pp. 363–380. *In*: Venema, K. and A.P. do Carmo (eds.). Probiotics and Prebiotics: Current Research and Future Trends. Caister Academic Press, U.K.

Brahe, L.K., A. Astrup and L.H. Larsen. 2013. Is butyrate the link between diet, intestinal microbiota and obesity-related metabolic diseases? Obes. Rev. 14: 950–959.

Browning, K.N. and R.A. Travagli. 2014. Central nervous system control of gastrointestinal motility and secretion and modulation of gastrointestinal functions. Compr. Physiol. 4: 1339–1368.

Caesar, R., V. Tremaroli and F. Bäckhed. 2015. Crosstalk between gut microbiota and dietary lipids aggravates WAT inflammation through TLR signaling. Cell Metab. 22: 658–668.

Canfora, E.E., J.W. Jocken and E.E. Blaak. 2015. Short-chain fatty acids in control of body weight and insulin sensitivity. Nat. Rev. Endocrinol. 11: 577–591.

Cani, P.D. and N.M. Delzenne. 2009. The role of the gut microbiota in energy metabolism and metabolic disease. Curr. Pharm. Des. 15: 1546–1558.

Carabotti, M., A. Scirocco, M.A. Maselli and C. Severi. 2015. The gut-brain axis: Interactions between enteric microbiota, central and enteric nervous systems. Ann. Gastroenterol. 28: 203–209.

Chaikham, P. and P. Rattanasena. 2017. Food bioscience combined effects of low-fat ice cream supplemented with probiotics on colon micro floral communities and their metabolites during fermentation in a human gut reactor. Food Biosci. 17: 35–41.

Chakraborti, C.K. 2015. New-found link between microbiota and obesity. World J. Gastrointest. Pathophysiol. 6: 110–119.

Champaneri, S., X. Xu, M.R. Carnethon, A.G. Bertoni, T. Seeman, A.S. DeSantis, A.D. Roux, S. Shrager and S.H. Golden. 2013. Diurnal salivary cortisol is associated with body mass index and waist circumference: The multi-ethnic study of atherosclerosis. Obes. (Silver Spring) 21: 1–18.

Chen, J., Y. Guo, Y. Gui and D. Xu. 2018a. Physical exercise, gut, gut microbiota, and atherosclerotic cardiovascular diseases. Lipids Health Dis. 17: 1–7.

Chen, Y., N. Yang, Y. Lin, S. Ho, K. Li, J. Lin, J. Liu and M. Chen. 2018b. A combination of *Lactobacillus mali* APS1 and dieting improved the efficacy of obesity treatment via manipulating gut microbiome in mice. Sci. Rep. 8(1): 6153.

Chiang, J.Y.L. 2009. Bile acids: regulation of synthesis. J. Lipid Res. 50: 1955–1966.

Chiang, J.Y.L., P. Pathak, H. Liu, A. Donepudi, J. Ferrell and S. Boehme. 2017. Intestinal FXR and TGR5 signaling in metabolic regulation John. Dig. Dis. 35: 241–245.

Christensen, L., H.M. Roager, A. Astrup and M.F. Hjorth. 2018. Microbial enterotypes in personalized nutrition and obesity management. Am. J. Clin. Nutr. 108(4): 645–651.

Christiansen, C.B., M. Buur, N. Gabe, B. Svendsen, L.O. Dragsted, M.M. Rosenkilde and J.J. Holst. 2018. The impact of short-chain fatty acids on GLP-1 and PYY secretion from the isolated perfused rat colon. Am. J. Physiol. Gastrointest. Liver Physiol. 315: 53–65.

Chu, D.T., N.T. Minh Nguyet, T.C. Dinh, N.V. Thai Lien, K.H. Nguyen, V.T. Nhu Ngoc, Y. Tao, L.H. Son, D.H. Le, V.B. Nga, A. Jurgoński, Q.H. Tran, P. Van Tu and V.H. Pham. 2018. An update on physical health and economic consequences of overweight and obesity. Diabetes Metab. Syndr. Clin. Res. Rev. 12: 1095–1100.

Chung, H., J.G. Yu, I. Lee, M. Liu, Y. Shen, S.P. Sharma, M.A.H.M. Jamal, J. Yoo, H. Kim and S. Hong. 2016. Intestinal removal of free fatty acids from hosts by *Lactobacilli* for the treatment of obesity. FEBS Open Bio. 6: 64–76.

Clarke, S.F., E.F. Murphy, O. O'Sullivan, R.P. Ross, P.W. O'Toole, F. Shanahan and P.D. Cotter. 2013. Targeting the microbiota to address diet-induced obesity: A time dependent challenge. PLoS One 8(6): e65790.

Clemente, J.C., L.K. Ursell, L.W. Parfrey and R. Knight. 2012. The impact of the gut microbiota on human health: An integrative view. Cell 148: 1258–1270.

Dahiya, D.K., Renuka, M. Puniya, U.K. Shandilya, T. Dhewa, N. Kumar, S. Kumar, A.K. Puniya and P. Shukla. 2017. Gut microbiota modulation and its relationship with obesity using prebiotic fibers and probiotics: A review. Front. Microbiol. 8: 563.

Daly, D.M., S.J. Park, W.C. Valinsky and M.J. Beyak. 2011. Impaired intestinal afferent nerve satiety signalling and vagal afferent excitability in diet induced obesity in the mouse. J. Physiol. 11: 2857–2870.

Dao, M.C., A. Everard, K. Clément and P.D. Cani. 2016. Losing weight for a better health: Role for the gut microbiota. Clin. Nutr. Exp. 6: 39–58.

Dave, M., P.D. Higgins, S. Middha and K.P. Rioux. 2012. The human gut microbiome: Current knowledge, challenges, and future directions. Transl. Res. 160: 246–257.

David, L.A., C.F. Maurice, R.N. Carmody, D.B. Gootenberg, J.E. Button, B.E. Wolfe, A.V. Ling, A.S. Devlin, Y. Varma, M.A. Fischbach, S.B. Biddinger, R.J. Dutton and P.J. Turnbaugh. 2014. Diet rapidly and reproducibly alters the human gut microbiome. Nature 505: 559–563.

Devi, S.M., S. Manjulata, N.K. Kurrey and P.M. Halami. 2018. *In vitro* anti-inflammatory activity among probiotic *Lactobacillus* species isolated from fermented foods. J. Funct. Foods 47: 19–27.

DiBaise, J.K., D.N. Frank and R. Mathur. 2012. Impact of the gut microbiota on the development of obesity: Current concepts. Am. J. Gastroenterol. Suppl. 1: 22–27.

Doré, J. and G. Corthier. 2010. Le microbiote intestinal. Gastroentérologie Clin. Biol. 34: 7–15.

Everard, A., S. Matamoros, L. Geurts, N.M. Delzenne and P.D. Cani. 2014. *Saccharomyces boulardii* administration changes gut microbiota and reduces hepatic steatosis, low-grade inflammation, and fat mass in obese and type 2 diabetic db/db mice. MBio. 5(3): e01011–01014.

de Fabiani, E., N. Mitro, F. Gilardi, D. Caruso, G. Galli and M. Crestani. 2003. Coordinated control of cholesterol catabolism to bile acids and of gluconeogenesis via a novel mechanism of transcription regulation linked to the fasted-to-fed cycle. J. Biol. Chem. 278: 39124–39132.

Fleissner, C.K., N. Huebel, M.M. Abd El-Bary, G. Loh, S. Klaus and M. Blaut. 2010. Absence of intestinal microbiota does not protect mice from diet-induced obesity. Br. J. Nutr. 104: 919–929.

Fooks, L.J. and G.R. Gibson. 2002. Probiotics as modulators of the gut flora. Br. J. Nutr. 88: 39–49.

Foster, J.A., L. Rinaman and J.F. Cryan. 2017. Neurobiology of stress & the gut-brain axis: Regulation by the microbiome. Neurobiol. Stress 7: 124–136.

Gibson, G.R. and M.B. Roberfroid. 1995. Dietary modulation of the human colonic microbiota: Introducing the concept of prebiotics. J. Nutr. 125: 1401–1412.

Gill, S., M. Pop, R. DeBoy and P. Eckburg. 2006. Metagenomic analysis of the human distal gut microbiome. Science 312: 1355–1359.

Go, G., S. Oh, M. Park, G. Gang, D. McLean, H. Yang, M.-H. Song and Y. Kim. 2013. t10,c12 conjugated linoleic acid upregulates hepatic *de novo* lipogenesis and triglyceride synthesis via mTOR pathway activation. J. Microbiol. Biotechnol. 23: 1569–1576.

Goodrich, J., J. Waters, A. Poole, J. Sutter, O. Koren, R. Blekhman, M. Beaumont, W. Van Treuren, R. Knight, J. Bell, T. Spector, A. Clark and R.E. Ley. 2014. Human genetics shape the gut microbiome. Cell 159: 789–799.

Gosalbes, M.J., S. Lop, Y. Vallès, A. Moya, F. Ballester and M.P. Francino. 2013. Meconium microbiota types dominated by lactic acid or enteric bacteria are differentially associated with maternal eczema and respiratory problems in infants. Clin. Exp. Allergy 43: 198–211.

Graham, C., A. Mullen and K. Whelan. 2015. Obesity and the gastrointestinal microbiota: A review of associations and mechanisms. Nutr. Rev. 73: 376–385.

Grasset, E., A. Puel, J. Charpentier, X. Collet and J.E. Christensen. 2017. A specific gut microbiota dysbiosis of type 2 diabetic mice induces GLP-1 resistance through an enteric no-dependent and gut-brain axis mechanism. Cell Metab. 25: 1075–1090.

Guillemot-legris, O. and G.G. Muccioli. 2017. Obesity-induced neuroinflammation: Beyond the Hypothalamus. Trends Neurosci. 40: 237–243.
Guo, X., J. Li, R. Tang, G. Zhang, H. Zeng, R.J. Wood and Z. Liu. 2017. High fat diet alters gut microbiota and the expression of paneth cell-antimicrobial peptides preceding changes of circulating inflammatory cytokines. Mediat. Inflamm. 2017: 9474896.
Harris, K., A. Kassis, G. Major and C.J. Chou. 2012. Is the gut microbiota a new factor contributing to obesity and its metabolic disorders? J. Obes. 2012: 879151.
Haro, C., M. Montes-borrego, O.A. Rangel-zúñiga, J.F. Alcalá-díaz, F. Gómez-delgado, P. Pérez-martínez, J. Delgado-lista, G.M. Quintana-navarro, F.J. Tinahones, B.B. Landa, J. López-miranda, A. Camargo and F. Pérez-jiménez. 2016. Community improving insulin sensitivity in a human obese population. J. Clin. Endocrinol. Metab. 101: 233–242.
Hartstra, A.V., K.E.C. Bouter, F. Bäckhed and M. Nieuwdorp. 2015. Insights into the role of the microbiome in obesity and type 2 diabetes. Diabetes Care 38: 159–165.
Hill, C., F. Guarner, G. Reid, G.R. Gibson, D.J. Merenstein, B. Pot, L. Morelli, R.B. Canani, H.J. Flint, S. Salminen, P.C. Calder and M.E. Sanders. 2014. The International Scientific Association for probiotics and prebiotics consensus statement on the scope and appropriate use of the term probiotic. Nat. Rev. Gastroenterol. Hepatol. 11: 506–515.
Hippe, B., M. Remely, E. Aumueller, A. Pointner, U. Magnet and A.G. Haslberger. 2016. *Faecalibacterium prausnitzii* phylotypes in type two diabetic, obese, and lean control subjects. Benef. Microbes 7: 511–517.
Hjorth, M.F., H.M. Roager, T.M. Larsen, S.K. Poulsen, T.R. Licht, M.I. Bahl, Y. Zohar and A. Astrup. 2018. Pre-treatment microbial *Prevotella*-to-*Bacteroides* ratio, determines body fat loss success during a 6-month randomized controlled diet intervention. Int. J. Obes. 42: 580–583.
Hotamisligil, G.S. 2006. Inflammation and metabolic disorders. Nature 444: 860–867.
Hsiao, E.Y., S.W. Mcbride, S. Hsien, G. Sharon, E.R. Hyde, T. Mccue, J.A. Codelli, J. Chow, S.E. Reisman, J.F. Petrosino, P.H. Patterson and S.K. Mazmanian. 2013. The microbiota modulates gut physiology and behavioral abnormalities associated with autism. Cell 155: 1451–1463.
Iepsen, E.W., J. Lundgren, C. Dirksen, J. Jensen, O. Pedersen, T. Hansen, S. Madsbad, J.J. Holst and S.S. Torekov. 2015. Treatment with a GLP-1 receptor agonist diminishes the decrease in free plasma leptin during maintenance of weight loss. Int. J. Obes. 39: 834–841.
Janik, R., L.A.M. Thomason, A.M. Stanisz, P. Forsythe, J. Bienenstock and G.J. Stanisz. 2016. Magnetic resonance spectroscopy reveals oral *Lactobacillus* promotion of increases in brain GABA, N-acetyl aspartate and glutamate. Neuroimage 125: 988–995.
Ji, J., D. Shu, M. Zheng, J. Wang, C. Luo, Y. Wang, F. Guo, X. Zou, X. Lv, Y. Li, T. Liu and H. Qu. 2016. Microbial metabolite butyrate facilitates M2 macrophage polarization and function. Sci. Rep. 6: 24838.
Jung, S., K. Lee, J. Kang, S. Yun, H. Park, Y. Moon and J. Kim. 2013. Effect of *Lactobacillus gasseri* BNR17 on overweight and obese adults: A randomized, double-blind clinical trial seung-pil. Korean J. Fam. Med. 34: 80–89.
Jung, S., Y. Ju, M. Kim, M. Kim and J. Hyun. 2015. Supplementation with two probiotic strains, *Lactobacillus curvatus* HY7601 and *Lactobacillus plantarum* KY1032, reduced body adiposity and Lp-PLA 2 activity in overweight subjects. J. Funct. Foods 19: 744–752.
Kaji, I., S. Karaki and A. Kuwahara. 2014. Short-chain fatty acid receptor and its contribution to glucagon-like peptide-1 release. Digestion 89: 31–36.
Kasai, C., K. Sugimoto, I. Moritani, J. Tanaka, Y. Oya, H. Inoue, M. Tameda, K. Shiraki, M. Ito, Y. Takei and K. Takase. 2015. Comparison of the gut microbiota composition between obese and non-obese individuals in a Japanese population, as analyzed by terminal restriction fragment length polymorphism and next-generation sequencing. BMC Gastroenterol. 15: 100.
Kerry, R.G., J.K. Patra, S. Gouda, Y. Park, H.S. Shin and G. Das. 2018. Benefaction of probiotics for human health: A review. J. Food Drug Anal. 26: 927–939.
Kim, D., H. Kim, D. Jeong, I. Kang, W. Chon, H. Kim, K. Song and K. Seo. 2017. Kefir alleviates obesity and hepatic steatosis in high-fat diet-fed mice by modulation of gut microbiota and mycobiota: Targeted and untargeted community analysis with correlation of biomarkers. J. Nutr. Biochem. 44: 35–43.

Kim, N., M. Yun, Y.J. Oh and H. Choi. 2018. Mind-altering with the gut: Modulation of the gut-brain axis with probiotics. J. Microbiol. 56: 172–182.
Kootte, R.S., A. Vrieze, F. Holleman, G.M. Dallinga-Thie, E.G. Zoetendal, W.M. de Vos, A.K. Groen, J.B.L. Hoekstra, E.S. Stroes and M. Nieuwdorp. 2012. The therapeutic potential of manipulating gut microbiota in obesity and type 2 diabetes mellitus. Diabetes, Obes. Metab. 14: 112–120.
Kootte, R.S., E. Levin, E.S.G. Stroes, A.K. Groen, M. Nieuwdorp, L.P. Smits, A.V. Hartstra and S.D. Udayappan. 2017. Clinical and translational report improvement of insulin sensitivity after lean donor feces in metabolic syndrome is driven by baseline intestinal microbiota composition. Cell Metab. 26: 611–619.
Kumari, M. and A.L. Kozyrskyj. 2017. Gut microbial metabolism defines host metabolism: An emerging perspective in obesity and allergic inflammation. Obes. Rev. 18: 18–31.
Lew, L., Y. Hor, N.A.A. Yusoff, S. Choi, M. Liong, L. Kwok and H. Zhang. 2018. Probiotic *Lactobacillus plantarum* P8 alleviated stress and anxiety while enhancing memory and cognition in stressed adults: A randomised, double-blind, placebo-controlled study. Clin. Nutr. pii: S0261–5614(18)32448-8.
Ley, R.E., P.J. Turnbaugh, S. Klein and J.I. Gordon. 2006. Microbial ecology: Human gut microbes associated with obesity. Nature 444: 1022–1023.
Lim, S., H. Jang, J. Jeong, M. Joo and D. Kim. 2017. *Lactobacillus johnsonii* CJLJ103 attenuates colitis and memory impairment in mice by inhibiting gut microbiota lipopolysaccharide production and NF-kB activation. J. Funct. Foods 34: 359–368.
Lin, H.V., A. Frassetto, E.J. Kowalik, A.R. Nawrocki, M.M. Lu, J.R. Kosinski, J.A. Hubert, D. Szeto, X. Yao, G. Forrest and D.J. Marsh. 2012. Butyrate and propionate protect against diet-induced obesity and regulate gut hormones via free fatty acid receptor 3-independent mechanisms. PLoS One 7(4): e35240.
Louis, P., P. Young, G. Holtrop and H.J. Flint. 2010. Diversity of human colonic butyrate-producing bacteria revealed by analysis of the butyryl-CoA: Acetate CoA-transferase gene. Environ. Microbiol. 12: 304–314.
Lozupone, C., J. Stomabaugh, J. Gordon, J. Jansson and R. Knight. 2012. Diversity, stability and resilience of the human gut microbiota. Nature 489: 220–230.
Lynch, S.V. and O. Pedersen. 2016. The human intestinal microbiome in health and disease. N. Engl. J. Med. 375: 2369–2379.
Macfarlane, G.T. and S. Macfarlane. 2012. Molecular methods to measure intestinal bacteria: A review. J. AOAC Int. 95: 5–23.
Marchesin, J. de C., L.S. Celiberto, A.B. Orlando, A.I. de Medeiros, R.A. Pinto, J.A.S. Zuanon, L.C. Spolidorio, A. dos Santos, M.P. Tarantod and D.C.U. Cavallini. 2018. A soy-based probiotic drink modulates the microbiota and reduces body weight gain in diet-induced obese mice. J. Funct. Foods 48: 302–313.
Mazloom, Z. 2013. Effect of probiotics on lipid profile, glycemic control, insulin action, oxidative stress, and inflammatory markers in patients with type 2 diabetes: A clinical trial. Iran J. Med. Sci. 38: 38–43.
Miyoshi, M., A. Ogawa, S. Higurashi and Y. Kadooka. 2014. Anti-obesity effect of *Lactobacillus gasseri* SBT2055 accompanied by inhibition of pro-inflammatory gene expression in the visceral adipose tissue in diet-induced obese mice. Eur. J. Nutr. 53: 599–606.
Moloney, R.D., L. Desbonnet, G. Clarke, T.G. Dinan and J.F. Cryan. 2014. The microbiome: stress, health and disease. Mamm. Genome 25: 49–74.
Moreira, A.P.B., T.F.S. Texeira, A.B. Ferreira, M. Do Carmo Gouveia Peluzio and R. De Cássia Gonçalves Alfenas. 2012. Influence of a high-fat diet on gut microbiota, intestinal permeability and metabolic endotoxaemia. Br. J. Nutr. 108: 801–809.
Morrison, D.J. and T. Preston. 2016. Formation of short chain fatty acids by the gut microbiota and their impact on human metabolism. Gut Microbes 7: 189–200.
Murphy, E.F., P.D. Cotter, A. Hogan, O. O'Sullivan, A. Joyce, F. Fouhy, S.F. Clarke, T.M. Marques, P.W. O'Toole, C. Stanton, E.M.M. Quigley, C. Daly, P.R. Ross, R.M. O'Doherty and F. Shanahan. 2013. Divergent metabolic outcomes arising from targeted manipulation of the gut microbiota in diet-induced obesity. Gut 62: 220–226.

Murugesan, S., K. Nirmalkar, C. Hoyo-Vadillo, M. García-Espitia, D. Ramírez-Sánchez and J. García-Mena. 2018. Gut microbiome production of short-chain fatty acids and obesity in children. Eur. J. Clin. Microbiol. Infect. Dis. 37: 621–625.

Nabavi, S., M. Rafraf and M. Somi. 2015. Probiotic yogurt improves body mass index and fasting insulin levels without affecting serum leptin and adiponectin levels in non-alcoholic fatty liver disease (NAFLD). J. Funct. Foods 18: 684–691.

Ness-Abramof, R. and C.M. Apovian. 2005. Drug-induced weight gain. Drugs Today (Barc) 41: 547–555.

O'Connor, S., S. Chouinard-castonguay, C. Gagnon and I. Rudkowska. 2017. Prebiotics in the management of components of the metabolic syndrome. Maturitas 104: 11–18.

O'Hara, A.M. and F. Shanahan. 2006. The gut flora as a forgotten organ. EMBO Rep. 7: 688–693.

Oliveira, C.D., C.M. Oliveira, I.C. de Macedo, A.S. Quevedo, PR. Filho, F.R. Silva, R. Vercelino, I.C. de Souza, W. Caumo and I.L. Torres. 2015. Hypercaloric diet modulates effects of chronic stress: A behavioral and biometric study on rats. Stress 18: 514–523.

Ozdal, T., D.A. Sela, J. Xiao, D. Boyacioglu, F. Chen and E. Capanoglu. 2016. The reciprocal interactions between polyphenols and gut microbiota and effects on bioaccessibility. Nutrients 8(2): 78.

Park, D.Y., Y.T. Ahn, S.H. Park, C.S. Huh, S.R. Yoo, R. Yu, M.K. Sung, R.A. McGregor and M.S. Choi. 2013. Supplementation of *Lactobacillus curvatus* KY1032 in diet-induced obese mice is associated with gut microbial changes and reduction in obesity. PLoS One 8(3): e59470.

Park, K.Y., B. Kim and C.K. Hyun. 2015. *Lactobacillus rhamnosus* GG improves glucose tolerance through alleviating ER stress and suppressing macrophage activation in db/db mice. J. Clin. Biochem. Nutr. 56: 240–246.

Park, S., Y. Ji, H.Y. Jung, H. Park, J. Kang, S.H. Choi, H. Shin, C.K. Hyun, K.T. Kim and W.H. Holzapfel. 2017. *Lactobacillus plantarum* HAC01 regulates gut microbiota and adipose tissue accumulation in a diet-induced obesity murine model. Appl. Microbiol. Biotechnol. 101: 1605–1614.

Parvova, I., N. Danchev and E. Hristov. 2011. Animal models of human diseases and their significance for clinical studies of new drugs. J. Clin. Med. 4: 19–29.

Payne, A.N., A. Zihler, C. Chassard and C. Lacroix. 2012. Advances and perspectives in *in vitro* human gut fermentation modeling. Trends Biotechnol. 30: 17–25.

Pendyala, S., J.M. Walker and P.R. Holt. 2012. A high-fat diet is associated with endotoxemia that originates from the gut. Gastroenterology 142: 1100–1101.

Peters, B.A., J.A. Shapiro, T.R. Church, G. Miller, C. Trinh-Shevrin, E. Yuen, C. Friedlander, R.B. Hayes and J. Ahn. 2018. A taxonomic signature of obesity in a large study of American adults. Sci. Rep. 8: 9749.

Pinto-sanchez, M.I., G.B. Hall, K. Ghajar, A. Nardelli, C. Bolino, J.T. Lau, O. Cominetti, C. Welsh, A. Rieder, J. Traynor, C. Gregory, G. De Palma, A.C. Ford, J. Macri, B. Berner, G. Bergonzelli, M.G. Surette and S.M. Collins. 2017. Probiotic *Bifidobacterium longum* NCC3001 reduces depression scores and alters brain activity: A pilot study in patients with irritable bowel syndrome. Gastroenterology 153: 448–459.

Possemiers, S., S. Bolca, W. Verstraete and A. Heyerick. 2011. The intestinal microbiome: A separate organ inside the body with the metabolic potential to influence the bioactivity of botanicals. Fitoterapia 82: 53–66.

Psichas, A., M.L. Sleeth, K.G. Murphy, L. Brooks, G.A. Bewick, A.C. Hanyaloglu, M.A. Ghatei, S.R. Bloom and G. Frost. 2015. The short chain fatty acid propionate stimulates GLP-1 and PYY secretion via free fatty acid receptor 2 in rodents. Int. J. Obes. 39: 424–429.

Puhl, R.M. and C.A. Heuer. 2010. Obesity stigma: Important considerations for public health. Am. J. Public Health 100: 1019–1028.

Rajoka, M.S.R., J. Shi, H.M. Mehwish, J. Zhu, Q. Li, D. Shao, Q. Huang and H. Yang. 2017. Interaction between diet composition and gut microbiota and its impact on gastrointestinal tract health. Food Sci. Hum. Wellness 6: 121–130.

Ramakrishna, B.S. 2013. Role of the gut microbiota in human nutrition and metabolism. J. Gastroenterol. Hepatol. 28: 9–17.

Rather, S.A., R. Pothuraju, R.K. Sharma, S. De, N.A. Mir and S. Jangra. 2014. Anti-obesity effect of feeding probiotic dahi containing *Lactobacillus casei* NCDC 19 in high fat diet-induced obese mice. Int. J. Dairy Technol. 67: 504–509.

Razmpoosh, E., A. Javadi, H. Sadat and P. Mirmiran. 2019. The effect of probiotic supplementation on glycemic control and lipid profile in patients with type 2 diabetes: A randomized placebo controlled trial. Diabetes Metab. Syndr. Clin. Res. Rev. 13: 175–182.

Remely, M., B. Hippe, J. Zanner, E. Aumueller, H. Brath and A.G. Haslberger. 2016. Gut microbiota of obese, type 2 diabetic individuals is enriched in *Faecalibacterium prausnitzii, Akkermansia muciniphila* and *Peptostreptococcus anaerobius* after weight loss. Endocrine, Metab. Immune Disord. - Drug Targets 16: 99–106.

Reyes, L.M., R.G. Vázquez, S.M.C. Arroyo, A.M. Avalos, P.A.R. Castillo, D.A.C. Pérez, I.R. Terrones, N.R. Ibáñez, M.M.R. Magallanes, P. Langella, L.B. Humarán and A.A. Espinosa. 2016. Correlation between diet and gut bacteria in a population of young adults. Int. J. Food Sci. Nutr. 67: 470–478.

Rhee, S.H., C. Pothoulakis and E.A. Mayer. 2009. Principles and clinical implications of the brain–gut–enteric microbiota axis. Nat. Rev. Gastroenterol. Hepatol. 6(5): 306–314.

Riaz, S. 2015. Modulation of immune system by taking probiotic bacteria: Especially focus on lactic acid bacteria. Asian J. Agr. Biol. 3: 74–77.

Ridaura, V.K., V.K. Ridaura, J.J. Faith, F.E. Rey, J. Cheng, A.E. Duncan, L. Kau, N.W. Griffi, V. Lombard, B. Henrissat, J.R. Bain, J. Michael, O. Ilkayeva, C.F. Semenkovich, K. Funai, D.K. Hayashi, J. Lyle, M.C. Martini, L.K. Ursell, J.C. Clemente, W. Van Treuren, A. William, R. Knight, C.B. Newgard, A.C. Heath and J.I. Gordon. 2013. Gut microbiota from twins discordant for obesity modulate metabolism in mice. Science 341: 1079–1090.

Robles-Alonso, V. and F. Guarner. 2013. Progreso en el conocimiento de la microbiota intestinal humana. Nutr. Hosp. 28: 553–557.

Roselli, M., A. Finamore, E. Brasili, R. Rami, F. Nobili, C. Orsi, A. Vittorio and E. Mengheri. 2018. Beneficial effects of a selected probiotic mixture administered to high fat-fed mice before and after the development of obesity. J. Funct. Foods 45: 321–329.

Rosenbaum, M., R. Knight and R.L. Leibel. 2015. The gut microbiota in human energy homeostasis and obesity. Trends Endocrinol. Metab. 26: 493–501.

Rouxinol-Dias, A.L., A.R. Pinto, C. Janeiro, D. Rodrigues, M. Moreira, J. Dias and P. Pereira. 2016. Probiotics for the control of obesity—Its effect on weight change. Porto Biomed. J. 1: 12–24.

Saad, S.M.I. 2006. Probióticos e prebióticos: O estado da arte. Rev. Bras. Ciências Farm. 42: 1–16.

Sanz, Y., A. Santacruz and P. Gauffin. 2010. Gut microbiota in obesity and metabolic disorders. Proc. Nutr. Soc. 69: 434–441.

Schachter, J., J. Martel, C. Lin, C. Chang, T. Wu, C. Lu, Y. Ko, H. Lai, D.M. Ojcius and J.D. Young. 2018. Effects of obesity on depression: A role for inflammation and the gut microbiota. Brain Behav. Immun. 69: 1–8.

Schneeberger, M., A. Everard, A.G. Gómez-Valadés, S. Matamoros, S. Ramírez, N.M. Delzenne, R. Gomis, M. Claret and P.D. Cani. 2015. *Akkermansia muciniphila* inversely correlates with the onset of inflammation, altered adipose tissue metabolism and metabolic disorders during obesity in mice. Sci. Rep. 5: 1–14.

Schwabe, R.F. and C. Jobin. 2013. The microbiome and cancer. Nat. Rev. Cancer 13: 800–812.

Schwiertz, A., D. Taras, K. Schäfer, S. Beijer, N.A. Bos, C. Donus and P.D. Hardt. 2010. Microbiota and SCFA in lean and overweight healthy subjects. Obesity 18: 190–195.

de Simone, C. 2018. The unregulated probiotic market. Clin. Gastroenterol. Hepatol. 1–9.

Simpson, H.L. and B.J. Campbell. 2015. Review article: Dietary fibre-microbiota interactions. Aliment. Pharmacol. Ther. 42: 158–179.

Singh, S., R.K. Sharma, S. Malhotra, R. Pothuraju and U.K. Shandilya. 2017. *Lactobacillus rhamnosus* NCDC17 ameliorates type-2 diabetes by improving gut function, oxidative stress and inflammation in high-fat-diet fed and streptozotocin treated rats. Benef. Microbes 8: 243–255.

Sommer, F. and F. Bäckhed. 2016. Know your neighbor: Microbiota and host epithelial cells interact locally to control intestinal function and physiology. BioEssays 38: 455–464.

Sonnenburg, E.D., S.A. Smits, M. Tikhonov, S.K. Higginbottom, N.S. Wingreen and J.L. Sonnenburg. 2016. Diet-induced extinction in the gut microbiota compounds over generations. Nature 529: 212–215.

Sperry, M.F., H.L.A. Silva, C.F. Balthazar, E.A. Esmerino, S. Verruck, E.S. Prudencio, R.P.C. Neto, M. Inês, B. Tavares, J.C. Peixoto, F. Nazzaro, R.S. Rocha, J. Moraes, A.S.G. Gomes, R.S.L. Raices, M.C. Silva, D. Granato, T.C. Pimentel, M.Q. Freitas and A.G. Cruz. 2018. Probiotic minas frescal cheese added with *L. casei* 01: Physicochemical and bioactivity characterization and effects on hematological/biochemical parameters of hypertensive overweighted women—A randomized double-blind pilot trial. J. Funct. Foods 45: 435–443.

Steenbergen, L., R. Sellaro, S. Van Hemert, J.A. Bosch and L.S. Colzato. 2015. A randomized controlled trial to test the effect of multispecies probiotics on cognitive reactivity to sad mood. Brain Behav. Immun. 48: 258–264.

Stenman, L.K., M.J. Lehtinen, N. Meland, J.E. Christensen, N. Yeung, M.T. Saarinen, M. Courtney, R. Burcelin, M. Lähdeaho, J. Linros, D. Apter, M. Scheinin, H. Kloster, A. Rissanen and S. Lahtinen. 2016. Probiotic with or without fiber controls body fat mass, associated with serum zonulin, in overweight and obese adults—Randomized controlled trial. EBIOM 13: 190–200.

Swann, J.R., E.J. Want, F.M. Geier, K. Spagou, I.D. Wilson, J.E. Sidaway, J.K. Nicholson and E. Holmes. 2011. Systemic gut microbial modulation of bile acid metabolism in host tissue compartments. Proc. Natl. Acad. Sci. 108: 4523–4530.

Swinburn, B.A., G. Sacks, K.D. Hall, K. McPherson, D.T. Finegood, M.L. Moodie and S.L. Gortmaker. 2011. The global obesity pandemic: Shaped by global drivers and local environments. Lancet 378: 804–814.

Szulinska, M., I. Łoniewski, S. Van Hemert, M. Sobieska and P. Bogdanski. 2018. Dose-dependent effects of multispecies probiotic supplementation on the lipopolysaccharide (LPS) level and cardiometabolic profile in obese postmenopausal women: A 12-week randomized clinical trial. Nutrients 10(6) pii: E773.

Thursby, E. and N. Juge. 2017. Introduction to the human gut microbiota. Biochem. J. 474: 1823–1836.

Torres-fuentes, C., H. Schellekens, T.G. Dinan and J.F. Cryan. 2017. The microbiota – gut–brain axis in obesity. Lancet Gastroenterol. Hepatol. 2: 747–756.

Tremmel, M., U.G. Gerdtham, P.M. Nilsson and S. Saha. 2017. Economic burden of obesity: A systematic literature review. Int. J. Environ. Res. Public Health 14(4). pii: E435.

Turnbaugh, P.J., R.E. Ley, M.A. Mahowald, V. Magrini, E.R. Mardis and J.I. Gordon. 2006. An obesity-associated gut microbiome with increased capacity for energy harvest. Nature 444: 1027–1031.

Vaghef-mehrabany, E., B. Alipour, A. Homayouni-rad, S. Sharif, M. Asghari-Jafarabadi and S. Zavvari. 2014. Probiotic supplementation improves inflammatory status in patients with rheumatoid arthritis. Nutrition 30: 430–435.

Vieira-Silva, S., G. Falony, Y. Darzi, G. Lima-Mendez, R. Garcia Yunta, S. Okuda, D. Vandeputte, M. Valles-Colomer, F. Hildebrand, S. Chaffron and J. Raes. 2016. Species-function relationships shape ecological properties of the human gut microbiome. Nat. Microbiol. 1(8): 16088.

Vrieze, A., E. Van Nood, F. Holleman, J. Salojärvi, R.S. Kootte, J.F.W.M. Bartelsman, G.M. Dallinga-Thie, M.T. Ackermans, M.J. Serlie, R. Oozeer, M. Derrien, A. Druesne, J.E. Van Hylckama Vlieg, V.W. Bloks, A.K. Groen, H.G.. Heilig, E.G. Zoetendal, E.S. Stroes, W.M. De Vos, J.B. Hoekstra and M. Nieuwdorp. 2012. Transfer of intestinal microbiota from lean donors increases insulin sensitivity in individuals with metabolic syndrome. Gastroenterology 143: 913–916.

Wanchai, K., S. Yasom, W. Tunaponga, T. Chunchai, S. Eaimworawuthikul, Parameth Thiennimitr, C. Chaiyasut, A. Pongchaidecha, V. Chatsudthipong, S. Chattipakorn, N. Chattipakorna and A. Lungkaphin. 2018. Probiotic *Lactobacillus paracasei* HII01 protects rats against obese-insulin resistance-induced kidney injury and impaired renal organic anion transporter 3 (Oat3) function. Clin. Sci. 132: 1545–1563.

Watanabe, M., S.M. Houten, C. Mataki, M.A. Christoffolete, B.W. Kim, H. Sato, N. Messaddeq, J.W. Harney, O. Ezaki, T. Kodama, K. Schoonjans, A.C. Bianco and J. Auwerx. 2006. Bile acids induce energy expenditure by promoting intracellular thyroid hormone activation. Nature 439: 484–489.

WHO/FAO. 2006. Probiotics in Food: Health and Nutritional Properties and Guidelines for Evaluation. FAO, Food and Nutrition Paper, Rome. http://www.fao.org/3/a-a0512e.pdf. Accessed April 2019.

WHO World health organization. 2018. Obesity and Overweight. http://www.who.int/news-room/factsheets/detail/obesity-and-overweight. Acessed April 2019.

Willson, K. and C. Situ. 2017. Systematic review on effects of diet on gut microbiota in relation to metabolic syndromes. J. Clin. Nutr. Metab. 1: 1–12.

Xiong, Y., N. Miyamoto, K. Shibata, M.A. Valasek, T. Motoike, R.M. Kedzierski and M. Yanagisawa. 2004. Short-chain fatty acids stimulate leptin production in adipocytes through the G protein-coupled receptor GPR41. Proc. Natl. Acad. Sci. USA 101: 1045–50.

Yadav, H., J.H. Lee, J. Lloyd, P. Walter and S.G. Rane. 2013. Beneficial metabolic effects of a probiotic via butyrate-induced GLP-1 hormone secretion. J. Biol. Chem. 288: 25088–25097.

Yang, J.L., D.X. Liu, H. Jiang, F. Pan, C.S.H. Ho and R.C.M. Ho. 2016. The effects of high-fat-diet combined with chronic unpredictable mild stress on depression-like behavior and leptin/lepRb in male rats. Sci. Rep. 6: 35239.

Yarandi, S.S., D.A. Peterson, G.J. Treisman, T.H. Moran and P.J. Pasricha. 2016. Modulatory effects of gut microbiota on the central nervous system: How gut could play a role in neuropsychiatric health and diseases. J. Neurogastroenterol. Motil. 22: 201–212.

Yatsunenko, T., F.E. Rey, M.J. Manary, I. Trehan, M.G. Dominguez-Bello, M. Contreras, M. Magris, G. Hidalgo, R.N. Baldassano, A.P. Anokhin, A.C. Heath, B. Warner, J. Reeder, J. Kuczynski, J.G. Caporaso, C.A. Lozupone, C. Lauber, J.C. Clemente, D. Knights, R. Knight and J.I. Gordon. 2012. Human gut microbiome viewed across age and geography Tanya. Nature 486: 222–227.

Zhang, L., M.I. Bahl, L.I. Hellgren, H.M. Roager and C.E. Fonvig. 2015. Obesity-associated fecal microbiota from human modulates body mass and metabolites in mice. Abstract from EMBL Conference, Heidelberg, Germany.

Zhang, M. and X. Yang. 2016. Effects of a high fat diet on intestinal microbiota and gastrointestinal diseases. World J. Gastroenterol. 22: 8905–8909.

Zhao, L. 2013. The gut microbiota and obesity: From correlation to causality. Nat. Rev. Microbiol. 11: 639–647.

10

Probiotics in the Management of Inflammatory Bowel Disease and Irritable Bowel Syndrome

Larissa S Celiberto,[1,*] *Bruce A Vallance*[1] and *Daniela CU Cavallini*[2]

Introduction

Diseases of the gastrointestinal tract, such as Inflammatory Bowel Diseases (IBD) and Irritable Bowel Syndrome (IBS), are highly prevalent worldwide, affecting people regardless of age, sex, or cultural background. These conditions can result in several physician visits per year, thus directly impacting the patient's quality of life as well as the local health care system (Sandler et al. 2002). The two major forms of IBD are ulcerative colitis (UC) and Crohn's disease (CD) and these conditions are characterized by chronic and relapsing intestinal inflammation. IBS is considered a functional gastrointestinal (GI) disorder associated with symptoms such as recurrent abdominal pain and overt changes in bowel habits (Lacy et al. 2016). Treatment for these diseases remains challenging, since patients often poorly adhere to conventional pharmacological treatments (Chan et al. 2017). The etiology of IBD and IBS remains elusive, although multifactorial aspects such as environmental factors (i.e., lifestyle, diet, stress, smoke, exposure to pathogens) and genetic predisposition seem to be involved. Moreover, there is increasing evidence linking alterations in the gut microbiota with the pathogenesis of IBD and IBS (Imhann et al. 2018). The human

[1] Division of Gastroenterology, Department of Pediatrics, BC Children's Hospital and the University of British Columbia, Vancouver, BC, Canada.
Email: bvallance@cw.bc.ca
[2] São Paulo State University (UNESP), School of Pharmaceutical Sciences, Department of Food and Nutrition, Araraquara, SP, Brazil.
Email: cavallinidc@fcfar.unesp.br
* Corresponding author: larissasbaglia@gmail.com

microbiota is a complex system, comprising thousands of species of bacteria and fungi, whose qualitative and quantitative alterations can interfere in several vital functions. Numerous strategies have been proposed to modulate the composition of the microbiota, such as the use of antibiotics, fecal microbiota transplantation and administration of probiotic microorganisms (Cammarota et al. 2017, Fodor et al. 2019). Therapeutic manipulation of the gut microbiota by probiotics has been shown to improve the clinical symptoms of GI disorders; however, the results are still inconsistent among studies (Marteau et al. 2006). The aim of this chapter is to provide scientific based evidence on the role of probiotics in IBD and IBS as well as to discuss the possible mechanisms underlying their protection. A deeper understanding of the microbial-host relationship is essential to assist in the design of more effective probiotics that could act as adjuvant therapies for these GI conditions.

The Intestinal Microbiota and its Role in Gastrointestinal Diseases

The intestinal microbiota is a collective term used to describe all the microorganisms within the gut, and this microbial community along with their genes are known as the intestinal microbiome. Although the human gut microbiome is a very diverse ecosystem with hundreds of different microbial species, it has been suggested that most individuals share a dominant core of species, thus indicating considerable dominance and inter-individual stability of microbes across humans (Qin et al. 2010). Moreover, microbiome sequencing data from different subjects around the world suggests that people are stratified into three major clusters of bacteria referred to as enterotypes (i.e., *Bacteroides*, *Prevotella*, and *Ruminococcus*) (Arumugam et al. 2011). The distinct enterotypes and their metabolic activity may help explain the different responses to dietary and drug interventions often found in humans.

The resident microbiome composition is established during the first years of life where both microbes and the immune system mature together in a symbiotic relationship [reviewed in (Belkaid and Hand 2014, Caballero and Pamer 2015)]. This complex and diverse ecosystem plays an important role in health and disease and its composition is strongly influenced by several factors such as mode of delivery (vaginal *versus* cesarean section), breast feeding, diet, lifestyle, hygiene conditions, infections, antibiotic use in early infancy, among others (Martin et al. 2010, Dominguez-bello et al. 2010). Usually the host and its resident microbiome coexist in homeostasis, thereby keeping the host immune system in a hypo-responsive state. While the host provides the necessary nutrients to promote microbial survival and development, the indigenous microbiota protects the host against potential pathogens through competition for space and nutrients. Moreover, resident microbes aid in food digestion and absorption and also release important metabolites such as vitamins, bacteriocins and other antimicrobial substances, as well as short chain fatty acids (SCFA) (Furusawa et al. 2013, LeBlanc et al. 2013, LeBlanc et al. 2017, Umu et al. 2017, Heeney et al. 2018, Hols et al. 2019). After the establishment of a "mature" microbiome, its composition seems to be relatively stable over time in healthy adults, and most often returns to a similar initial composition after exogenous perturbations such as antibiotic administration or infections with pathogenic microorganisms (Spor et al. 2011). This microbiome resilience is an important mechanism that protects the host against pathogen competition with the resident microbes, however it also represents an obstacle to engrafting exogenous beneficial microbes into the microbiota, and thus providing long term effects. However, if somehow this dynamic ecosystem becomes perturbed, a detrimental imbalance called dysbiosis may occur.

Considering how different each person's microbiome is, the concept of dysbiosis itself does not necessarily reflect a disease. Nevertheless, quantitative and qualitative alterations in the host microbiome, especially as compared to its baseline, have been implicated in several chronic GI conditions such as colon cancer, celiac disease, IBD, and IBS, as well as extra-intestinal diseases such as diabetes, obesity, and asthma (Ott et al. 2004, Ley et al. 2006, Wu et al. 2010, Andoh et al. 2012, Ohigashi et al. 2013, Walters et al. 2014, Arrieta et al. 2015). Although a dysbiotic microbiota seems to be involved in the pathogenesis of several GI diseases, it remains unclear whether the changes in microbiota composition truly play a causative role. While there is strong evidence that alterations in the gut microbial community play a key role in the development of diseases such as *Clostridium difficile* infection and small intestinal bacterial overgrowth (SIBO), it remains unclear for other conditions such as IBD and IBS. Even though microbial dysbiosis is often identified in IBD patients (Fava and Danese 2011), it remains elusive whether the characteristic intestinal inflammation results primarily from an abnormal immune response to an unbalanced microbiota or if the dysbiotic microbiota are simply a reflection of the inflammatory process during the course of disease. The answer probably lies in between these two options, since there are studies supporting both possibilities (Celiberto et al. 2018). The potential that an altered gut microbiota promotes the pathogenesis of IBS also represents a challenge in the scientific community. Since there is no sole pattern for IBS and its symptoms are often non-specific, associations between specific groups of microbes and IBS are difficult to be predicted (Shanahan and Quigley 2014).

Inflammatory Bowel Disease and Irritable Bowel Syndrome

The exact etiology of IBD remains unknown; however studies have indicated that genetics, the immune system, and environmental factors are involved in both the initiation and the progression of the disease [reviewed in Celiberto et al. (2018)]. While UC leads to superficial mucosal inflammation that primarily affects the terminal portion of the colon and the rectum, CD is characterized by transmural inflammation that may be localized to any part of the GI tract, from the mouth to the anus. Some risk factors associated with IBD are smoking, hygiene, appendectomy, use of medications (i.e., antibiotics, oral contraceptives, and nonsteroidal anti-inflammatory drugs), nutrition, and stress. Nonetheless, the mechanisms involved in these associations are still poorly understood (Loftus et al. 2004, Ye et al. 2015).

Europe and North America have the highest prevalence of IBD in the world, although its incidence is significantly increasing in developing countries, potentially due to their adoption of more Western type lifestyles (Ye et al. 2015). UC and CD may occur at any age, however around 25 percent of IBD patients are diagnosed in late adolescence and early adulthood, followed by a second incidence peak in later decades (Loftus et al. 2004, Ye et al. 2015). Likewise, the incidence of pediatric IBD and Very Early Onset IBD (VEO-IBD), diagnosis in the first six years of life, has significantly increased worldwide (Moran 2017). Gender distribution in IBD is highly dependent on the disease subtype (i.e., CD or UC). Shah et al. (2018) found that despite having a lower risk of CD during childhood, female patients had a higher risk of CD thereafter (e.g., 25–29 years and older than 35 years) in Western countries. Regarding UC, there was no difference in the incidence between female and male patients until age 45, except for children between 5 and 9 years old. Thereafter, men had a higher incidence of UC as compared to women (Shah et al. 2018). In Canadian adults, women are more affected by CD (1.3 females for every male), whereas no gender difference is described in UC patients (Bernstein et al. 2006). Interestingly, in Canadian children, CD is the most prevalent subtype of IBD, with boys displaying a 1.25:1 ratio over girls (Benchimol et al. 2009).

IBS is a highly prevalent functional GI disorder characterized by the association of recurrent abdominal pain and changes in bowel habits (Lacy et al. 2016). Symptoms such as abdominal bloating/distention and excessive gas are often present, as well as a change in stool consistency or frequency towards a pattern of diarrhea (IBS-D), constipation (IBS-C), or mixed bowel habits (IBS-M). These symptoms vary in intensity and frequency, with some patients describing it as a chronic condition. In addition, a subset of IBS patients is classified as unsubtyped (IBS-U), since although symptoms are present they do not follow a specific or consistent pattern (Lacy et al. 2017). Therefore, even though IBS is characterized as a relatively benign disease, it is also a source of considerable discomfort with a remarkable strong negative impact on the quality of life and work productivity of patients (Chey et al. 2015, Martinez et al. 2015, Lacy et al. 2016). The pathogenesis of IBS, as well as its phenotype, is heterogenous and likely encompasses multiple aspects such as changes in gut motility, altered pain perception, visceral hypersensitivity, gut-brain interactions, and psychosocial distress. Moreover, recent studies have highlighted the possible involvement of an increased gut mucosal immune activation, impaired intestinal permeability, and the gut microbiome as potential drivers in the disease initiation and progression (Simrén et al. 2013, DuPont 2014, Chey et al. 2015). Early life stressors, food intolerance and antibiotic usage are described as risk factors for IBS (Chey et al. 2015). Additionally, it is well documented that bacterial gastroenteritis significantly increases the risk of subsequently developing IBS; with this subset of patients referred to as suffering post-infectious IBS (PI-IBS) (Rodríguez and Ruigómez 1999, Klem et al. 2017). Unlike typical IBS, which is considered by many patients as a chronic and relapsing disease, the PI-IBS subtype tends to settle within 6–8 years following the infection in approximately 50% of patients (Spiller and Lam 2012).

The prevalence of IBS is estimated to range between 7 and 21% worldwide (Chey et al. 2015). However, an accurate estimate of prevalence and incidence of IBS is difficult since disease diagnosis is defined by subjective criteria rather than a specific biomarker. Moreover, multiple methods and criteria (i.e., Manning, Rome I–IV, and others) have been used to assess the prevalence of IBS in different countries, which ultimately leads to under or over-estimation worldwide (Chey et al. 2015, Sperber et al. 2017). A meta-analysis conducted by Lovell and Ford (2012) showed that the prevalence of IBS in North America ranges around 12%, with South America (21%) and Southeast Asia (7%) being the most and least prevalent regions, respectively. Regarding gender-related differences, IBS symptoms are 1.5 to 2 times more prevalent in women than men in North America and Israel. Conversely, prevalence in Asia is described as more equally distributed. Female patients commonly report abdominal pain and constipation, whereas male patients usually describe a diarrhea-predominant subtype of IBS.

Alteration of the gut microbiota in IBD and IBS

Patients with IBD often display low diversity in their gut microbiota as compared to healthy individuals (Ott et al. 2004, Andoh et al. 2012), with reduced beneficial bacteria such as *Bifidobacterium* spp. (Joossens et al. 2011, Andoh et al. 2012), *Lactobacillus* spp. (Ott et al. 2004), and *Faecalibacterium prausnitzii* (Sokol et al. 2009, Joossens et al. 2011, Andoh et al. 2012), together with an increased abundance of pathobionts such as certain strains of *Escherichia coli* (Darfeuille-Michaud et al. 2004, Sokol et al. 2006) as well as true pathogens such as *C. difficile* (Rodemann et al. 2007). Besides *F. prausnitzii*, Takahashi et al. (2016) showed that butyrate-producers such as *Blautia faecis, Roseburia inulinivorans, Ruminococcus torques, Clostridium lavalense, Bacteroides uniformis* were more likely to be significantly reduced in CD patients as compared to healthy individuals. Additionally, studies have shown that UC patients display an increased abundance of *C. perfringens* (Li et al. 2016) as well as a decrease in *Fusicatenibacter saccharivorans* (Takeshita et al. 2016), which represents a single species of the *Clostridium* subcluster XIVa that supresses intestinal inflammation and was inversely correlated with UC activity. These changes in the gut microbiota composition of IBD patients highlights the hypothesis that microbial dysbiosis is an important mechanism underlying disease pathogenesis (Chassaing and Darfeuille-Michaud 2011). However, considering that inflammation on its own favours the depletion of *Firmicutes* and the expansion of *Enterobacteriaceae* such as *E. coli*, it is not clear if the observed alterations in microbiota composition play a role in the development of the disease or are instead a consequence of the typical inflammatory environment found in IBD patients (Chassaing et al. 2017, Celiberto et al. 2018).

Alterations in the gut microbiota composition and its temporal stability have also been described in IBS patients [reviewed in Matto et al. (2005), Maukonen et al. (2006), Collins (2014), and Rodiño-Janeiro et al. (2018)], especially in the diarrhea-predominant subtype (IBS-D). In contrast to IBD, studies have described an increase in the abundance of *Firmicutes* (*Ruminococcaceae* spp. and *Clostridium* cluster XIVa) in patients with IBS, along with a lower relative abundance of *Bacteroides* spp. (Salonen et al. 2010, Rajilić-Stojanović et al. 2011, Jeffery et al. 2012). An increase of bacterial species from the *Enterobacteriaceae* family has also been described in the microbiota of IBS patients, while the abundance of beneficial microbes such as *Lactobacillus* spp. and *Bifidobacterium* spp. often show mixed results depending on the IBS subtype and the geographic region investigated (Zhuang et al. 2017, Rodiño-Janeiro et al. 2018). Nonetheless, no species-specific microbial enhancement or decrease has been consistently reported in IBS that could act as a biomarker of dysbiosis for this condition (Collins 2014). Interestingly, fecal samples from IBS patients can induce gut dysfunction when inoculated into germ-free animals, reflecting similar responses to those seen in IBS patients (e.g., higher visceral pain response (Crouzet et al. 2013), faster intestinal transit when receiving IBD-D samples (De Palma et al. 2017), and lower abundance of *Bifidobacterium* spp. and higher levels of *Enterobacteriaceae* and sulphate-reducing bacteria (Chassard et al. 2012, Crouzet et al. 2013). These results strongly suggest an important role for the gut microbiota in IBS, although the underlying mechanisms of action still need to be elucidated.

Aside from changes in the gut microbiota composition, IBD and IBS are also associated with changes in bacterial metabolites such as SCFA. Acetate, propionate and butyrate are the major SCFA found in the gut and are mainly produced by bacteria that are able to ferment dietary fibers (Gonçalves et al. 2018). SCFAs, particularly butyrate, are important metabolites in maintaining intestinal homeostasis through their anti-inflammatory actions. They improve intestinal barrier function by serving as an energy source for colonocytes and by increasing regulatory T cell (Treg) function (Celiberto et al. 2018, Parada Venegas et al. 2019). Furthermore, SCFA act as ligands for specific G protein-coupled receptors, thus promoting epithelial cell production of cytokines and chemokines and consequently activating anti-inflammatory signaling pathways (Kim et al. 2013, Celiberto et al. 2018, Parada Venegas et al. 2019). Beneficial bacteria that ferment fibers and produce SCFA (e.g., *F. prausnitzii, Roseburia hominis*) are commonly reduced in the mucosal biopsies and feces of IBD patients in comparison with healthy individuals (Huda-Faujan et al. 2010, Kumari et al. 2013, Machiels et al. 2014, Parada Venegas et al. 2019). Therefore, not only do the microbial communities within the gut seem to be involved in IBD pathogenesis, but so too is the loss of key metabolites produced by strict anaerobic bacteria commonly associated with the microbiota of healthy individuals. In contrast, increased levels of SCFA have been associated with IBS (Collins 2014), with enemas of butyrate being used as a model to study colonic hypersensitivity in rats, to model IBS (Bourdu et al. 2005). The influence of SCFA on GI motility might be related to the resulting visceral pain response, bloating and other GI symptoms described by IBS patients. However, considering the different subtypes of IBS, and their specific dietary restrictions, the different levels of SCFA in the gut may reflect the changes in the microbiome as a consequence of each intervention (diet, drug therapy, psychological and alternative therapies) rather than playing a causative role in the disease.

Current therapies for IBD and IBS

At present, the traditional therapies for IBD follow a step-up approach that starts with immunosuppresives such as 5-aminosalicylic acid (5-ASA), moving to immunomodulators (azathioprine and methotrexate), and finally to biologics (infliximab, adalimumab, vedolizumab, etrolizumab, ustekinumab) [reviewed in (Burger and Travis (2011), and Celiberto et al. (2018)]. The drug type, dosage and possible combination of therapies must be managed through an individualized approach for each patient, considering important factors such as age, clinical background and/or associated co-morbidities, as well as disease severity. At present, conventional treatments for IBD focus on suppressing the pathological immune response rather than targeting potential root causes of the disease. Moreover, several limitations and side effects have been reported when using these approaches (Stallmach et al. 2010). For instance, the efficacy of biologics and steroids often reduces over time in most patients, while some patients do not respond at all to these treatments (Renna et al. 2014, Ben-Horin et al. 2015, Roda et al. 2016). Notably, a resulting increase in susceptibility to cancer and infectious disease has been described following steroid and biologic regimes (Renna et al. 2014, Ben-Horin et al. 2015), which represents a significant concern especially for pediatric patients, since they will be on medications for many years following diagnosis. Therefore, studies investigating potential causal factors of IBD, as well as therapies targeting them, are extremely important to improve the quality of life for these patients.

Considering the heterogeneity of IBS among patients, the current treatments for this condition focus on relieving the symptoms associated with each IBS subtype. The initial steps in treating IBS usually consist of supportive treatments to alleviate gastrointestinal symptoms through dietary adjustments, regular exercise, and stress control (Grundmann and Yoon 2014). Studies also indicate that the patient-practitioner relationship plays a key role in the treatment outcome, thus supporting the psychological aspect of the disease as well as the importance of personalized treatments focusing on each patient's individual needs (Halpert and Godena 2011). Patients commonly describe that certain foods (e.g., dairy, cereals, citrus fruits, potatoes, caffeine, alcohol, additives, and preservatives) exacerbate their IBS symptoms. However, studies investigating exclusion diets in IBS often lack proper controls, leading to questionable results with high placebo response rates. Likewise, several studies have proposed the use of dietary fiber, bulking agents, and laxatives in the management of IBS (Peyton and Greene 2014). Again, with the exception of the fiber psyllium hydrophilic mucilloid that is used clinically as a bulk laxative and seems to confer modest benefits to IBS patients, the results have been extremely inconsistent, with most studies presenting insufficient details of their methodological design. Therefore, the American College of Gastroenterology - IBS Task Force makes only a weak recommendation to support the use of psyllium hydrophilic mucilloid, and recommends against food-allergy testing, strict exclusion diets, and routine use of other fibers or bulking agents (Peyton and Greene 2014). Interestingly, one of the most successful dietary approaches to manage IBS symptoms is the administration of a low FODMAP diet (Marsh et al. 2016, McKenzie et al. 2016). FODMAP stands for "Fermentable Oligo-, Di-, and Mono-saccharides and Polyols" and is a group of short-chain carbohydrates that undergo rapid fermentation in the colon due to its incomplete absorption in the small intestine (Ooi et al. 2019). A high FODMAP diet is also associated with alterations in gut microbiota composition, increased inflammation, impaired intestinal barrier function and visceral hypersensitivity (Khan et al. 2015, Farzaei et al. 2016, Staudacher and Whelan 2017). Clinical intervention for IBS involves restricting FODMAP intake for 4–8 weeks to test whether this reduces clinical symptoms. If a positive response is identified, then FODMAP food sources are slowly reintroduced into the diet to test for specific intolerances for each patient (Ooi et al. 2019). Despite the positive impact of a low FODMAP diet on IBS symptoms, especially abdominal pain and bloating, studies have raised concerns regarding the long-term effects of such restricted diets in comparison with traditional IBS diets (i.e., reduction in fiber intake, regularly scheduled meals, elimination of trigger foods—that differ from patient to patient) (Altobelli et al. 2017, Krogsgaard et al. 2017). Therefore, future studies assessing the long-term outcomes of FODMAP diets are necessary to investigate its therapeutic potential for IBS patients. Moreover, FODMAP interventions should only be managed by a dietitian specialized in IBS/gut health to avoid patients suffering nutritional deficiencies.

Regarding the medications currently used to treat IBS, prokinetics and antispasmodics are the most commonly used for IBS-C, whereas opioid agonists, anticholinergics, and 5-hydroxytryptamin (5-HT3) antagonists are frequently prescribed for patients with IBS-D (Grundmann and Yoon 2014). Since psychiatric disorders such as depression and anxiety have been increasingly described as co-morbidities to IBS, the use of selective serotonin reuptake inhibitors and other antidepressants have shown to yield significant improvements in a subset of patients with IBS (Halper et al. 2005, Ford et al. 2009, Rahimi et al. 2009). Lastly, several complementary and alternative therapies (i.e., dietary supplements, acupuncture, yoga, hypnotherapy, cognitive behavioral therapy) have also been suggested for the treatment of IBS as options to manage disease with reduced side effects as compared to pharmacological approaches.

Over the last several decades, the use of beneficial microbes known as probiotics have been seen as a potential strategy to favorably modulate the intestinal microbiota to promote health and treat microbial dysbiosis-related diseases. Probiotics are defined by the WHO as live microorganisms that confer a health benefit to the host when administered in adequate amounts (FAO/WHO 2002, Hill et al. 2014), and these beneficial microbes may positively influence microbial interactions with the immune system and the gut epithelium (DuPont and DuPont 2011). Several microorganisms classified as probiotics belong to the lactic acid bacteria (LAB) group (e.g., *Lactobacillus* spp., *Lactococcus* spp., *Enterococcus* spp.). LAB produce lactic acid as their major end product of carbohydrate fermentation. Moreover, besides having microbicidal activity against enteric pathogens, some LAB strains compete for cell surface and mucin binding sites, thus offering protection to the host by strengthening the intestinal mucosal barrier (Ljungh and Wadström 2006). Additionally, lactic acid also contributes to the sensory and texture profile of food products, making this group of bacteria highly investigated as potential probiotics.

Microorganisms from the genera *Bifidobacterium*, as well as *E. coli* Nissle 1917 and the yeast *Saccharomyces boulardii* are also among the most common probiotic candidates used in current clinical practice for GI diseases (Celiberto et al. 2018). However, the common genera used in probiotic foods and supplements are often minor components of the gut microbiota in adults (Xiao et al. 2015), thus making their incorporation and persistence within the gut very difficult to be achieved over the long term.

Probiotics in Gastrointestinal Diseases

Probiotics in IBD

Preclinical studies to elucidate possible mechanisms underlying the beneficial effects of probiotics in IBD have been performed both *in vitro* as well as *in vivo* through animal studies. Several probiotics have been shown to promote anti-inflammatory activity and help maintain intestinal barrier function in *in vitro* experiments (Meijerink and Wells 2010, Wells 2011). Some of these beneficial microorganisms, especially *Lactobacillus* spp., are able to interact with mucosal dendritic cells (DC) within the lamina propria as well as with the epithelial cells lining the intestinal mucosa. Binding of microbe-associated molecular patterns with innate pattern-recognition receptors [Toll like receptors (TLRs), nucleotide-binding oligomerization domain-like receptors, or NOD-like receptors (NLRs), and C-type lectin receptors (CLRs)] can activate antigen presenting cells and modulate their function through the expression of surface receptors, secreted cytokines and chemokines (Meijerink and Wells 2010). For example, DC from patients with UC had their normal stimulatory capacity restored after treatment with *Lactobacillus casei* Shirota, and this effect was attributed to the probiotic causing a reduction in their expression of TLR2 and TLR4 (Mann et al. 2013, Mann et al. 2014). Another strain of the *Lactobacillus* genera, *L. plantarum* CGMCC1258, was found to protect against epithelial barrier disruption, when the intestinal epithelial cell line IPEC-J2 was challenged by the pathogen *E. coli* K88. Besides increasing tight junction protein levels and decreasing intestinal permeability, the probiotic reduced the expression of the pro-inflammatory cytokines IL-8 and TNF-α, possibly through a combination of several mechanisms including causing a decrease in TLR expression as well as nuclear factor-κB (NF-κB) activation, and mitogen-activated protein kinase pathways (Wu et al. 2016). Lastly, *Lactobacillus casei* DN-114001 was found to inhibit the adhesion and invasion of human intestinal epithelial cells (Intestine-407 and Caco-2) by the pathobiont adherent-invasive *E. coli* (AIEC) isolated from Crohn's patients (Ingrassia et al. 2005).

In the past decades, a variety of animal models have been developed to study the pathogenesis of IBD and intestinal inflammation. These models can be broadly divided into those involving chemically induced colitis, bacterial induced colitis, spontaneous colitis (i.e., congenital and genetically engineered), and adoptive transfer models (Perše and Cerar 2012). Although animal models cannot fully represent human disease, they do allow us to study different aspects of GI inflammation thus being extremely helpful in understanding the pathogenesis of IBD and in the development and testing of novel therapeutic approaches. At present, many studies have demonstrated the beneficial effects of probiotics in murine models of IBD and intestinal inflammation. The mechanisms of action involved in this protection are diverse, including pathogen exclusion, enhancement of the epithelial barrier function, and activation of anti-inflammatory and regulatory immune effects via adaptive immunity (Gareau et al. 2010). A mix of two probiotic strains, *Lactobacillus acidophilus* Bar 13 and *Bifidobacterium longum* Bar 33, promoted the expansion of Treg cells and reduced the number of intraepithelial lymphocytes in the 2,4,6-trinitrobenzene sulfonic acid (TNBS) induced colitis model (Roselli et al. 2009). In the infectious-colitis model caused by the murine pathogen *Citrobacter rodentium*, the mix of *Lactobacillus helveticus* R0052 and *Lactobacillus rhamnosus* R0011 (Johnson-Henry et al. 2005) as well as the yeast *Saccharomyces boulardii* (Wu et al. 2008) were able to attenuate disease activity by reducing the attachment of the pathogen to colonic epithelial cells. Finally, the combination of eight distinct probiotic strains (VSL#3) has repeatedly shown positive results in animal models of colitis. This commercial probiotic mixture is known to effectively reduce disease activity and colonic inflammation by significantly reducing inflammatory markers such as IL-1β, NF-κB activation, and neutrophil myeloperoxidase (MPO) in mice (Dai et al. 2013, Salim et al. 2013, Talero et al. 2015). Similarly, administration of *Lactobacillus plantarum* 299V prevented spontaneous colitis development in IL-10 deficient ($IL\text{-}10^{-/-}$) mice (Schultz et al. 2002), while treatment with VSL#3 also ameliorates colitis and overall disease activity in $IL\text{-}10^{-/-}$ mice.

In comparison with the number of successful probiotic studies performed in animal models of colitis, far fewer clinical studies have demonstrated promising results in patients with IBD (Veerappan et al. 2012, Whelan and Quigley 2013, DuPont et al. 2014). The considerable array of probiotics available in the market, as well as all the possible multistrain combinations, together with the several disease phenotypes of IBD, have often yielded mixed results in clinical trials (Isaacs and Herfarth 2008). The commercial multistrain VSL#3 is the probiotic most commonly shown to exert beneficial effects in IBD patients, mostly in those with UC or pouchitis (Isaacs and Herfarth 2008, Celiberto et al. 2015). Pronio et al. (2008) demonstrated that patients undergoing ileal-anal anastomosis for UC showed a reduction in their pouchitis activity index when treated with VSL#3. The protective role of the probiotics in this particular study appeared to be mediated, at least in part, by reducing IL-1β mRNA levels as well as increasing Foxp3 mRNA expression by Treg cells (Pronio et al. 2008). Moreover, VSL#3 and *Lactobacillus rhamnosus* GG (Zocco et al. 2006) seem to be effective in maintaining remission in UC patients (Bibiloni et al. 2005). Interestingly, VSL#3 was also shown to effectively maintain disease remission in pediatric patients with UC when compared to children that received a placebo (Huynh et al. 2009, Miele et al. 2009). A pilot study conducted by Kato et al. (2004) showed that UC patients who ingested a commercial bifidobacteria-fermented milk (BFM—*Bifidobacterium breve* strain Yakult, *Bifidobacterium bifidum* strain Yakult, and a *Lactobacillus acidophillus* strain) combined with conventional treatments displayed a lower clinical activity index as compared to a non-supplemented group. Besides highlighting the characteristic strain and dose specific effects of probiotics, this study also indicated a possible benefit of the food matrix carrying the probiotic strains. Despite multiple studies showing the beneficial effects of probiotics in UC patients, other trials have failed to show any positive outcome (Prantera et al. 2002, Marteau et al. 2006, Van Gossum et al. 2007, Wildt et al. 2011, Petersen et al. 2014). Most systematic reviews and meta-analysis studies find that the single or multi-strain probiotics most commonly used in medical practice are ineffective in inducing or maintaining remission in CD patients (Shen et al. 2009, Jonkers et al. 2012, Bjarnason et al. 2019), although administration of synbiotics (probiotics plus prebiotics) has been associated with improved clinical status (Saez-Lara et al. 2015). To date, there is insufficient data to recommend probiotics for CD patients (DuPont et al. 2014), probably due to the severe, yet patchy inflammation typical of the disease, and the lack of long-term effective studies.

Probiotics in IBS

Similar to IBD, the use of probiotics to treat and/or prevent IBS remains controversial. The mechanisms of action of probiotics in IBS seem to involve effects on epithelial barrier function as well as visceral hyperalgesia and motility (Gareau et al. 2010). Animal models of early life stress, acute psychological stress, and intestinal hypersensitivity are often used to study typical symptoms of IBS such as microbial dysbiosis, abdominal pain, and altered visceral perception (Söderholm et al. 2002). Oral administration of *Lactobacillus helveticus* R0052 combined with *Lactobacillus rhamnosus* R0011 was shown to prevent bacterial translocation and improved intestinal epithelial barrier function in rats subjected to chronic psychological stress (Zareie et al. 2006). However, no mechanistic basis for this protection was determined in the study. Another study conducted in a mouse model of chronic colonic hypersensitivity demonstrated that treatment with *Lactobacillus acidophilus* NCMF positively modulated the composition of the gut microbiota. The probiotic was able to induce opioid and cannabinoid receptor expression by intestinal epithelial cells (IECs), thus restoring normal perceptions of visceral pain (Rousseaux et al. 2007). Furthermore, the probiotic yeast *Saccharomyces boulardii* CNCM I-745 was shown to attenuate GI neuromuscular anomalies in mice suffering from gut dysfunction typically observed in patients suffering from IBS (Brun et al. 2017).

Regarding the use of probiotics in patients with IBS, several clinical studies and meta-analyses have investigated different probiotics and their effects. A meta-analysis of 15 studies involving multiple probiotic strains belonging to different genera (*Lactobacillus* spp., *Lactococcus* spp., *Propionibacterium* spp., *Bifidobacterium* spp., *Enterococcus* spp., *Escherichia coli*) showed that probiotics reduce pain and symptom severity scores in IBS patients as compared to placebo controls (Didari et al. 2015). Another meta-analysis concluded that supplementation with *Saccharomyces cerevisiae* CNCM I-3856 ameliorates GI symptoms in a general IBS population and in the IBS-C subpopulation (Cayzeele-Decherf et al. 2017). In a multicentre, double-blind, controlled trial, a *Bifidobacterium* strain proved to be beneficial in adults with constipation-predominant IBS (IBS-C). The consumption of fermented milk containing *Bifidobacterium animalis* DN-173 010 for more than 6 weeks led to an improvement in health-related quality of life, a specific marker to evaluate the effect of probiotics in IBS (Guyonnet et al. 2007). Furthermore, different studies have also demonstrated that VSL#3 is effective at ameliorating symptoms and improving the quality of life of children (Guandalini et al. 2010) and adults (Kim et al. 2005) affected by IBS.

Taken together, numerous studies suggest that certain probiotic strains are effective in IBS animal models, with their ability to promote epithelial barrier function as well as normalize altered gut motility as well as visceral pain sensitivity identified as potential underlying mechanisms of action. While there is a role for probiotics in the management of IBS, the exact mechanisms underlying this protection in humans are difficult to determine. Particular combinations of probiotics, as well as single strains, appear to have positive effects on the global symptoms of IBS along with abdominal pain (Corbitt et al. 2018, Ford et al. 2018). Moreover, several probiotic strains have proven to be safe in IBS, thus reflecting the high compliance by patients seeking complementary and alternative therapies. However, given the heterogeneity of IBS and its numerous subtypes, studies testing the effects of probiotics focusing on specific subgroups of patients and their particular IBS-related symptoms may result in more promising recommendations.

Controversies and challenges regarding probiotics in IBD and IBS

Several issues may underlie the sometimes limited and contradictory effects of probiotics in IBD and IBS patients. These divergent results may be due to the strain specific effects of probiotics as well as methodological differences between studies (doses, sole probiotic use *versus* symbiotics, given in combination with medication, etc.). Furthermore, one issue may stem from the "one size fits all" approach that has been commonly employed with probiotics. Considering that the resident gut microbiota is established in the early infancy, the host's immune system is programmed to recognize indigenous microbes as part of the human body. In contrast, new (exogenous) microbes encountered later in life are usually identified as harmful stimuli that need to be rapidly expelled. Thus, providing probiotics isolated from exogenous sources to patients without defining whether there are environmental/immunological niches for these gut microbes, may mean they will be seen as foreign and unable to take up permanent residence in the GI tracts of treated patients. Moreover, the microbial dysbiosis found in complex diseases such as IBD and IBS may require treatment with a diverse combination of microbes as similar as possible to the baseline microbiota of the host (i.e., before disease onset) rather than with a combination of a few commercial isolates. In addition, the environment in the intestines of IBD and IBS patients is often inhospitable to probiotic microbes due to the exaggerated inflammatory and antimicrobial responses seen during disease. These responses are aimed at rapidly clearing microbes from the intestine, including probiotics and commensal bacteria, often before they have the opportunity to exhibit their beneficial effects.

Considering our growing knowledge of the structure and function of the human gut microbiome as well as the increasing availability of cutting-edge microbiological techniques, there is growing interest in isolating novel probiotic strains from the dominant bacterial groups of the adult intestinal microbiota. Therefore, species such as *F. prausnitzii* and *Akkermansia muciniphila* along with several *Bacteroides* spp. have been considered as potential next-generation probiotics (NGPs) (O'Toole et al. 2017, Chang et al. 2019). More rigorous studies need to be published supporting the efficacy and safety of these agents before they become a mainstay for IBD and IBS medical treatments. Additionally, some studies over the past two decades have changed the focus of medical treatment from a disease-specific approach towards a patient-specific approach, favoring the stratification of subpopulations and precision/personalized strategies to promote accuracy and cost-effective medical treatments (Rebbeck et al. 2004, Jameson et al. 2015, Zmora et al. 2016). In the same line of thought, our research group has recently proposed a personalized strategy for the use of probiotics in IBD, with preliminary studies performed in the DSS-colitis mouse model. The results show that commensal microbes isolated from the host's microbiota are more effective at preventing DSS-colitis in mice as compared with a commercial probiotic strain (Celiberto et al. 2018). While commercial and novel probiotics have proven promising in treating conditions where there is a specific target to be modulated, single or even multi-strain probiotics appear to exert only modest benefits in diseases such as IBD and IBS, since each patient seems to have a distinct dysbiotic profile. The individual characteristics of a person's microbiota, as well as how it changes during their lifetime or in response to environmental stimuli, are the only factors known regarding what controls microbiota composition. Therefore, strategies focusing on personalized probiotics based on each individual's microbiota may provide advantages over the use of specific isolates from an exogenous source.

Finally, it is now clear that the use of probiotics is increasing in popularity for both the prevention and treatment of a variety of diseases. This popularity has a direct impact on the growing number of pharmaceutical and food companies seeking profits through the development of new probiotic supplements and food products. While a growing number of well-conducted human studies are emerging and investigations of the underlying mechanisms of action of probiotics are being undertaken, questions still remain with respect to the specific immune and physiological effects of these microorganisms in health and disease. The minimal requirements for the evaluation of an effective probiotic strain should be strictly followed by researchers in the field. Compliance with these rules avoids the misuse of the term "probiotic" that generally causes confusion among the public and clinical professionals, as well as leads to inadequate conclusions based on improperly conducted intervention studies (Reid et al. 2019).

Conclusion

Although several probiotics have been prescribed for IBD and IBS patients in the past years, additional human studies are required to clearly define their precise roles and potential recommendation to treat these conditions. Research determining which probiotics will benefit patients with IBD and IBS requires the proper identification and characterization of these microorganisms following strict compliance to the scientific definition of the term "probiotic". In addition, recent advances in cutting-edge techniques will help scientists and health professionals better understand the pathogenesis of IBD and IBS, and thus help identify the most promising probiotic candidates to be used for preventive and therapeutic strategies. Lastly, considering the heterogenicity of IBD and IBS, personalized approaches addressing each patient's microbiome may benefit the treatment of such complex diseases and their diverse symptoms.

References

Arumugam, M., J. Raes, E. Pelletier, D. Le Paslier, T. Yamada, D.R. Mende, G.R. Fernandes, J. Tap, T. Bruls, J.-M. Batto, M. Bertalan, N. Borruel, F. Casellas, L. Fernandez, L. Gautier, T. Hansen, M. Hattori, T. Hayashi, M. Kleerebezem, K. Kurokawa, M. Leclerc, F. Levenez, C. Manichanh, H.B. Nielsen, T. Nielsen, N. Pons, J. Poulain, J. Qin, T. Sicheritz-Ponten, S. Tims, D. Torrents, E. Ugarte, E.G. Zoetendal, J. Wang, F. Guarner, O. Pedersen, W.M. de Vos,S. Brunak, J. Doré, M. Antolín, F. Artiguenave, H.M. Blottiere, M. Almeida, C. Brechot, C. Cara, C. Chervaux, A. Cultrone, C. Delorme, G. Denariaz, R. Dervyn, K.U. Foerstner, C. Friss, M. van de Guchte, E. Guedon, F. Haimet, W. Huber, J. van Hylckama-Vlieg, A. Jamet, C. Juste, G. Kaci, J. Knol, O. Lakhdari, S. Layec, K. Le Roux, E. Maguin, A. Mérieux, R. Melo Minardi, C. M'rini, J. Muller, R. Oozeer, J. Parkhill, P. Renault, M. Rescigno, N. Sanchez, S. Sunagawa, A. Torrejon, K. Turner, G. Vandemeulebrouck, E. Varela, Y. Winogradsky, G. Zeller, J. Weissenbach, S.D. Ehrlich, P. Bork, S.D. Ehrlich and P. Bork. 2011. Enterotypes of the human gut microbiome. Nature. 473: 174–180.

Andoh, A., H. Kuzuoka, T. Tsujikawa, S. Nakamura, F. Hirai, Y. Suzuki, T. Matsui, Y. Fujiyama and T. Matsumoto. 2012. Multicenter analysis of fecal microbiota profiles in Japanese patients with Crohn's disease. J. Gastroenterol. 47: 1298–1307.

Arrieta, M.-C., L.T. Stiemsma, P.A. Dimitriu, L. Thorson, S. Russell, S. Yurist-Doutsch, B. Kuzeljevic, M.J. Gold, H.M. Britton, D.L. Lefebvre, P. Subbarao, P. Mandhane, A. Becker, K.M. McNagny, M.R. Sears, T. Kollmann, CHILD Study Investigators, W.W. Mohn, S.E. Turvey and B.B. Finlay. 2015. Early infancy microbial and metabolic alterations affect risk of childhood asthma. Sci. Transl. Med. 7: 307ra152.

Altobelli, E., V. Del Negro, P. Angeletti, G. Latella, E. Altobelli, V. Del Negro, P.M. Angeletti and G. Latella. 2017. Low-FODMAP diet improves irritable bowel syndrome symptoms: A meta-analysis. Nutrients. 9: E940.

Belkaid, Y. and T.W. Hand. 2014. Role of the microbiota in immunity and inflammation. Cell. 157: 121–41.

Bernstein, C.N., A. Wajda, L.W. Svenson, A. MacKenzie, M. Koehoorn, M. Jackson, R. Fedorak, D. Israel and J.F. Blanchard. 2006. The epidemiology of inflammatory bowel disease in Canada: A population-based study. Am. J. Gastroenterol. 101: 1559–1568.

Benchimol, E.I., A. Guttmann, A.M. Griffiths, L. Rabeneck, D.R. Mack, H. Brill, J. Howard, J. Guan and T. To. 2009. Increasing incidence of paediatric inflammatory bowel disease in Ontario, Canada: evidence from health administrative data. Gut. 58: 1490–1497.

Bourdu, S., M. Dapoigny, E. Chapuy, F. Artigue, M.-P. Vasson, P. Dechelotte, G. Bommelaer, A. Eschalier and D. Ardid. 2005. Rectal instillation of butyrate provides a novel clinically relevant model of noninflammatory colonic hypersensitivity in rats. Gastroenterology 128: 1996–2008.

Burger, D. and S. Travis. 2011. Conventional medical management of inflammatory bowel disease. Gastroenterology 140: 1827–1837.

Ben-Horin, S., R. Mao and M. Chen. 2015. Optimizing biologic treatment in IBD: objective measures, but when, how and how often? BMC Gastroenterol. 15: 178.

Bibiloni, R., R.N. Fedorak, G.W. Tannock, K.L. Madsen, P. Gionchetti, M. Campieri, C. De Simone and R.B. Sartor. 2005. VSL#3 probiotic-mixture induces remission in patients with active ulcerative colitis. Am. J. Gastroenterol. 100: 1539–1546.

Bjarnason, I., G. Sission and B. Hayee. 2019. A randomised, double-blind, placebo-controlled trial of a multi-strain probiotic in patients with asymptomatic ulcerative colitis and Crohn's disease. Inflammopharmacology 27: 465–473.

Brun, P., M. Scarpa, C. Marchiori, G. Sarasin, V. Caputi, A. Porzionato, M.C. Giron, G. Palù and I. Castagliuolo. 2017. *Saccharomyces boulardii* CNCM I-745 supplementation reduces gastrointestinal dysfunction in an animal model of IBS. PLoS One 12: e0181863.

Caballero, S. and E.G. Pamer. 2015. Microbiota-mediated inflammation and antimicrobial defense in the intestine. Annu. Rev. Immunol. 33: 227–256.

Cammarota, G., G. Ianiro, H. Tilg, M. Rajilić-Stojanović, P. Kump, R. Satokari, H. Sokol, P. Arkkila, C. Pintus, A. Hart, J. Segal, M. Aloi, L. Masucci, A. Molinaro, F. Scaldaferri, G. Gasbarrini, A. Lopez-Sanroman, A. Link, P. de Groot, W.M. de Vos, C. Högenauer, P. Malfertheiner, E. Mattila, T. Milosavljević, M. Nieuwdorp, M. Sanguinetti, M. Simren and A. Gasbarrini. 2017. European FMT Working Group, European consensus conference on faecal microbiota transplantation in clinical practice. Gut. 66: 569–580.

Cayzeele-Decherf, A., F. Pélerin, S. Leuillet, B. Douillard, B. Housez, M. Cazaubiel, G.K. Jacobson, P. Jüsten and P. Desreumaux. 2017. *Saccharomyces cerevisiae* CNCM I-3856 in irritable bowel syndrome: An individual subject meta-analysis. World J. Gastroenterol. 23: 336–344.

Celiberto, L.S., R. Bedani, E. Rossi and D.C.U. Cavallini. 2017. Probiotics: the scientific evidence in the context of inflammatory bowel disease. Crit. Rev. Food Sci. Nutr. 57: 1759–1768.

Celiberto, L.S., F.A. Graef, G.R. Healey, E.S. Bosman, K. Jacobson, L.M. Sly and B.A. Vallance. 2018. Inflammatory Bowel Disease and immunonutrition: Novel therapeutic approaches through modulation of diet and the gut microbiome. Immunology 155: 36–52.

Celiberto, L.S., R.A. Pinto, E.A. Rossi, B.A. Vallance and D.C.U. Cavallini. 2018. Isolation and characterization of potentially probiotic bacterial strains from mice: proof of concept for personalized probiotics. Nutrients 10: 1684.

Chan, W., A. Chen, D. Tiao, C. Selinger and R. Leong. 2017. Medication adherence in inflammatory bowel disease. Intest. Res. 15: 434–445.

Chang, C.-J., T.-L. Lin, Y.-L. Tsai, T.-R. Wu, W.-F. Lai, C.-C. Lu and H.-C. Lai. 2019. Next generation probiotics in disease amelioration. J. Food Drug Anal. 27(3): 615–622.

Chassaing, B. and A. Darfeuille–Michaud. 2011. The commensal microbiota and enteropathogens in the pathogenesis of inflammatory bowel diseases. Gastroenterology 140: 1720–1728.

Chassaing, B., T. Van de Wiele, J. De Bodt, M. Marzorati and A.T. Gewirtz. 2017. Dietary emulsifiers directly alter human microbiota composition and gene expression *ex vivo* potentiating intestinal inflammation. Gut. 66: 1414–1427.

Chassard, C., M. Dapoigny, K.P. Scott, L. Crouzet, C. Del'homme, P. Marquet, J.C. Martin, G. Pickering, D. Ardid, A. Eschalier, C. Dubray, H.J. Flint and A. Bernalier-Donadille. 2012. Functional dysbiosis within the gut microbiota of patients with constipated-irritable bowel syndrome. Aliment. Pharmacol. Ther. 35: 828–838.

Chey, W.D., J. Kurlander and S. Eswaran. 2015. Irritable bowel syndrome. JAMA 313: 949.

Collins, S.M. 2014. A role for the gut microbiota in IBS. Nat. Rev. Gastroenterol. Hepatol. 11: 497–505.

Conboy, L.A., E. Macklin, J. Kelley, E. Kokkotou, A. Lembo and T. Kaptchuk. 2010. Which patients improve: Characteristics increasing sensitivity to a supportive patient–practitioner relationship. Soc. Sci. Med. 70: 479–484.

Corbitt, M., N. Campagnolo, D. Staines and S. Marshall-Gradisnik. 2018. A systematic review of probiotic interventions for gastrointestinal symptoms and irritable bowel syndrome in chronic fatigue syndrome/myalgic encephalomyelitis (CFS/ME). Probiotics Antimicrob. Proteins. 10: 466–477.

Crouzet, L., E. Gaultier, C. Del'Homme, C. Cartier, E. Delmas, M. Dapoigny, J. Fioramonti and A. Bernalier-Donadille. 2013. The hypersensitivity to colonic distension of IBS patients can be transferred to rats through their fecal microbiota. Neurogastroenterol. Motil. 25: e272–e282.

Dai, C., C.-Q. Zheng, F. Meng, Z. Zhou, L. Sang and M. Jiang. 2013. VSL#3 probiotics exerts the anti-inflammatory activity via PI3k/Akt and NF-κB pathway in rat model of DSS-induced colitis. Mol. Cell. Biochem. 374: 1–11.

Darfeuille-Michaud, A., J. Boudeau, P. Bulois, C. Neut, A.L. Glasser, N. Barnich, M.A. Bringer, A. Swidsinski, L. Beaugerie and J.F. Colombel. 2004. High prevalence of adherent-invasive *Escherichia coli* associated with ileal mucosa in Crohn's disease. Gastroenterology 127: 412–421.

De Palma, G., M.D.J. Lynch, J. Lu, V.T. Dang, Y. Deng, J. Jury, G. Umeh, P.M. Miranda, M. Pigrau Pastor, S. Sidani, M.I. Pinto-Sanchez, V. Philip, P.G. McLean, M.-G. Hagelsieb, M.G. Surette, G.E. Bergonzelli, E.F. Verdu, P. Britz-McKibbin, J.D. Neufeld, S.M. Collins and P. Bercik. 2017. Transplantation of fecal microbiota from patients with irritable bowel syndrome alters gut function and behavior in recipient mice. Sci. Transl. Med. 9: eaaf6397.

Didari, T., S. Mozaffari, S. Nikfar and M. Abdollahi. 2015. Effectiveness of probiotics in irritable bowel syndrome: Updated systematic review with meta-analysis. World J. Gastroenterol. 21: 3072–3084.

Dominguez-bello, M.G., E.K. Costello, M. Contreras, M. Magris and G. Hidalgo. 2010. Delivery mode shapes the acquisition and structure of the initial microbiota across multiple body habitats in newborns. Proc. Natl. Sci. USA 107: 11971–11975.

DuPont, H.L. 2014. Review article: evidence for the role of gut microbiota in irritable bowel syndrome and its potential influence on therapeutic targets. Aliment. Pharmacol. Ther. 39: 1033–1042.

DuPont, A.W. and H.L. DuPont. 2011. The intestinal microbiota and chronic disorders of the gut. Nat. Rev. Gastroenterol. Hepatol. 8: 523–531.

DuPont, A., D.M. Richards, K.A. Jelinek, J. Krill, E. Rahimi and Y. Ghouri. 2014. Systematic review of randomized controlled trials of probiotics, prebiotics, and synbiotics in inflammatory bowel disease. Clin. Exp. Gastroenterol. 7: 473–487.

FAO/WHO. 2002. Guidelines for the Evaluation of Probiotics in Food. Report of a Joint FAO/WHO Working Group on Drafting Guidelines for the Evaluation of Probiotics in Food. London, Ontario, Canada. Available from: https://www.who.int/foodsafety/fs_management/en/probiotic_guidelines.pdf.

Farzaei, M.H., R. Bahramsoltani, M. Abdollahi and R. Rahimi. 2016. The role of visceral hypersensitivity in irritable bowel syndrome: Pharmacological targets and novel treatments. J. Neurogastroenterol. Motil. 22: 558–574.

Fava, F. and S. Danese. 2011. Intestinal microbiota in inflammatory bowel disease: Friend of foe? World J. Gastroenterol. 17: 557–566.

Fodor, A.A., M. Pimentel, W.D. Chey, A. Lembo, P.L. Golden, R.J. Israel and I.M. Carroll. 2019. Rifaximin is associated with modest, transient decreases in multiple taxa in the gut microbiota of patients with diarrhoea-predominant irritable bowel syndrome. Gut Microbes. 10: 22–33.

Ford, A.C., N.J. Talley, P.S. Schoenfeld, E.M.M. Quigley and P. Moayyedi. 2009. Efficacy of antidepressants and psychological therapies in irritable bowel syndrome: systematic review and meta-analysis. Gut. 58: 367–378.

Ford, A.C., L.A. Harris, B.E. Lacy, E.M.M. Quigley and P. Moayyedi. 2018. Systematic review with meta-analysis: the efficacy of prebiotics, probiotics, synbiotics and antibiotics in irritable bowel syndrome. Aliment. Pharmacol. Ther. 48: 1044–1060.

Furusawa, Y., Y. Obata, S. Fukuda, T.A. Endo, G. Nakato, D. Takahashi, Y. Nakanishi, C. Uetake, K. Kato, T. Kato, M. Takahashi, N.N. Fukuda, S. Murakami, E. Miyauchi, S. Hino, K. Atarashi, S. Onawa, Y. Fujimura, T. Lockett, J.M. Clarke, D.L. Topping, M. Tomita, S. Hori, O. Ohara, T. Morita, H. Koseki, J. Kikuchi, K. Honda, K. Hase and H. Ohno. 2013. Commensal microbe-derived butyrate induces the differentiation of colonic regulatory T cells. Nature. 504: 446–450.

Gareau, M.G., P.M. Sherman and W.A. Walker. 2010. Probiotics and the gut microbiota in intestinal health and disease. Nat. Rev. Gastroenterol. Hepatol. 7: 503–514.

Gonçalves, P., J.R. Araújo and J.P.A. Di Santo. 2018. Cross-talk between microbiota-derived short-chain fatty acids and the host mucosal immune system regulates intestinal homeostasis and inflammatory bowel disease. Inflamm. Bowel Dis. 24: 558–572.

Grundmann, O. and S.L. Yoon. 2014. Complementary and alternative medicines in irritable bowel syndrome: an integrative view. World J. Gastroenterol. 20: 346–362.

Guandalini, S., G. Magazzù, A. Chiaro, V. La Balestra, G. Di Nardo, S. Gopalan, A. Sibal, C. Romano, R.B. Canani, P. Lionetti and M. Setty. 2010. VSL#3 improves symptoms in children with irritable bowel syndrome: a multicenter, randomized, placebo-controlled, double-blind, crossover study. J. Pediatr. Gastroenterol. Nutr. 51: 24–30.

Guyonnet, D., O. Chassany, P. Ducrotte, C. Picard, M. Mouret, C.-H. Mercier and C. Matuchansky. 2007. Effect of a fermented milk containing *Bifidobacterium animalis* DN-173 010 on the health-related quality of life and symptoms in irritable bowel syndrome in adults in primary care: a multicentre, randomized, double-blind, controlled trial. Aliment. Pharmacol. Ther. 26: 475–486.

Halpert, A., C.B. Dalton, N.E. Diamant, B.B. Toner, Y. Hu, C.B. Morris, S.I. Bangdiwala, W.E. Whitehead and D.A. Drossman. 2005. Clinical response to tricyclic antidepressants in functional bowel disorders is not related to dosage. Am. J. Gastroenterol. 100: 664–671.

Halpert, A. and E. Godena. 2011. Irritable bowel syndrome patients' perspectives on their relationships with healthcare providers. Scand. J. Gastroenterol. 46: 823–830.

Heeney, D.D., Z. Zhai, Z. Bendiks, J. Barouei, A. Martinic, C. Slupsky and M.L. Marco. 2018. *Lactobacillus plantarum* bacteriocin is associated with intestinal and systemic improvements in diet-induced obese mice and maintains epithelial barrier integrity *in vitro*. Gut Microbes. 10(3): 382–397.

Hill, C., F. Guarner, G. Reid, G.R. Gibson, D.J. Merenstein, B. Pot, L. Morelli, R.B. Canani, H.J. Flint, S. Salminen, P.C. Calder and M.E. Sanders. 2014. The International Scientific Association for Probiotics and Prebiotics consensus statement on the scope and appropriate use of the term probiotic. Nat. Rev. Gastroenterol. Hepatol. 11: 506–514.

Hols, P., L. Ledesma-García, P. Gabant and J. Mignolet. 2019. Mobilization of microbiota commensals and their bacteriocins for therapeutics. Trends Microbiol. 27(8): 690–702.

Huda-Faujan, N., A.S. Abdulamir, A.B. Fatimah, O.M. Anas, M. Shuhaimi, A.M. Yazid and Y.Y. Loong. 2010. The impact of the level of the intestinal short chain fatty acids in inflammatory bowel disease patients versus healthy subjects. Open Biochem. J. 4: 53–58.

Huynh, H.Q., J. deBruyn, L. Guan, H. Diaz, M. Li, S. Girgis, J. Turner, R. Fedorak and K. Madsen. 2009. Probiotic preparation VSL#3 induces remission in children with mild to moderate acute ulcerative colitis: A pilot study. Inflamm. Bowel Dis. 15: 760–768.

Imhann, F., A. Vich Vila, M.J. Bonder, J. Fu, D. Gevers, M.C. Visschedijk, L.M. Spekhorst, R. Alberts, L. Franke, H.M. van Dullemen, R.W.F. Ter Steege, C. Huttenhower, G. Dijkstra, R.J. Xavier, E.A.M. Festen, C. Wijmenga, A. Zhernakova and R.K. Weersma. 2018. Interplay of host genetics and gut microbiota underlying the onset and clinical presentation of inflammatory bowel disease. Gut. 67: 108–119.

Ingrassia, I., A. Leplingard and A. Darfeuille-Michaud. 2005. *Lactobacillus casei* DN-114 001 inhibits the ability of adherent-invasive *Escherichia coli* isolated from Crohn's disease patients to adhere to and to invade intestinal epithelial cells. Appl. Environ. Microbiol. 71: 2880–2887.

Isaacs, K. and H. Herfarth. 2008. Role of probiotic therapy in IBD. Inflamm. Bowel Dis. 14: 1597–1605.

Jeffery, I.B., P.W. O'Toole, L. Öhman, M.J. Claesson, J. Deane, E.M.M. Quigley and M. Simrén. 2012. An irritable bowel syndrome subtype defined by species-specific alterations in faecal microbiota. Gut. 61: 997–1006.

Johnson-Henry, K.C., M. Nadjafi, Y. Avitzur, D.J. Mitchell, B. Ngan, E. Galindo-Mata, N.L. Jones and P.M. Sherman. 2005. Amelioration of the effects of *Citrobacter rodentium* infection in mice by pretreatment with probiotics. J. Infect. Dis. 191: 2106–2117.

Jameson, J.L. and D.L. Longo. 2015. Precision medicine—Personalized, problematic, and promising. N. Engl. J. Med. 372: 2229–2234.

Jonkers, D., J. Penders, A. Masclee and M. Pierik. 2012. Probiotics in the management of inflammatory bowel disease. Drugs. 72: 803–823.

Joossens, M., G. Huys, M. Cnockaert, V. De Preter, K. Verbeke, P. Rutgeerts, P. Vandamme and S. Vermeire. 2011. Dysbiosis of the faecal microbiota in patients with Crohn's disease and their unaffected relatives. Gut. 60: 631–637.

Kato, K., S. Mizuno, Y. Umesaki, Y. Ishii, M. Sugitani, A. Imaoka, M. Otsuka, O. Hasunuma, R. Kurihara, A. Iwasaki and Y. Arakawa. 2004. Randomized placebo-controlled trial assessing the effect of bifidobacteria-fermented milk on active ulcerative colitis. Aliment. Pharmacol. Ther. 20: 1133–1141.

Klem, F., A. Wadhwa, L.J. Prokop, W.J. Sundt, G. Farrugia, M. Camilleri, S. Singh and M. Grover. 2017. Prevalence, risk factors, and outcomes of irritable bowel syndrome after infectious enteritis: A systematic review and meta-analysis. Gastroenterology 152: 1042–1054.

Khan, M.A., S. Nusrat, M.I. Khan, A. Nawras and K. Bielefeldt. 2015. Low-FODMAP diet for irritable bowel syndrome: Is it ready for prime time? Dig. Dis. Sci. 60: 1169–1177.

Kim, H.J., M.I. Vazquez Roque, M. Camilleri, D. Stephens, D.D. Burton, K. Baxter, G. Thomforde and A.R.A. Zinsmeister. 2005. Randomized controlled trial of a probiotic combination VSL# 3 and placebo in irritable bowel syndrome with bloating. Neurogastroenterol. Motil. 17: 687–696.

Kim, M.H., S.G. Kang, J.H. Park, M. Yanagisawa and C.H. Kim. 2013. Short-chain fatty acids activate GPR41 and GPR43 on intestinal epithelial cells to promote inflammatory responses in mice. Gastroenterology 145: 396–406.

Krogsgaard, L.R., M. Lyngesen and P. Bytzer. 2017. Systematic review: quality of trials on the symptomatic effects of the low FODMAP diet for irritable bowel syndrome. Aliment. Pharmacol. Ther. 45: 1506–1513.

Kumari, R., V. Ahuja and J. Paul. 2013. Fluctuations in butyrate-producing bacteria in ulcerative colitis patients of North India. World J. Gastroenterol. 19: 3404.

Lacy, B.E., F. Mearin, L. Chang, W.D. Chey, A.J. Lembo, M. Simren and R. Spiller. 2016. Bowel disorders. Gastroenterology 150: 1393–1407.

Lacy, B.E. and N.K. Patel. 2017. Rome criteria and a diagnostic approach to irritable bowel syndrome. J. Clin. Med. 6(11): E99.

LeBlanc, J.G., C. Milani, G.S. de Giori, F. Sesma, D. van Sinderen and M. Ventura. 2013. Bacteria as vitamin suppliers to their host: a gut microbiota perspective. Curr. Opin. Biotechnol. 24: 160–168.

LeBlanc, J.G., F. Chain, R. Martín, L.G. Bermúdez-Humarán, S. Courau and P. Langella. 2017. Beneficial effects on host energy metabolism of short-chain fatty acids and vitamins produced by commensal and probiotic bacteria. Microb. Cell Fact. 16: 79.

Ley, R.E., P.J. Turnbaugh, S. Klein and J.I. Gordon. 2006. Microbial ecology: Human gut microbes associated with obesity. Nature. 444: 1022–1023.

Li, K.-Y., J.-L. Wang, J.-P. Wei, S.-Y. Gao, Y.-Y. Zhang, L.-T. Wang and G. Liu. 2016. Fecal microbiota in pouchitis and ulcerative colitis. World J. Gastroenterol. 22: 8929–8939.

Ljungh, A. and T. Wadström. 2006. Lactic acid bacteria as probiotics. Curr. Issues Intest. Microbiol. 7: 73–89.

Loftus, E.V. 2004. Clinical epidemiology of inflammatory bowel disease: incidence, prevalence, and environmental influences. Gastroenterology 126: 1504–1517.

Lovell, R.M. and A.C. Ford. 2012. Effect of gender on prevalence of irritable bowel syndrome in the community: Systematic review and meta-analysis. Am. J. Gastroenterol. 107: 991–1000.

Machiels, K., M. Joossens, J. Sabino, V. De Preter, I. Arijs, V. Eeckhaut, V. Ballet, K. Claes, F. Van Immerseel, K. Verbeke, M. Ferrante, J. Verhaegen, P. Rutgeerts and S. Vermeire. 2014. A decrease of the butyrate-producing species *Roseburia hominis* and *Faecalibacterium prausnitzii* defines dysbiosis in patients with ulcerative colitis. Gut. 63: 1275–1283.

Mann, E.R., J. You, V. Horneffer-van der Sluis, D. Bernardo, H. Omar Al-Hassi, J. Landy, S.T. Peake, L.V. Thomas, C.T. Tee, G.H. Lee, A.L. Hart, P. Yaqoob and S.C. Knight. 2013. Dysregulated circulating dendritic cell function in ulcerative colitis is partially restored by probiotic strain *Lactobacillus casei* Shirota. Mediators Inflamm. 2013: 573576.

Mann, E.R., D. Bernardo, S.C. Ng, R.J. Rigby, H.O. Al-Hassi, J. Landy, S.T.C. Peake, H. Spranger, N.R. English, L.V. Thomas, A.J. Stagg, S.C. Knight and A.L. Hart. 2014. Human gut dendritic cells drive aberrant gut-specific t-cell responses in ulcerative colitis, characterized by increased IL-4 production and loss of IL-22 and IFNγ. Inflamm. Bowel Dis. 20: 2299–2307.

Marsh, A., E.M. Eslick and G.D. Eslick. 2016. Does a diet low in FODMAPs reduce symptoms associated with functional gastrointestinal disorders? A comprehensive systematic review and meta-analysis. Eur. J. Nutr. 55: 897–906.

Marteau, P., M. Lémann, P. Seksik, D. Laharie, J.F. Colombel, Y. Bouhnik, G. Cadiot, J.C. Soulé, A. Bourreille, E. Metman, E. Lerebours, F. Carbonnel, J.L. Dupas, M. Veyrac, B. Coffin, J. Moreau, V. Abitbol, S. Blum-Sperisen and J.Y. Mary. 2006. Ineffectiveness of *Lactobacillus johnsonii* LA1 for prophylaxis of postoperative recurrence in Crohn's disease: a randomised, double blind, placebo controlled GETAID trial. Gut. 55: 842–847. doi:10.1136/gut.2005.076604.

Martin, R., A.J. Nauta, K. Ben Amor, L.M.J. Knippels, J. Knol and J. Garssen. 2010. Early life: gut microbiota and immune development in infancy. Benef Microbes. 1: 367–382.

Martinez, R.C.R., R. Bedani and S.M.I. Saad. 2015. Scientific evidence for health effects attributed to the consumption of probiotics and prebiotics: an update for current perspectives and future challenges. Br. J. Nutr. 114: 1993–2015.

Matto, J., L. Maunuksela, K. Kajander, A. Palva, R. Korpela, A. Kassinen and M. Saarela. 2005. Composition and temporal stability of gastrointestinal microbiota in irritable bowel syndrome—a longitudinal study in IBS and control subjects. FEMS Immunol. Med. Microbiol. 43: 213–222.

Maukonen, J., R. Satokari, J. Mättö, H. Söderlund, T. Mattila-Sandholm and M. Saarela. 2006. Prevalence and temporal stability of selected clostridial groups in irritable bowel syndrome in relation to predominant faecal bacteria. J. Med. Microbiol. 55: 625–633.

McKenzie, Y.A., R.K. Bowyer, H. Leach, P. Gulia, J. Horobin, N.A. O'Sullivan, C. Pettitt, L.B. Reeves, L. Seamark, M. Williams, J. Thompson and M.C.E. Lomer. 2016. British Dietetic Association systematic review and evidence-based practice guidelines for the dietary management of irritable bowel syndrome in adults (2016 update). J. Hum. Nutr. Diet. 29: 549–575.

Meijerink, M. and J. Wells. 2010. Probiotic modulation of dendritic cells and T cell responses in the intestine. Benef. Microbes. 1: 317–326.

Miele, E., F. Pascarella, E. Giannetti, L. Quaglietta, R.N. Baldassano and A. Staiano. 2009. Effect of a probiotic preparation (VSL#3) on induction and maintenance of remission in children with ulcerative colitis. Am. J. Gastroenterol. 104: 437–443.

Moran, C.J. 2017. Very early onset inflammatory bowel disease. Semin. Pediatr. Surg. 26: 356–359.

Ohigashi, S., K. Sudo, D. Kobayashi, O. Takahashi, T. Takahashi, T. Asahara, K. Nomoto and H. Onodera. 2013. Changes of the intestinal microbiota, short chain fatty acids, and fecal pH in patients with colorectal cancer. Dig. Dis. Sci. 58: 1717–1726.

Ooi, S.L., D. Correa and S.C. Pak. 2019. Probiotics, prebiotics, and low FODMAP diet for irritable bowel syndrome—What is the current evidence? Complement. Ther. Med. 43: 73–80.

O'Toole, P.W., J.R. Marchesi and C. Hill. 2017. Next-generation probiotics: the spectrum from probiotics to live biotherapeutics. Nat. Microbiol. 2: 17057.

Ott, S.J., M. Musfeldt, D.F. Wenderoth, J. Hampe, O. Brant, U.R. Fölsch, K.N. Timmis and S. Schreiber. 2004. Reduction in diversity of the colonic mucosa associated bacterial microflora in patients with active inflammatory bowel disease. Gut. 53: 685–93.

Parada Venegas, D., M.K. De la Fuente, G. Landskron, M.J. González, R. Quera, G. Dijkstra, H.J.M. Harmsen, K.N. Faber and M.A. Hermoso. 2019. Short chain fatty acids (SCFAs)-mediated gut epithelial and immune regulation and its relevance for inflammatory bowel diseases. Front. Immunol. 10: 277.

Perše, M. and A. Cerar. 2012. Dextran sodium sulphate colitis mouse model: traps and tricks. J. Biomed. Biotechnol. 2012: 718617.

Petersen, A.M., H. Mirsepasi, S.I. Halkjær, E.M. Mortensen, I. Nordgaard-Lassen and K.A. Krogfelt. 2014. Ciprofloxacin and probiotic *Escherichia coli* Nissle add-on treatment in active ulcerative colitis: A double-blind randomized placebo controlled clinical trial. J. Crohn's Colitis 8: 1498–1505.

Peyton, L. and J. Greene. 2014. Irritable bowel syndrome: current and emerging treatment options. P T. 39: 567–578.

Prantera, C. 2002. Ineffectiveness of probiotics in preventing recurrence after curative resection for Crohn's disease: a randomised controlled trial with *Lactobacillus* GG. Gut. 51: 405–409.

Pronio, A., C. Montesani, C. Butteroni, S. Vecchione, G. Mumolo, A. Vestri, D. Vitolo and M. Boirivant. 2008. Probiotic administration in patients with ileal pouch–anal anastomosis for ulcerative colitis is associated with expansion of mucosal regulatory cells. Inflamm. Bowel Dis. 14: 662–668.

Qin, J., R. Li, J. Raes, M. Arumugam, K.S. Burgdorf, C. Manichanh, T. Nielsen, N. Pons, F. Levenez, T. Yamada, D.R. Mende, J. Li, J. Xu, S. Li, D. Li, J. Cao, B. Wang, H. Liang, H. Zheng, Y. Xie, J. Tap, P. Lepage, M. Bertalan, J.-M. Batto, T. Hansen, D. Le Paslier, A. Linneberg, H.B. Nielsen, E. Pelletier, P. Renault, T. Sicheritz-Ponten, K. Turner, H. Zhu, C. Yu, S. Li, M. Jian, Y. Zhou, Y. Li, X. Zhang, S. Li, N. Qin, H. Yang, J. Wang, S. Brunak, J. Doré, F. Guarner, K. Kristiansen, O. Pedersen, J. Parkhill, J. Weissenbach, P. Bork, S.D. Ehrlich, J. Wang and J. Wang. 2010. A human gut microbial gene catalogue established by metagenomic sequencing. Nature. 64: 59–65.

Rahimi, R., S. Nikfar, A. Rezaie and M. Abdollahi. 2009. Efficacy of tricyclic antidepressants in irritable bowel syndrome: A meta-analysis. World J. Gastroenterol. 15: 1548–1553.

Rajilić–Stojanović, M., E. Biagi, H.G.H.J. Heilig, K. Kajander, R.A. Kekkonen, S. Tims and W.M. de Vos. 2011. Global and deep molecular analysis of microbiota signatures in fecal samples from patients with irritable bowel syndrome. Gastroenterology 141: 1792–1801.

Rebbeck, T.R., T. Friebel, H.T. Lynch, S.L. Neuhausen, L. van 't Veer, J.E. Garber, G.R. Evans, S.A. Narod, C. Isaacs, E. Matloff, M.B. Daly, O.I. Olopade and B.L. Weber. 2004. Bilateral prophylactic mastectomy reduces breast cancer risk in BRCA1 and BRCA2 mutation carriers: the PROSE study group. J. Clin. Oncol. 22: 1055–1062.

Reid, G., A.A. Gadir and R. Dhir. 2019. Probiotics: Reiterating what they are and what they are not. Front. Microbiol. 10: 424.

Renna, S., M. Cottone and A. Orlando. 2014. Optimization of the treatment with immunosuppressants and biologics in inflammatory bowel disease. World J. Gastroenterol. 20: 9675–9690.

Roda, G., B. Jharap, N. Neeraj and J.-F. Colombel. 2016. Loss of response to anti-TNFs: Definition, epidemiology, and management. Clin. Transl. Gastroenterol. 7: e135.

Rodemann, J.F., E.R. Dubberke, K.A. Reske, D.H. Seo and C.D. Stone. 2007. Incidence of *Clostridium difficile* infection in inflammatory bowel disease. Clin. Gastroenterol. Hepatol. 5: 339–344.

Rodiño-Janeiro, B.K., M. Vicario, C. Alonso-Cotoner, R. Pascua-García and J. Santos. 2018. A Review of microbiota and irritable bowel syndrome: Future in therapies. Adv. Ther. 35: 289–310.

Rodríguez, L.A. and A. Ruigómez. 1999. Increased risk of irritable bowel syndrome after bacterial gastroenteritis: cohort study. BMJ 318: 565–566.

Roselli, M., A. Finamore, S. Nuccitelli, P. Carnevali, P. Brigidi, B. Vitali, F. Nobili, R. Rami, I. Garaguso and E. Mengheri. 2009. Prevention of TNBS-induced colitis by different *Lactobacillus* and *Bifidobacterium* strains is associated with an expansion of γδT and regulatory T cells of intestinal intraepithelial lymphocytes. Inflamm. Bowel Dis. 15: 1526–1536.

Rousseaux, C., X. Thuru, A. Gelot, N. Barnich, C. Neut, L. Dubuquoy, C. Dubuquoy, E. Merour, K. Geboes, M. Chamaillard, A. Ouwehand, G. Leyer, D. Carcano, J.-F. Colombel, D. Ardid and P. Desreumaux. 2007. *Lactobacillus acidophilus* modulates intestinal pain and induces opioid and cannabinoid receptors. Nat. Med. 13: 35–37.

Saez-Lara, M.J., C. Gomez-Llorente, J. Plaza-Diaz and A. Gil. 2015. The role of probiotic lactic acid bacteria and bifidobacteria in the prevention and treatment of inflammatory bowel disease and other related diseases: A systematic review of randomized human clinical trials. Biomed. Res. Int. 2015: 505878.

Salim, S.Y., P.Y. Young, C.M. Lukowski, K.L. Madsen, B. Sis, T.A. Churchill and R.G. Khadaroo. 2013. VSL#3 probiotics provide protection against acute intestinal ischaemia/reperfusion injury. Benef. Microbes. 4: 357–365.

Salonen, A., W.M. de Vos and A. Palva. 2010. Gastrointestinal microbiota in irritable bowel syndrome: present state and perspectives. Microbiology 156: 3205–3215.

Sandler, R.S., J.E. Everhart, M. Donowitz, E. Adams, K. Cronin, C. Goodman, E. Gemmen, S. Shah, A. Avdic and R. Rubin. 2002. The burden of selected digestive diseases in the United States. Gastroenterology 122: 1500–1511.

Schultz, M., C. Veltkamp, L.A. Dieleman, W.B. Grenther, P.B. Wyrick, S.L. Tonkonogy and R.B. Sartor. 2002. *Lactobacillus plantarum* 299V in the treatment and prevention of spontaneous colitis in interleukin-10-deficient mice. Inflamm. Bowel Dis. 8: 71–80.

Shah, S.C., H. Khalili, C. Gower-Rousseau, O. Olen, E.I. Benchimol, E. Lynge, K.R. Nielsen, P. Brassard, M. Vutcovici, A. Bitton, C.N. Bernstein, D. Leddin, H. Tamim, T. Stefansson, E.V. Loftus, B. Moum, W. Tang, S.C. Ng, R. Gearry, B. Sincic, S. Bell, B.E. Sands, P.L. Lakatos, Z. Végh, C. Ott, G.G. Kaplan, J. Burisch and J.-F. Colombel. 2018. Sex-based differences in incidence of inflammatory bowel diseases-pooled analysis of population-based studies from Western countries. Gastroenterology 155: 1079–1089.e3.

Shanahan, F. and E.M.M. Quigley. 2014. Manipulation of the microbiota for treatment of IBS and IBD—challenges and controversies. Gastroenterology 146: 1554–1563.

Shen, J., H.Z. Ran, M.H. Yin, T.X. Zhou and D.S. Xiao. 2009. Meta-analysis: the effect and adverse events of *Lactobacilli* versus placebo in maintenance therapy for Crohn disease. Intern. Med. J. 39: 103–109.

Simrén, M., G. Barbara, H.J. Flint, B.M.R. Spiegel, R.C. Spiller, S. Vanner, E.F. Verdu, P.J. Whorwell and E.G. Zoetendal. 2013. Rome Foundation Committee Intestinal microbiota in functional bowel disorders: a Rome foundation report. Gut. 62: 159–176.

Söderholm, J.D., D.A. Yates, M.G. Gareau, P.-C. Yang, G. MacQueen and M.H. Perdue. 2002. Neonatal maternal separation predisposes adult rats to colonic barrier dysfunction in response to mild stress. Am. J. Physiol. Liver Physiol. 283: G1257–G1263. doi:10.1152/ajpgi.00314.2002.

Sokol, H., P. Lepage, P. Seksik, J. Doré and P. Marteau. 2006. Temperature gradient gel electrophoresis of fecal 16S rRNA reveals active *Escherichia coli* in the microbiota of patients with ulcerative colitis. J. Clin. Microbiol. 44: 3172–3177.

Sokol, H., P. Seksik, J.P. Furet, O. Firmesse, I. Nion-Larmurier, L. Beaugerie, J. Cosnes, G. Corthier, P. Marteau and J. Doraé. 2009. Low counts of *Faecalibacterium prausnitzii* in colitis microbiota. Inflamm. Bowel Dis. 15: 1183–1189.

Sperber, A.D., D. Dumitrascu, S. Fukudo, C. Gerson, U.C. Ghoshal, K.A. Gwee, A.P.S. Hungin, J.-Y. Kang, C. Minhu, M. Schmulson, A. Bolotin, M. Friger, T. Freud and W. Whitehead. 2017. The global prevalence of IBS in adults remains elusive due to the heterogeneity of studies: a Rome Foundation working team literature review. Gut. 66: 1075–1082.

Spiller, R. and C. Lam. 2012. An update on post-infectious irritable bowel syndrome: Role of genetics, immune activation, serotonin and altered microbiome. J. Neurogastroenterol. Motil. 18: 258–268.

Spor, A., O. Koren and R. Ley. 2011. Unravelling the effects of the environment and host genotype on the gut microbiome. Nat. Rev. Microbiol. 9: 279–290.

Takahashi, K., A. Nishida, T. Fujimoto, M. Fujii, M. Shioya, H. Imaeda, O. Inatomi, S. Bamba, A. Andoh and M. Sugimoto. 2016. Reduced abundance of butyrate-producing bacteria species in the fecal microbial community in Crohn's disease. Digestion 93: 59–65.

Takeshita, K., S. Mizuno, Y. Mikami, T. Sujino, K. Saigusa, K. Matsuoka, M. Naganuma, T. Sato, T. Takada, H. Tsuji, A. Kushiro, K. Nomoto and T.A. Kanai. 2016. Single species of *Clostridium* subcluster XIVa decreased in ulcerative colitis patients. Inflamm. Bowel Dis. 22: 2802–2810.

Talero, E., S. Bolivar, J. Ávila-Román, A. Alcaide, S. Fiorucci and V. Motilva. 2015. Inhibition of chronic ulcerative colitis-associated adenocarcinoma development in mice by VSL#3. Inflamm. Bowel Dis. 21: 1027–1037.

Umu, Ö.C.O., K. Rudi and D.B. Diep. 2017. Modulation of the gut microbiota by prebiotic fibres and bacteriocins. Microb. Ecol. Health Dis. 28: 1348886.

Van Gossum, A., O. Dewit, E. Louis, G. de Hertogh, F. Baert, F. Fontaine, M. DeVos, M. Enslen, M. Paintin and D. Franchimont. 2007. Multicenter randomized-controlled clinical trial of probiotics (*Lactobacillus johnsonii* LA1) on early endoscopic recurrence of Crohn's disease after ileo-caecal resection. Inflamm. Bowel Dis. 13: 135–142.

Veerappan, G.R., J. Betteridge and P.E. Young. 2012. Probiotics for the treatment of inflammatory bowel disease. Curr. Gastroenterol. Rep. 14: 324–333.

Walters, W.A., Z. Xu and R. Knight. 2014. Meta-analyses of human gut microbes associated with obesity and IBD. FEBS Lett. 588: 4223–4233.

Wells, J.M. 2011. Immunomodulatory mechanisms of lactobacilli. Microb. Cell Fact. 10Suppl 1: S17.

Whelan, K. and E.M. Quigley. 2013. Probiotics in the management of irritable bowel syndrome and inflammatory bowel disease. Curr. Opin. Gastroenterol. 29: 184–189.

Wildt, S., I. Nordgaard, U. Hansen, E. Brockmann and J.J. Rumessen. 2011. A randomised double-blind placebo-controlled trial with *Lactobacillus acidophilus* La-5 and *Bifidobacterium animalis* subsp. *lactis* BB-12 for maintenance of remission in ulcerative colitis. J. Crohn's Colitis 5: 115–121.

Wu, X., B.A. Vallance, L. Boyer, K.S.B. Bergstrom, J. Walker, K. Madsen, J.R. O'Kusky, A.M. Buchan and K. Jacobson. 2008. *Saccharomyces boulardii* ameliorates *Citrobacter rodentium*-induced colitis through actions on bacterial virulence factors. Am. J. Physiol. Liver Physiol. 294: G295–G306.

Wu, X., C. Ma, L. Han, M. Nawaz, F. Gao, X. Zhang, P. Yu, C. Zhao, L. Li, A. Zhou, J. Wang, J.E. Moore, B. Cherie Millar and J. Xu. 2010. Molecular characterisation of the faecal microbiota in patients with type II diabetes. Curr. Microbiol. 61: 69–78.

Wu, Y., C. Zhu, Z. Chen, Z. Chen, W. Zhang, X. Ma, L. Wang, X. Yang and Z. Jiang. 2016. Protective effects of *Lactobacillus plantarum* on epithelial barrier disruption caused by enterotoxigenic *Escherichia coli* in intestinal porcine epithelial cells. Vet. Immunol. Immunopathol. 172: 55–63.

Xiao, L., Q. Feng, S. Liang, S.B. Sonne, Z. Xia, X. Qiu, X. Li, H. Long, J. Zhang, D. Zhang, C. Liu, Z. Fang, J. Chou, J. Glanville, Q. Hao, D. Kotowska, C. Colding, T.R. Licht, D. Wu, J. Yu, J.J.Y. Sung, Q. Liang, J. Li, H. Jia, Z. Lan, V. Tremaroli, P. Dworzynski, H.B. Nielsen, F. Bäckhed, J. Doré, E. Le Chatelier, S.D. Ehrlich, J.C. Lin, M. Arumugam, J. Wang, L. Madsen and K. Kristiansen. 2015. A catalog of the mouse gut metagenome. Nat. Biotechnol. 33: 1103–1108.

Ye, Y., Z. Pang, W. Chen, S. Ju and C. Zhou. 2015. The epidemiology and risk factors of inflammatory bowel disease. Int. J. Clin. Exp. Med. 8: 22529–22542.

Zareie, M., K. Johnson-Henry, J. Jury, P.-C. Yang, B.-Y. Ngan, D.M. McKay, J.D. Soderholm, M.H. Perdue and P.M. Sherman. 2006. Probiotics prevent bacterial translocation and improve intestinal barrier function in rats following chronic psychological stress. Gut. 55: 1553–1560.

Zhuang, X., L. Xiong, L. Li, M. Li and M. Chen. 2017. Alterations of gut microbiota in patients with irritable bowel syndrome: A systematic review and meta-analysis. J. Gastroenterol. Hepatol. 32: 28–38.

Zmora, N., D. Zeevi, T. Korem, E. Segal and E. Elinav. 2016. Taking it Personally: personalized utilization of the human microbiome in health and disease. Cell Host Microbe 19: 12–20.

Zocco, M.A., L.Z. Dal Verme, F. Cremonini, A.C. Piscaglia, E.C. Nista, M. Candelli, M. Novi, D. Rigante, I.A. Cazzato, V. Ojetti, A. Armuzzi, G. Gasbarrini and A. Gasbarrini. 2006. Efficacy of *Lactobacillus* GG in maintaining remission of ulcerative colitis. Aliment. Pharmacol. Ther. 23: 1567–1574.

11

The Potential Use of Lactic Acid Bacteria in Neurodegenerative Pathologies

Daiana E Perez Visñuk,[1] *María del Milagro Teran,*[1] *Graciela Savoy de Giori,*[1,2] *Jean Guy LeBlanc*[1] and *Alejandra de Moreno de LeBlanc*[1,*]

Introduction

Aging is a common factor that leads to the development of neurodegenerative diseases such as Parkinson's and Alzheimer's. Probiotics have the potential to act against some of the degenerative effects associated with aging and these diseases, like the decrease of neurotransmitter levels, chronic inflammation, oxidative stress and neuronal apoptosis. They can also exert benefits against gastrointestinal disorders that accompany these pathologies such as diarrhea, constipation, vitamin deficiency, etc. In the present chapter, some mechanisms related to neurological disorders that can be prevented or improved by probiotics as well as the importance of the gut-brain axis in order to understand the potential of probiotics against neurodegenerative diseases will be discussed.

Some Mechanisms Involved in Neurodegenerative Diseases

Neurodegenerative diseases can be triggered by an accumulation of harmful and random interactions from the environment. For example, Parkinson's disease (PD)

[1] Centro de Referencia para Lactobacilos (CERELA-CONICET), Chacabuco 145, (T4000ILC), San Miguel de Tucumán, Tucumán, Argentina.
Emails: dperez@cerela.org.ar; mteran@cerela.org.ar; leblanc@cerela.org.ar, leblancjeanguy@gmail.com
[2] Cátedra de Microbiología Superior, Facultad de Bioquímica, Química y Farmacia, UNT, San Miguel de Tucumán, Tucumán, Argentina.
Email: gsavoy@cerela.org.ar
* Corresponding author: demoreno@cerela.org.ar

and Alzheimer's disease (AD) are both linked to exposure to environmental toxins such as herbicides, fungicides and pesticides (Baltazar et al. 2014) or to lifestyle changes such as diets or stress (Virmani et al. 2013). It was also reported that many of the early symptoms of neurodegenerative diseases originate in the gastrointestinal tract (Goldman and Postuma 2014).

Regarding oxidative stress, the generation of oxygen and nitrogen reactive species (RNOS) are normally controlled by the antioxidant defense mechanisms; however, aging can be associated with an imbalance between RONS and the anti-oxidant defenses with progressive ROS (reactive oxygen species)-induced cell damages that can lead to cell death (Dias et al. 2013). In this sense, oxidative stress is involved in many age-related disorders including neurological diseases (Liguori et al. 2018). The brain is particularly sensitive to oxidative damage since neurons have high energy demand and membranes rich in polyunsaturated fatty acids that are more sensitive to damages induced by ROS (Wang and Michaelis 2010).

The slow accumulation of ROS in the neurons stimulates the release of cytokines with the consequent microglial activation and neuroinflammation. In addition, chronic, low-grade inflammation (inflammaging) observed in aging, characterized by an increase in cytokines and inflammatory mediators, contribute to neuronal damage/degeneration (Franceschi and Campisi 2014).

Many studies have also established the relationship between the neuroendocrine pathways, the gut microbiota and the immune system. These systems are communicated by complex molecular networks; systemic imbalances can cause neurodegeneration (Westfall et al. 2017).

The Intestine-Brain Axis

There is a bidirectional communication between the intestine and the brain mediated by neurological connections, hormonal signals and immunological factors (Forsythe et al. 2014, Burokas et al. 2015, Bauer et al. 2016), establishing an association between the gut microbiota and its metabolites with the brain. These interactions enable gastrointestinal functions such as mobility, production and secretion of mucin and immunological functions including the production of cytokines by intestinal and mucosal immune system's cells (Tracey 2009, Mayer 2011). According to these findings, it has been shown that emotional factors such as stress or depression can worsen symptoms and/or induce recurrence of chronic gastrointestinal diseases such as Crohn's disease and irritable bowel syndrome (IBS) (Mawdsley and Rampton 2005, Mayer 2011). They can also alter the integrity and mobility of the intestinal epithelium, and modify the composition and/or activity of the microbiota (Collins and Bercik 2009). It has also been shown that the release of catecholamines could interfere with inter-bacteria signals as well as with the expression of virulence genes in the intestine (Kaper and Sperandio 2005, Freestone et al. 2008).

The neuroendocrine system is responsible for the production of hormones and neuropeptides (Toni 2004, Prevot 2010). An important axis of this system is represented by the Hypothalamus Pituitary Adrenal axis (HPA), considered the main regulator of psychological and physical stress (Smith et al. 2014). The corticotrophin release factor (CRF) is responsible for regulating the HPA axis. This is generated in the paraventricular nucleus of the hypothalamus (PVN) and in response to stress releases the adrenocorticotropic hormone (ACTH) in the bloodstream which in turn induces the secretion of glucocorticoids (cortisol in humans and corticosterone in rodents). These glucocorticoids inhibit the synthesis and release of CRF and ACTH by binding to their receptor in the hypothalamus, and the body regulates the stress response. Chronic high cortisol blood levels negatively affect the brain activating the HPA pathway without regulation (Liu 2017). This activation can later affect the composition of the microbiota (dysbiosis) (de Punder and Pruimboom 2015, Kelly et al. 2015).

Signals transmitted through the vagus nerve are part of an "inflammatory reflex" in order to maintain immunological homeostasis; and an imbalance produced in it could lead to the development of diseases such as atherosclerosis, obesity, cancer, inflammatory bowel disease, neurodegeneration, rheumatoid arthritis, among others. Many nerve fibers from the gut are constantly providing information to the Central Nervous System (CNS), and this communication allows the CNS to know what is happening in the intestine. Local changes in the levels of cytokines or other molecules, or changes in pH in the gut can be communicated in this way from the enteric nervous system (ENS) to the CNS (Andersson and Tracey 2012).

The importance of the gut-brain axis in neurodegenerative diseases has been recently reviewed demonstrating that its dysfuntionality is observed in patients suffering PD (Chapelet et al. 2018). These authors summarized results that demonstrated intestinal inflammation and changes in gut microbiota composition are risk factors for PD.

Studies performed in mice showed that gut microbiota has an essential role for the development of the ENS during the early postnatal life, similar to its role in the maturation of the mucosal immune system (De Vadder et al. 2018). These authors demonstrated that serotonin release and activation of the 5-hydroxytryptamine receptor 4 (5-HT4) are mechanisms of communication between the microbiota and enteric neurons and suggested that the modulation of gut microbiota (e.g., by diet) has the potential to be used against neurogastroenterological pathologies. It has been demonstrated that alterations in the microbiota contribute to inflammatory and functional bowel disorders with the development of depressive and anxiety episodes. The alteration of the microbiota of mice by administering three antibiotics for 7 days were associated to an increase in BDNF, a protein present in the hippocampus, the cerebral cortex and the cerebellum, whose excess negatively affects memory and motivation, generating anxiety and depression (Bercik et al. 2011a).

Another recent review article showed the role of the gut microbiota not only in the development and function of the ENS, but also its influence on the CNS (Heiss and Olofsson 2019). The importance of this communication between ENS and CNS was recently analysed in patients suffering Parkinson's diseases. Aberrant expression of α-synuclein was observed in the gastrointestinal nervous system of PD patients and these α-synuclein inclusions in gastrointestinal mucous biopsy has the potential to be used for the early-stage diagnostic of PD (Yan et al. 2018). All these observations suggest that interventions to maintain a healthy microbiota or to avoid dysbiosis can be used against neurological disorders. In this sense, probiotics have an important potential.

Probiotics and the Gut-Brain Axis

Beneficial effects of probiotics against neurological disorders

Recent studies present probiotics as microorganisms that alter the neurochemistry of the brain and protect it from damage that could be caused by the generation of free radicals and/or inflammatory cytokines. Preclinical and clinical studies indicates the important role of probiotics in the control of anxiety, depression, cognitive loss, chronic pain and inflammatory diseases (Luo et al. 2014, Abhari et al. 2016, Wang and Wang 2016, Giannetti et al. 2017). These studies were carried out using *B. longum* subsp. *infantis* 35624, *B. longum* NCC3001, *Lactobacillus* (*L.*) *rhamnosus* R0011 and *L. helveticus* R0052 allowing to establish that the interaction between them with the gut microbiota and the gut-brain axis is a strain specific characteristic. In addition, the response carried out by probiotics depends on the parameter of behavior under study and the model used to disrupt it (Collins et al. 2012).

It has been reported that regular feeding with *Lactobacillus* would be able to alter the expression of the receptors for the BDNF and the neurotransmitters of serotonin, and gamma-aminobutyric acid (GABA) (Bercik et al. 2011b, Tillisch 2014, Liang et al. 2015). In addition, several LAB can also produce GABA during their metabolism becoming an important bioactive component of pharmaceutical products and food. These LAB included strains of *L. fermentum* (Ramos et al. 2013), *L. futsaii* (Sanchart et al. 2016), *L. brevis* (Binh et al. 2014, Hsueh et al. 2017, Wu et al. 2018), and *L. senmaizukei* (Hiraga et al. 2008). Other mechanisms involved in the benefits of probiotics include their anti-inflammatory properties. In this sense, it has been reported that *L. helveticus* NS8 prevented cognitive decline and anxiety-like behavior in rats by reducing the level of inflammatory markers (inducible nitric oxide synthase, prostaglandin E2, and interleukin-1 β) in the brain (Luo et al. 2014). Oral administration of *Enterococcus faecium* CFR 3003 and *L. rhamnosus* GG MTCC 1408 decreased oxidative markers in the brain of young mice and enhanced antioxidant enzymes activities, with increased GABA and dopamine levels (Divyashri et al. 2015).

It was also shown that preventive treatment with a mixture of 4 potential probiotic bacteria strains (*B. breve*, *L. casei*, *L. delbrueckii* subsp. *bulgaricus*, and *L. acidophilus*) reduced the levels of TNF-α (tumor necrosis factor-alpha) and the stress biomarker MDA (malondialdehyde) in the ischemic brain tissue by using a mouse model. These authors suggested that based on their results, probiotics should be evaluated for use in patients at risk of stroke (Akhoundzadeh et al. 2018).

The role of adaptive immune cells in maintaining intestinal and brain health was reported by using $Rag1^{-/-}$ mice. In this model, the probiotic mixture composed of *L. rhamnosus* R0011 and *L. helveticus* R0052 administered at weaning normalized behavior alterations, intestinal physiology and microbiota, demonstrating that probiotics can improve the gut-brain-microbiota axis even when there is an immune deficiency (Smith et al. 2014). Furthermore, it was revealed that the administration of the probiotic *B. longum* subsp. *infantis* 35624 to rats has potential antidepressant properties. This effect was associated with the modulation of the systemic inflammatory immune response and to the reduction of dopamine and 5-hyroxyindole acetic acid (5-HIAA) in the frontal cortex of rats (Desbonnet et al. 2008). In addition, the same probiotic was evaluated in newborn rats stressed by separation from their mothers;

the results showed that probiotic administration normalized the immune response, reversed behavioral deficits, and decreased noradrenalin (NA) concentrations in the brainstem (Desbonnet et al. 2010).

The chronic administration of *L. rhamnosus* JB1 to mice modulated the GABA system in specific regions of the brain and promoted an exploratory behavior in the animals suggesting its potential for the treatment of depression and anxiety associated with GABA alterations. In addition, it has been demonstrated that the vagus nerve was involved in the observed effects and was responsible to maintain the communication between bacteria, gut and brain (Bravo et al. 2011).

B Group Vitamins and Neurological Disorders

Potential role of selected vitamin-producing lactic acid bacteria

There is increasing evidence to suggest that certain vitamins could provide anti-inflammatory effects and it has been shown that certain strains of LAB can produce biologically active forms of certain vitamins (LeBlanc et al. 2011, 2013).

The potential benefits of vitamins produced by LAB are discussed not only because they could be used to decrease inflammation and oxidative stress associated with some neurological disorders, but also because they could provide the host with essential nutrients that are normally deficient in aging patients or patients suffering from neurodegenerative diseases such as PD and AD. These micronutrients could also be beneficial for the intestinal microbiome (Sechi et al. 2016).

The commensal microbiota is capable of producing a variety of neuroactive molecules that can act directly or indirectly on the CNS. Among the molecules produced by microorganisms and representing attractive complementary strategies in the treatment of mental disorders, vitamin D and B group vitamins were evaluated by different research groups.

Vitamin D

Vitamin D is associated with the functioning and development of the brain (Cui et al. 2007). Recently the role of this vitamin in neurological diseases has been reviewed (Moretti et al. 2018). Even when it is known that low levels of vitamin D are related to many neurological disorders, these authors remarked that, at the moment, the results for this vitamin supplementation against neurological diseases are inconclusive. This steroid vitamin has anti-inflammatory properties and the ability to regulate the immune system. The neuronal cells and those of the glia present receptors for vitamin D (Kalueff and Tuohimaa 2007). A recent article showed that vitamin D has an essential role for the normal development of dopaminergic neurons (Pertile et al. 2018). It was also reported that intestinal microbiota regulates vitamin D metabolism (Bora et al. 2018). In this sense it was also reported that vitamin D deficiency was linked to changes of intestinal microbiome and can reduce the production of B vitamins in the gut. These changes are also associated with many diseases including sleep disorders. These authors also discussed the pro-inflammatory state associated to this dysbiosis and the importance of supplementing with B group vitamins in addition to vitamin D when there are disorders related to the deficiency of this last vitamin (Gominak 2016).

At this level, probiotics can exert benefits by modifying the intestinal microbiota, by increasing the vitamin D receptor (VDR) expression, and VDR activity in the host (Shang and Sun 2017) or as producers of B group vitamins (LeBlanc et al. 2013, de Moreno de LeBlanc et al. 2018).

Riboflavin (vitamin B2)

Recent evidences suggest that riboflavin (vitamin B2) can exert a neuroprotective effect and offer protection against some neurological disorders, such as Parkinson's disease, migraines, and multiple sclerosis. The mechanisms involved in these effects are related to the antioxidant/anti-inflammatory properties of this vitamin. It can prevent oxidative stress, neuroinflammation, and mitochondrial dysfunction associated with PD and other neurological disorders (Marashly and Bohlega 2017). Riboflavin deficiency is also a risk factor for multiple sclerosis because vitamin B2 is associated to myelin formation. It was demonstrated that increased levels of BDNF in the CNS are associated with the beneficial effects of riboflavin in an animal model of multiple sclerosis (Naghashpour et al. 2017). In this sense, the production of riboflavin can be another mechanism by which lactic acid bacteria have the potential to exert beneficial effects against different neurological disorders. LAB capable of producing riboflavin were isolated, and the bioavailability of the vitamin produced by selected LAB have been shown by using animal models of vitamin deficiency (Juarez Del Valle et al. 2016). The administration of *L. plantarum* CRL 2130, a riboflavin overproducer strain, showed anti-oxidant/anti-inflammatory effects *in vivo,* by using murine models of colitis (Levit et al. 2017, 2018a) and mucositis (Levit et al. 2018c). The anti-oxidant/anti-inflammatory properties of *L. plantarum* CRL 2130 were also associated to *in vitro* and *in vivo* neuroprotection (unpublished data) demonstrating the potential of this strain to protect against neurological diseases related to oxidant stress and neuroinflammation.

Folates (vitamin B9)

Folates and cobalamin (B group vitamins) participate in metabolic processes as coenzymes, essential for the development and functioning of the nervous system, since their activity is to donate methyl groups for the synthesis of neurotransmitters and hormones (Mitchell et al. 2014). Vitamins B6, B12 and B9 are involved in the methionine-homocysteine metabolism and their deficiencies produce an increase in homocysteine levels. High levels of this amino acid are negative for brain functioning due to the inhibition of methylation and the increase of oxidative stress generating DNA damage and neurotoxicity (Kennedy 2016). Hyperhomocysteinemia is associated with depressive disorders and cognitive deterioration (Mitchell et al. 2014), such as schizophrenia (Nishi et al. 2014).

It has been demonstrated that folate deficiency on C57Bl/6 mice caused damage by ischemia in the sensorimotor cortex with high levels of plasma homocysteine (Jadavji et al. 2017), and the addition of a diets rich in folate and choline improved motor function after ischemic damage. Mice that received folate and choline showed increased proliferation, neuroplasticity and antioxidant activity within the perilesional cortex.

Folate deficiency and hyperhomocysteinemia have also been associated with neurodegenerative diseases such as PD and AD, especially in aging (Sharma et al. 2015, Ostrakhovitch and Tabibzadeh 2019).

Folate producing LAB could exert beneficial effects against pathologies associated to hyperhomocysteinemia and folate deficiency such as some neurological diseases. *Streptococcus thermophillus* CRL 808 was selected as a bacterium capable of producing high levels of folates intracellularly, and using *in vivo* and *in vitro* evaluations it was demonstrated that exerted beneficial effects against mucositis caused by an anti-cancer drug (Levit et al. 2018b). The neuroprotective potential of this bacterial strain was recently evaluated; it allows recovering the motor deficits in a neurodegenerative disease mouse model as well as exerted a neuroprotective effect in neuronal cell cultures (unpublished data).

Thiamine (vitamin B1)

Thiamine is another essential nutrient for human growth and development due to its multilateral participation in key biochemical and physiological processes. In addition to its function as a cofactor during metabolic processes (Hirsch and Parrott 2012), thiamine plays a neuro-modulatory role in the acetylcholine neurotransmitter system and contributes to the structure and function of cellular membranes, including neurons and neuroglia (Ba 2008).

Decreased thiamine levels are correlated with the aging process (Zhang et al. 2013), and insufficient levels of thiamine cause neurological and metabolic disorders. The CNS is particularly vulnerable to thiamine deficiency (TD), because of its dependence on TPP-mediated metabolic pathways involved in energy metabolism and neurotransmitter synthesis (Vernau et al. 2015). In the brain, TD causes a cascade of events including mild impairment of oxidative metabolism, neuroinflammation, and neurodegeneration, which are commonly observed in majority of the neurodegenerative diseases (Kennedy 2016, Liu et al. 2017). The human neurological disorder most clearly associated with TD is Wernicke-Korsakoff syndrome (WKS), characterized by behavioral, cognitive, and neuropathological deficits (Dror et al. 2014). TD is also involved in several neurodegenerative diseases, such as Alzheimer's disease (AD), Parkinson's disease (PD), Huntington's disease (HD) and alcohol-induced dementia (Tanev et al. 2008). For instance, Pan et al. (2017) found that the contents of thiamine diphosphate (TDP) in blood were significantly lower in patients with AD compared with control subjects, and that also thiamine diphosphatase (TDPase) and thiamine monophosphatase (TMPase) activities were significantly enhanced in the first group.

Until now, many studies have shown that the administration of thiamine has therapeutic effects in some neurodegenerative diseases and alcohol-induced dementia (Lu'o'ng and Nguyen 2011, Costantini et al. 2013, Costantini et al. 2015). Thiamine metabolites may serve as promising biomarkers for neurodegenerative diseases, and thiamine supplementation has potential as therapy for patients suffering from neurodegenerative diseases. Moreover, the use of LAB capable of producing this vitamin could be an innovative source for thiamine supplementation, used as a bio-strategy to avoid and/or to treat their deficiency, similar to other B group vitamins.

Some bifidobacteria strains were described as capable of producing thiamine (Deguchi et al. 1985), and other authors investigated the potential of different LAB

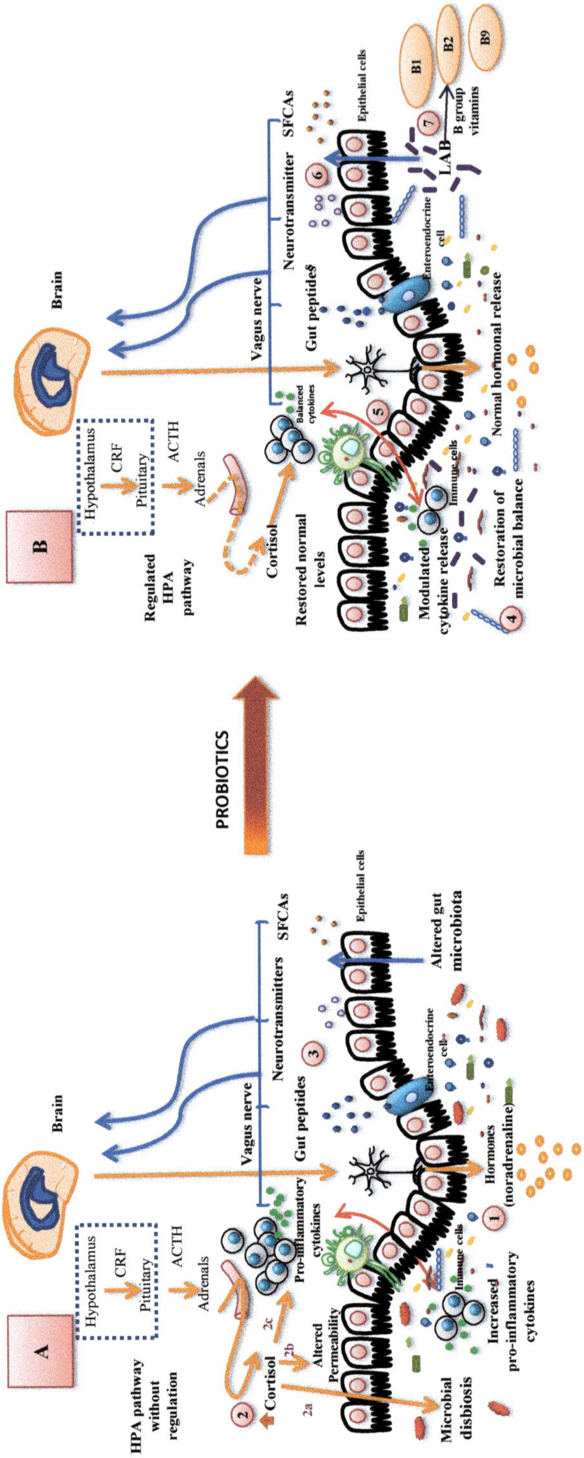

Figure 1 contd. ...

to produce this and other B group vitamins; however no bacterial strain under study produced high levels of thiamine (Masuda et al. 2012). Recent studies by our group selected LAB that were able to produce thiamine. One of these strains was associated to neuroprotective effect *in vitro* by using a neuronal model and also *in vivo* by reducing the motor deficits in a PD animal model (unpublished data).

Conclusion

Considering the current knowledge about the microbiota gut-brain axis, and the relationship between its unbalance and the development of neurological disorders, probiotics are an alternative to prevent or improve symptoms related to these pathologies. The administration of lactic acid bacteria with specific characteristics (anti-inflammatory properties, B group vitamin producers) can have a beneficial effect on the intestinal microbiota and on the neuro-immunological alterations associated with the aging process and neurodegenerative pathologies such as PD. Figure 1 shows a schematic representation of potential mechanisms by which probiotics can exert beneficial effects against neurological diseases that affect the CNS.

Probiotic microorganisms could have a promising effect, mainly for elderly individuals, in early stages of the disease or for individuals suffering neurodegenerative disorders and receiving conventional treatments by improving their quality of life. Studies with human patients only demonstrated the potential of probiotics to prevent some intestinal deficiencies associated with certain neurological diseases. The properties demonstrated for certain probiotics against other pathologies

...Figure 1 contd.

Figure 1: Schematic representation of potential mechanisms by which probiotics can exert benefits in the Central Nervous System (CNS). Participation in the Gut-Brain Axis. Figure **(A)** represents the findings described in patients with central neurological disorders; the neuroendocrine system is responsible for the production of hormones and neuropeptides. (1) The abnormal release of hormones (noradrenaline) alters intestinal physiology and microbiota, favoring the growth of certain specific microorganisms. (2) Elevated chronic blood cortisol levels adversely affect the brain by activating Hypothalamus Pituitary Adrenal axis (HPA) pathway without regulation. This activation can later affect the composition of the microbiota (dysbiosis) (2a), the permeability and barrier function of the intestine (2b), and the immune cells to secrete pro-inflammatory cytokines locally in the intestine and also systemically (2c). Intestinal inflammation and changes in the composition of the intestinal microbiota are risk factors for several neurological diseases. (3) Many nerve fibers of the intestine constantly provide information to the CNS. Local changes in the levels of cytokines, neurotransmitters or other molecules (such as short-chain fatty acids, SCFA) can be communicated in this way from the enteric nervous system (ENS) to the CNS. Figure **(B)** shows possible mechanisms by which probiotics can beneficially affect the gut-brain axis, altered in neurological diseases. (4) Probiotics can restore the microbial balance in the gut microbiota that affects the intestinal nervous and immune systems. (5) LAB with anti-inflammatory properties can modulate cytokine profiles and/or exert antioxidant activities at the intestinal or systemic level. These anti-inflammatory effects can exert benefits indirectly through the ENS, or directly the anti-oxidant/anti-inflammatory compounds can reach the CNS. (6) In addition, some LAB can produce neuroactive molecules such as gamma aminobutyric acid (GABA), serotonin, acetylcholine and also SCFAs that have neuroactive properties; they can modulate the immune system and affect the CNS from the ENS. (7) Finally, the bacteria that produce B vitamins can also exert benefits due to the antioxidant/anti-inflammatory properties of these vitamins, described as diminished in certain neurological diseases.

show the potential of these microorganisms to act against disorders associated with neurological diseases. However, specific pre-clinical studies (with adequate *in vitro* and *in vivo* models) are needed in order to understand the mechanisms involved in the neuroprotective potential of probiotics, especially for aging and people with predisposition to develop neurodegenerative diseases.

Acknowledgments

Authors thank the financial support of Consejo Nacional de Investigaciones Científicas y Técnicas (CONICET, project PIP0697), and Agencia Nacional de Promoción Científica y Tecnológica (ANPCyT, projects 2554 and 0301).

References

Abhari, K., S.S. Shekarforoush, S. Hosseinzadeh, S. Nazifi, J. Sajedianfard and M.H. Eskandari. 2016. The effects of orally administered *Bacillus coagulans* and inulin on prevention and progression of rheumatoid arthritis in rats. Food Nutr. Res. 60: 30876.

Akhoundzadeh, K., A. Vakili, M. Shadnoush and J. Sadeghzadeh. 2018. Effects of the oral ingestion of probiotics on brain damage in a transient model of focal cerebral ischemia in mice. Iran J. Med. Sci. 43(1): 32–40.

Andersson, U. and K.J. Tracey. 2012. Neural reflexes in inflammation and immunity. J. Exp. Med. 209(6): 1057–1068.

Ba, A. 2008. Metabolic and structural role of thiamine in nervous tissues. Cell Mol. Neurobiol. 28(7): 923–931.

Baltazar, M.T., R.J. Dinis-Oliveira, M. de Lourdes Bastos, A.M. Tsatsakis, J.A. Duarte and F. Carvalho. 2014. Pesticides exposure as etiological factors of Parkinson's disease and other neurodegenerative diseases—a mechanistic approach. Toxicol. Lett. 230(2): 85–103.

Bauer, K.C., K.E. Huus and B.B. Finlay. 2016. Microbes and the mind: emerging hallmarks of the gut microbiota-brain axis. Cell Microbiol. 18(5): 632–644.

Bercik, P., E. Denou, J. Collins, W. Jackson, J. Lu, J. Jury, Y. Deng, P. Blennerhassett, J. Macri, K.D. McCoy, E.F. Verdu and S.M. Collins. 2011a. The intestinal microbiota affect central levels of brain-derived neurotropic factor and behavior in mice. Gastroenterology 141(2): 599–609, 609 e591–593.

Bercik, P., A.J. Park, D. Sinclair, A. Khoshdel, J. Lu, X. Huang, Y. Deng, P.A. Blennerhassett, M. Fahnestock, D. Moine, B. Berger, J.D. Huizinga, W. Kunze, P.G. McLean, G.E. Bergonzelli, S.M. Collins and E.F. Verdu. 2011b. The anxiolytic effect of *Bifidobacterium longum* NCC3001 involves vagal pathways for gut-brain communication. Neurogastroenterol. Motil. 23(12): 1132–1139.

Binh, T.T., W.T. Ju, W.J. Jung and R.D. Park. 2014. Optimization of gamma-amino butyric acid production in a newly isolated *Lactobacillus brevis*. Biotechnol. Lett. 36(1): 93–98.

Bora, S.A., M.J. Kennett, P.B. Smith, A.D. Patterson and M.T. Cantorna. 2018. The gut microbiota regulates endocrine vitamin D metabolism through fibroblast growth factor 23. Front. Immunol. 9: 408.

Bravo, J.A., P. Forsythe, M.V. Chew, E. Escaravage, H.M. Savignac, T.G. Dinan, J. Bienenstock and J.F. Cryan. 2011. Ingestion of *Lactobacillus* strain regulates emotional behavior and central GABA receptor expression in a mouse via the vagus nerve. Proc. Natl. Acad. Sci. USA 108(38): 16050–16055.

Burokas, A., R.D. Moloney, T.G. Dinan and J.F. Cryan. 2015. Microbiota regulation of the Mammalian gut-brain axis. Adv. Appl. Microbiol. 91: 1–62.

Chapelet, G., L. Leclair-Visonneau, T. Clairembault, M. Neunlist and P. Derkinderen. 2018. Can the gut be the missing piece in uncovering PD pathogenesis? Parkinsonism Relat. Disord.

Collins, S.M. and P. Bercik. 2009. The relationship between intestinal microbiota and the central nervous system in normal gastrointestinal function and disease. Gastroenterology 136(6): 2003–2014.

Collins, S.M., M. Surette and P. Bercik. 2012. The interplay between the intestinal microbiota and the brain. Nat. Rev. Microbiol. 10(11): 735–742.

Costantini, A., M.I. Pala, L. Compagnoni and M. Colangeli. 2013. High-dose thiamine as initial treatment for Parkinson's disease. BMJ Case Rep. 2013.

Costantini, A., M.I. Pala, E. Grossi, S. Mondonico, L.E. Cardelli, C. Jenner, S. Proietti, M. Colangeli and R. Fancellu. 2015. Long-term treatment with high-dose thiamine in Parkinson Disease: An open-label pilot study. J. Altern. Complement. Med. 21(12): 740–747.

Cui, X., J.J. McGrath, T.H. Burne, A. Mackay-Sim and D.W. Eyles. 2007. Maternal vitamin D depletion alters neurogenesis in the developing rat brain. Int. J. Dev. Neurosci. 25(4): 227–232.

de Moreno de LeBlanc, A., R. Levit, G.S. de Giori and J.G. LeBlanc. 2018. Vitamin producing lactic acid bacteria as complementary treatments for intestinal inflammation. Antiinflamm. Antiallergy Agents Med. Chem. 17(1): 50–56.

de Punder, K. and L. Pruimboom. 2015. Stress induces endotoxemia and low-grade inflammation by increasing barrier permeability. Front. Immunol. 6: 223.

De Vadder, F., E. Grasset, L. Manneras Holm, G. Karsenty, A.J. Macpherson, L.E. Olofsson and F. Backhed. 2018. Gut microbiota regulates maturation of the adult enteric nervous system via enteric serotonin networks. Proc. Natl. Acad. Sci. USA 115(25): 6458–6463.

Deguchi, Y., T. Morishita and M. Mutai. 1985. Comparative studies on synthesis of water-soluble vitamins among human species of bifidobacteria. Agric. Biol. Chem. 49(1): 13–19.

Desbonnet, L., L. Garrett, G. Clarke, J. Bienenstock and T.G. Dinan. 2008. The probiotic *Bifidobacteria infantis*: An assessment of potential antidepressant properties in the rat. J. Psychiatr. Res. 43(2): 164–174.

Desbonnet, L., L. Garrett, G. Clarke, B. Kiely, J.F. Cryan and T.G. Dinan. 2010. Effects of the probiotic Bifidobacterium infantis in the maternal separation model of depression. Neuroscience 170(4): 1179–1188.

Dias, V., E. Junn and M.M. Mouradian. 2013. The role of oxidative stress in Parkinson's Disease. J. Parkinsons Dis. 3(4): 461–491.

Divyashri, G., G. Krishna, Muralidhara and S.G. Prapulla. 2015. Probiotic attributes, antioxidant, anti-inflammatory and neuromodulatory effects of *Enterococcus faecium* CFR 3003: *in vitro* and *in vivo* evidence. J. Med. Microbiol. 64(12): 1527–1540.

Dror, V., M. Rehavi, I.E. Biton and S. Eliash. 2014. Rasagiline prevents neurodegeneration in thiamine deficient rats-a longitudinal MRI study. Brain Res. 1557: 43–54.

Forsythe, P., J. Bienenstock and W.A. Kunze. 2014. Vagal pathways for microbiome-brain-gut axis communication. Adv. Exp. Med. Biol. 817: 115–133.

Franceschi, C. and J. Campisi. 2014. Chronic inflammation (inflammaging) and its potential contribution to age-associated diseases. J. Gerontol. A. Biol. Sci. Med. Sci. 69 Suppl 1: S4–9.

Freestone, P.P., S.M. Sandrini, R.D. Haigh and M. Lyte. 2008. Microbial endocrinology: How stress influences susceptibility to infection. Trends Microbiol. 16(2): 55–64.

Giannetti, E., M. Maglione, A. Alessandrella, C. Strisciuglio, D. De Giovanni, A. Campanozzi, E. Miele and A. Staiano. 2017. A mixture of 3 Bifidobacteria decreases abdominal pain and improves the quality of life in children with irritable bowel syndrome: A multicenter, randomized, double-blind, placebo-controlled, crossover trial. J. Clin. Gastroenterol. 51(1): e5–e10.

Goldman, J.G. and R. Postuma. 2014. Premotor and nonmotor features of Parkinson's disease. Curr. Opin. Neurol. 27(4): 434–441.

Gominak, S.C. 2016. Vitamin D deficiency changes the intestinal microbiome reducing B vitamin production in the gut. The resulting lack of pantothenic acid adversely affects the immune system, producing a "pro-inflammatory" state associated with atherosclerosis and autoimmunity. Med. Hypotheses. 94: 103–107.

Heiss, C.N. and L.E. Olofsson. 2019. The role of the gut microbiota in development, function and disorders of the central nervous system and the enteric nervous system. J. Neuroendocrinol. 2019: e12684.

Hiraga, K., Y. Ueno, S. Sukontasing, S. Tanasupawat and K. Oda. 2008. *Lactobacillus senmaizukei* sp. nov., isolated from Japanese pickle. Int. J. Syst. Evol. Microbiol. 58(Pt 7): 1625–1629.

Hirsch, J.A. and J. Parrott. 2012. New considerations on the neuromodulatory role of thiamine. Pharmacology 89(1-2): 111–116.

Hsueh, Y.H., W.C. Liaw, J.M. Kuo, C.S. Deng and C.H. Wu. 2017. Hydrogel film-immobilized *Lactobacillus brevis* RK03 for gamma-aminobutyric acid production. Int. J. Mol. Sci. 18(11).

Jadavji, N.M., J.T. Emmerson, A.J. MacFarlane, W.G. Willmore and P.D. Smith. 2017. B-vitamin and choline supplementation increases neuroplasticity and recovery after stroke. Neurobiol. Dis. 103: 89–100.

Juarez Del Valle, M., J.E. Laino, A. de Moreno de LeBlanc, G. Savoy de Giori and J.G. LeBlanc. 2016. Soyamilk fermented with riboflavin-producing *Lactobacillus plantarum* CRL 2130 reverts and prevents ariboflavinosis in murine models. Br. J. Nutr. 116(7): 1229–1235.

Kalueff, A.V. and P. Tuohimaa. 2007. Neurosteroid hormone vitamin D and its utility in clinical nutrition. Curr. Opin. Clin. Nutr. Metab. Care. 10(1): 12–19.

Kaper, J.B. and V. Sperandio. 2005. Bacterial cell-to-cell signaling in the gastrointestinal tract. Infect. Immun. 73(6): 3197–3209.

Kelly, J.R., P.J. Kennedy, J.F. Cryan, T.G. Dinan, G. Clarke and N.P. Hyland. 2015. Breaking down the barriers: the gut microbiome, intestinal permeability and stress-related psychiatric disorders. Front. Cell. Neurosci. 9: 392.

Kennedy, D.O. 2016. B Vitamins and the Brain: Mechanisms, dose and efficacy. A review. Nutrients. 8(2): 68.

LeBlanc, J.G., J.E. Laino, M.J. del Valle, V. Vannini, D. van Sinderen, M.P. Taranto, G.F. de Valdez, G.S. de Giori and F. Sesma. 2011. B-group vitamin production by lactic acid bacteria—current knowledge and potential applications. J. Appl. Microbiol. 111(6): 1297–1309.

LeBlanc, J.G., C. Milani, G.S. de Giori, F. Sesma, D. van Sinderen and M. Ventura. 2013. Bacteria as vitamin suppliers to their host: a gut microbiota perspective. Curr. Opin. Biotechnol. 24(2): 160–168.

Levit, R., G.S. de Giori, A. de Moreno de LeBlanc and J.G. LeBlanc. 2017. Evaluation of the effect of soymilk fermented by a riboflavin-producing *Lactobacillus plantarum* strain in a murine model of colitis. Benef. Microbes. 8(1): 65–72.

Levit, R., G. Savoy de Giori, A. de Moreno de LeBlanc and J.G. LeBlanc. 2018a. Effect of riboflavin-producing bacteria against chemically induced colitis in mice. J. Appl. Microbiol. 124(1): 232–240.

Levit, R., G. Savoy de Giori, A. de Moreno de LeBlanc and J.G. LeBlanc. 2018b. Folate-producing lactic acid bacteria reduce inflammation in mice with induced intestinal mucositis. J. Appl. Microbiol. 125(5): 1494–1501.

Levit, R., G. Savoy de Giori, A. de Moreno de LeBlanc and J.G. LeBlanc. 2018c. Protective effect of the riboflavin-overproducing strain *Lactobacillus plantarum* CRL2130 on intestinal mucositis in mice. Nutrition 54: 165–172.

Liang, S., T. Wang, X. Hu, J. Luo, W. Li, X. Wu, Y. Duan and F. Jin. 2015. Administration of *Lactobacillus helveticus* NS8 improves behavioral, cognitive, and biochemical aberrations caused by chronic restraint stress. Neuroscience 310: 561–577.

Liguori, I., G. Russo, F. Curcio, G. Bulli, L. Aran, D. Della-Morte, G. Gargiulo, G. Testa, F. Cacciatore, D. Bonaduce and P. Abete. 2018. Oxidative stress, aging, and diseases. Clin. Interv. Aging. 13: 757–772.

Liu, D., Z. Ke and J. Luo. 2017. Thiamine deficiency and neurodegeneration: The interplay among oxidative stress, endoplasmic reticulum stress, and autophagy. Mol. Neurobiol. 54(7): 5440–5448.

Liu, R.T. 2017. The microbiome as a novel paradigm in studying stress and mental health. Am. Psychol. 72(7): 655–667.

Lu'o'ng, K. and L.T. Nguyen. 2011. Role of thiamine in Alzheimer's disease. Am. J. Alzheimers Dis. Other. Demen. 26(8): 588–598.

Luo, J., T. Wang, S. Liang, X. Hu, W. Li and F. Jin. 2014. Ingestion of *Lactobacillus* strain reduces anxiety and improves cognitive function in the hyperammonemia rat. Sci. China Life Sci. 57(3): 327–335.

Marashly, E.T. and S.A. Bohlega. 2017. Riboflavin has neuroprotective potential: Focus on Parkinson's Disease and migraine. Front. Neurol. 8: 333.

Masuda, M., M.I. de, H. Utsumi, T. Niiro, Y. Shimamura and M. Murata. 2012. Production potency of folate, vitamin B(12), and thiamine by lactic acid bacteria isolated from Japanese pickles. Biosci. Biotechnol. Biochem. 76(11): 2061–2067.

Mawdsley, J.E. and D.S. Rampton. 2005. Psychological stress in IBD: new insights into pathogenic and therapeutic implications. Gut. 54(10): 1481–1491.

Mayer, E.A. 2011. Gut feelings: the emerging biology of gut-brain communication. Nat. Rev. Neurosci. 12(8): 453–466.

Mitchell, E.S., N. Conus and J. Kaput. 2014. B vitamin polymorphisms and behavior: evidence of associations with neurodevelopment, depression, schizophrenia, bipolar disorder and cognitive decline. Neurosci. Biobehav. Rev. 47: 307–320.

Moretti, R., M.E. Morelli and P. Caruso. 2018. Vitamin D in neurological diseases: A rationale for a pathogenic impact. Int. J. Mol. Sci. 19(8).

Naghashpour, M., S. Jafarirad, R. Amani, A. Sarkaki and A. Saedisomeolia. 2017. Update on riboflavin and multiple sclerosis: a systematic review. Iran J. Basic. Med. Sci. 20(9): 958–966.

Nishi, A., S. Numata, A. Tajima, M. Kinoshita, K. Kikuchi, S. Shimodera, M. Tomotake, K. Ohi, R. Hashimoto, I. Imoto, M. Takeda and T. Ohmori. 2014. Meta-analyses of blood homocysteine levels for gender and genetic association studies of the MTHFR C677T polymorphism in schizophrenia. Schizophr. Bull. 40(5): 1154–1163.

Ostrakhovitch, E.A. and S. Tabibzadeh. 2019. Homocysteine and age-associated disorders. Ageing Res. Rev. 49: 144–164.

Pan, X., S. Sang, G. Fei, L. Jin, H. Liu, Z. Wang, H. Wang and C. Zhong. 2017. Enhanced activities of blood thiamine diphosphatase and monophosphatase in Alzheimer's disease. PLoS One 12(1): e0167273.

Pertile, R.A.N., X. Cui, L. Hammond and D.W. Eyles. 2018. Vitamin D regulation of GDNF/Ret signaling in dopaminergic neurons. FASEB J. 32(2): 819–828.

Prevot, V. 2010. Plasticity of neuroendocrine systems. Eur. J. Neurosci. 32(12): 1987–1988.

Ramos, C.L., L. Thorsen, R.F. Schwan and L. Jespersen. 2013. Strain-specific probiotics properties of *Lactobacillus fermentum*, *Lactobacillus plantarum* and *Lactobacillus brevis* isolates from Brazilian food products. Food Microbiol. 36(1): 22–29.

Sanchart, C., O. Rattanaporn, D. Haltrich, P. Phukpattaranont and S. Maneerat. 2016. Technological and safety properties of newly isolated GABA-producing *Lactobacillus futsaii* strains. J. Appl. Microbiol. 121(3): 734–745.

Sechi, G., E. Sechi, C. Fois and N. Kumar. 2016. Advances in clinical determinants and neurological manifestations of B vitamin deficiency in adults. Nutr. Rev. 74(5): 281–300.

Shang, M. and J. Sun. 2017. Vitamin D/VDR, Probiotics, and gastrointestinal diseases. Curr. Med. Chem. 24(9): 876–887.

Sharma, M., M. Tiwari and R.K. Tiwari. 2015. Hyperhomocysteinemia: Impact on neurodegenerative diseases. Basic Clin. Pharmacol. Toxicol. 117(5): 287–296.

Smith, C.J., J.R. Emge, K. Berzins, L. Lung, R. Khamishon, P. Shah, D.M. Rodrigues, A.J. Sousa, C. Reardon, P.M. Sherman, K.E. Barrett and M.G. Gareau. 2014. Probiotics normalize the gut-brain-microbiota axis in immunodeficient mice. Am. J. Physiol. Gastrointest. Liver. Physiol. 307(8): G793–802.

Tanev, K.S., M. Roether and C. Yang. 2008. Alcohol dementia and thermal dysregulation: a case report and review of the literature. Am. J. Alzheimers Dis. Other Demen. 23(6): 563–570.

Tillisch, K. 2014. The effects of gut microbiota on CNS function in humans. Gut Microbes. 5(3): 404–410.

Toni, R. 2004. The neuroendocrine system: organization and homeostatic role. J. Endocrinol. Invest. 27(6 Suppl): 35–47.

Tracey, K.J. 2009. Reflex control of immunity. Nat. Rev. Immunol. 9(6): 418–428.

Vernau, K., E. Napoli, S. Wong, C. Ross-Inta, J. Cameron, D. Bannasch, A. Bollen, P. Dickinson and C. Giulivi. 2015. Thiamine deficiency-mediated brain mitochondrial pathology in Alaskan huskies with mutation in SLC19A3.1. Brain Pathol. 25(4): 441–453.

Virmani, A., L. Pinto, Z. Binienda and S. Ali. 2013. Food, nutrigenomics, and neurodegeneration. Neuroprotection by what you eat! Mol. Neurobiol. 48(2): 353–362.

Wang, H.X. and Y.P. Wang. 2016. Gut microbiota-brain axis. Chin. Med. J. 129(19): 2373–2380.

Wang, X. and E.K. Michaelis. 2010. Selective neuronal vulnerability to oxidative stress in the brain. Front. Aging Neurosci. 2: 12.

Westfall, S., N. Lomis, I. Kahouli, S.Y. Dia, S.P. Singh and S. Prakash. 2017. Microbiome, probiotics and neurodegenerative diseases: deciphering the gut brain axis. Cell Mol. Life Sci. 74(20): 3769–3787.

Wu, C.H., Y.H. Hsueh, J.M. Kuo and S.J. Liu. 2018. Characterization of a potential probiotic *Lactobacillus brevis* RK03 and efficient production of gamma-aminobutyric acid in batch fermentation. Int. J. Mol. Sci. 19(1).

Yan, F., Y. Chen, M. Li, Y. Wang, W. Zhang, X. Chen and Q. Ye. 2018. Gastrointestinal nervous system alpha-synuclein as a potential biomarker of Parkinson's disease. Medicine (Baltimore) 97(28): e11337.

Zhang, G., H. Ding, H. Chen, X. Ye, H. Li, X. Lin and Z. Ke. 2013. Thiamine nutritional status and depressive symptoms are inversely associated among older Chinese adults. J. Nutr. 143(1): 53–58.

12

The Role of the Microbiota and the Application of Probiotics in Reducing the Risk of Cardiovascular Diseases

Raquel Bedani[1,2,*] and *Susana Marta Isay Saad*[1,2]

Introduction

Cardiovascular diseases (CVDs) are considered as a group of disorders of the heart and blood vessels, including coronary heart disease, brain-vessels disease, rheumatic heart disease and others (WHO 2018). According to World Health Organization, 17.9 million people die each year from CVDs, which corresponds to approximately 31% of all deaths worldwide and accounting for costs of approximately $ 1 trillion per year (WHO 2018, Battson et al. 2018). It is noteworthy that 85% of all CVD deaths are due to heart attacks and strokes, caused mainly by a blocking of the blood route to the heart and brain as a consequence of fatty deposits on the inner walls of these blood vessels. Thus, atherosclerosis could be appointed as the most common cause of cardiovascular complications (Singh et al. 2016). Individuals at risk of developing CVD may present elevated blood pressure, glucose, lipids, as well as overweight, and obesity (WHO 2018). In spite of the significant progress in the prevention and treatment of CVD, it is important to note that due to an increasing prevalence in obesity, type 2 diabetes, and metabolic syndrome worldwide, it is estimated that the number of people with chronic cardiometabolic diseases tends to continue to grow

[1] Department of Biochemical-Pharmaceutical Technology, School of Pharmaceutical Sciences, University of São Paulo, São Paulo, Brazil.
[2] Food Research Center, University of São Paulo, São Paulo, Brazil.
Email: susaad@usp.br
* Corresponding author: raquelbedani@yahoo.com.br

in the coming decades (Schiattarella et al. 2019). In this sense, the identification of novel therapeutic targets becomes extremely important in terms of both prevention and treatment of these diseases (Frankenfeld 2017, Battson et al. 2018).

Genetic factors (responsible for less than 20% CVD risk) and diet are known to influence the pathogenesis of cardiovascular diseases; however, studies in recent years have supported a promising relationship among nutrient intake, intestinal microbial metabolism, and the host to change the risk of developing CVD (Drosos et al. 2015, Brown and Hazen 2018). Therefore, the modulation of host-microbiota interaction in order to develop microbiota-targeted therapies with the aim of inducing beneficial changes in the composition and/or metabolism of the microbial community has become a promising approach for reducing, not only cardiometabolic risk factors, but also of other metabolic disorders (Frankenfeld 2017, Battson et al. 2018).

The gastrointestinal tract harbors at least 10 trillion microorganisms, collectively named intestinal microbiota (Marchesi et al. 2015, Battson et al. 2018). Bacteria, archea, fungi, and viruses are part of this intestinal ecosystem. The intestinal microbiota and its collective genetic material, termed intestinal microbiome, contains approximately 100 times more genes than the human genome (Marchesi et al. 2015, Battson et al. 2018) and are involved in the development and regulation of the immune system; protection against pathogens; elimination of exogenous toxins; regulation of intestinal function; and nutrient synthesis, absorption and metabolism (Battson et al. 2018).

The identification of a wide diversity of intestinal microbiota composition and the role of the intestinal microbiome on human physiology and disease processes have advanced significantly due to novel culture-independent molecular assays, based on the analysis of the 16S ribosomal RNA, and the use of experimental models, including germ-free mice and fecal microbiota transplantation (Zoetendal et al. 2008, Kuczynski et al. 2011, Battson et al. 2018). In general, most of the bacteria present in the intestinal microbiota from healthy individuals belong to three main phyla: Firmicutes, Bacteroidetes, and Actinobacteria (Zoetendal et al. 2008, Tap et al. 2009, Kasselman et al. 2018). The most abundant genera in the intestinal microbiota are: Gram-positive bacteria such as *Clostridium* spp., *Bifidobacterium* spp., *Lactobacillus* spp., *Ruminococcus* spp., *Streptococcus*, and Gram-negative bacteria like *Bacteroides* spp., *Prevotella* spp., and *Akkermansia* spp. (Kasselman et al. 2018). Nevertheless, as each person has a distinct and highly variable intestinal microbial composition, recognizing a "healthy" microbiome has been a challenge (Turnbaugh et al. 2009, Backhed et al. 2012, Battson et al. 2018). However, it has been accepted that the microbiome of healthy subjects is characterized by high overall microbial diversity, resilience against external pressures, and prevalence of Bacteroidetes and Firmicutes phyla (Battson et al. 2018). Disturbance or imbalance of the healthy gut microbiota (dysbiosis) is characterized by decreased microbial diversity and changes in the relative abundance of certain bacterial phyla (Petersen and Round 2014, Battson et al. 2018). Several studies have linked intestinal dysbiosis to obesity-related diseases, such as CVD (Battson et al. 2018). It is noteworthy that the intestinal microbiota composition is influenced by major risk factors for CVD, including aging, obesity, sedentary lifestyle, and diet (Battson et al. 2018).

On the other hand, current evidence has pointed that changes in bacterial composition through specific dietary components, for example, probiotic microorganisms, or regulation of microbial metabolite formation might have effect on the development of CVD (Jonsson and Bäckhed 2017). In this sense, it has been speculated that the beneficial modulation of the gut microbiota with probiotics might be a promising approach to prevent and/or treat CVD pathogenesis (Thushara et al. 2016). Therefore, this chapter presents scientific evidence about the relationship between microbiota and CVD, as well as the probiotic potential in modulating the intestinal microbiota with a focus on their effects on CVD-related risk factors.

Microbiota and its Influence on Cardiovascular Disease

Role of oral microbiota on cardiovascular disease

Findings suggest that atherosclerotic plaques are affected by a distant or by a direct infection of the vessel wall cells (Jonsson and Backhed 2016). This information is supported by the discovery of bacterial DNA in plaques (Koren et al. 2011, Jonsson and Backhed 2016). It is noteworthy that the bacteria detected in plaques are found predominantly in the oral cavity and the gut, which could be considered as reservoirs of these potentially pathogenic microorganisms (Jonsson and Backhed 2016).

Several studies have shown a relationship between the microbiota from the oral cavity and the atherosclerotic plaques of individuals with atherosclerosis (Kholy et al. 2015, Bui et al. 2019). The most common bacterial genera in the oral cavity are *Treponema* spp., *Porphyromonas* spp., *Prevotella* spp., *Capnocytophaga* spp., *Peptostreptococcus* spp., *Fusobacterium* spp., *Actinobacillus* spp., and *Eikenella* spp. (Bui et al. 2019). The oral microbiota is found in saliva, gingival epithelium, and other oral cavity surfaces, and concentrated in dental plaques (Bui et al. 2019).

Studies have suggested the microorganisms *Porphyromonas* (*P.*) *gingivalis*, *Actinobacilus* (*A.*) *actinomycetemcomitans*, and *Tanerella* (*T.*) *forsythia* as responsible for periodontal disease (Bui et al. 2019). Periodontitis is a chronic infectious and inflammatory disease caused by periodontal pathogens, and *P. gingivalis* is one of the main bacterial species associated to this pathology (Ao et al. 2014). Research on the pathogen *P. gingivalis* and the association of periodontal disease and CVD risk have supported the understanding of the role of oral microbiota on CVD pathogenesis (Brown and Hazen 2018). In this line, evidence has indicated that oral or systemic infection with *P. gingivalis* is able to accelerate the atherosclerosis development in animal models (Brown and Hazen 2018).

A meta-analysis of five prospective cohort studies with a total of 86,092 patients showed that individuals with periodontitis presented a 1.14 times higher risk of developing coronary heart disease compared to controls, suggesting that periodontitis may be a risk factor for this disease (Bahekar et al. 2007). High levels of inflammatory mediators in individuals with periodontitis indicate their role in atherothrombogenesis leading to coronary heart disease (Bahekar et al. 2007).

Armingohar et al. (2014) reported that 84 bacterial taxa were present in atherosclerotic plaques from patients with chronic periodontitis. However, 18 different taxa were identified in the vascular tissues from patients with vascular disease without chronic periodontitis, showing that a higher bacterial load and a more diverse colonization was verified in vascular disease lesions of subjects with periodontal disease when compared to subjects without periodontitis. In addition, Ao et al. (2014)

demonstrated that *P. gingivalis* from a dental infection was able to invade the aortal tissue through the circulation and increased the endothelial damage induced by a high fat diet in mice, in part by rising endothelial cell apoptosis. Immunohistochemical analyzes showed that *P. gingivalis* was detected deep in the smooth muscle of the aorta, and the aortal wall of high-fat diet mice had an increased number of this bacterial species compared to mice fed with a normal chow diet (control), suggesting that endothelial damage induced by high-fat diet can increase the *P. gingivalis* invasion rate.

An *in vitro* study showed that *P. gingivalis* could induce platelet aggregation, which could be responsible for thrombus development related to atherosclerosis and myocardial infarction (Sharma et al. 2000). On the other hand, oral species such as *A. actinomycetemcomitans*, *T. forsythia*, *Campylobacter* (*C.*) *rectus*, *Fusibacterium* (*F.*) *nucleatum*, *Prevotella* (*P.*) *intermedia*, and *Treponema* (*T.*) *denticola* did not promote platelet aggregation when tested for aggregation activity under similar conditions. This same study reported that outer membrane vesicles of *P. gingivalis* are potent inducers of mouse platelet aggregation *in vitro*, suggesting that these vesicles may be related to systemic virulence manifestations, including platelet aggregation and the development of atherosclerotic plaques.

Although the action mechanisms are unknown, it is believed that there are two ways of invasion of the cardiovascular tissues by periodontal microorganisms. Bacteria observed in the oral cavity might translocate to the vessels, affecting the growth and stability of the atherosclerotic plaques (Jonsson and Backhed 2016). Additionally, bacteremia-related bacteria can invade the endothelial layer producing pro-inflammatory cytokines (for example, monocyte chemoattractant protein 1, MCP-1) (Kholy et al. 2015). In the second way, transmigrating phagocytes harboring oral bacteria are proposed to disseminate to atherosclerotic plaques, contributing to the growth of atheroma (Kholy et al. 2015).

In general, evidence has shown that several oral pathogens are associated with increased risk of CVD in humans; however, further studies are required to support the relationship between infection by oral pathogens and the development of CVD, as well as the action mechanisms involved.

Intestinal microbiota and cardiovascular disease

Recent findings have shown that bacterial structural components as well as microbial metabolites are able to migrate from the intestine to the systemic circulation and influence the cardiovascular physiology towards the CVD development. The relationship between the bacterial components LPS and peptidoglycan and the metabolites TMAO, short-chain fatty acids, and bile acids on CVD will be discussed in the following sections.

Intestinal barrier function and the translocation of bacterial components

Several studies have shown a relationship between intestinal dysbiosis and changes in the intestinal epithelial barrier (Cani et al. 2007, Wang et al. 2014, Battson et al. 2018). The combination of reduced expression and reorganization of tight junction proteins (zonula occludens-1, claudin-1, and occludin) and an imbalance between intestinal epithelial cell death and proliferation may lead to increased intestinal permeability (Battson et al. 2018). It is noteworthy that barrier function is crucial in preventing translocation of the intestinal content, such as bacterial components, from entering the systemic circulation (Battson et al. 2018). The impairment of the barrier function allows that pathogen associated molecular patters (PAMPs) activate an immune response that results in systemic and tissue-specific inflammation (Battson et al. 2018). Therefore, damages in the barrier function related to dysbiosis could be considered as sources of the chronic inflammation present in several diseases linked to obesity, including CVD (Battson et al. 2018). The main PAMPs recognized by host immune receptors and associated to CVD risk are lipopolysaccharide (also referred as LPS or endotoxin) and peptidoglycan (Org et al. 2015).

The absorption of LPS from the intestine to the circulatory system may be explained through two mechanisms: (1) transport via chylomicron (transcellular transport) and (2) extracellular leakage through tight junctions in the epithelium (paracellular transport) (Caesar et al. 2010).

The translocation of LPS, a cell wall component of Gram-negative bacteria, to systemic circulation leads to a low-grade rise in plasma LPS named metabolic endotoxemia (Battson et al. 2018). This condition is associated with an elevated pro-inflammatory and oxidant environment often observed in obesity (Boutagy et al. 2016). The responsible factors for mediating pro-inflammatory effects of the microbiota are Toll-like receptors (TLRs) and NOD-like receptors (NLRs) that recognize bacterial components, including LPS and peptidoglycan (Org et al. 2015). Microbial LPS may be considered as a potent trigger of innate immune responses when recognized by TLR-4 (Battson et al. 2018). This binding (LPS/TLR-4) activates an intracellular signaling cascade that leads to the activation of NFκB (nuclear factor kappa B) and the up-regulation of several genes encoding pro-inflammatory mediators, such as cytokines, adhesion molecules, and pro-oxidant enzymes (Battson et al. 2018). TLR-4 is present on the cell surface of monocytes, other immune cells, and other cells types, including skeletal muscle, adipose tissue, and liver (Boutagy et al. 2016). Macrophages that accumulate in the walls of the blood vessels during hyperlipidemia conditions display a pro-inflammatory profile induced by TLRs (Caesar et al. 2010).

It is hypothesized that changes in the tight junction permeability by diet and intestinal microbiota composition could promote endotoxemia (Caesar et al. 2010). Studies suggest that the relationship among inflammation induced by high-fat diet, oxidative stress, metabolic disorders, and intestinal microbiota could be LPS-dependent (Cani et al. 2009). In this line, Cani et al. (2008) reported that the modifications of intestinal bacteria following a high-fat diet increase intestinal permeability by reducing the expression of genes encoding tight junction proteins (ZO-1 and occludin). On the other hand, the authors observed that obese and high-fat fed diabetic mice treated with antibiotic recovered normal intestinal epithelial integrity, which shows the influence of intestinal bacteria on this process (Cani et al. 2008).

Some evidence has shown that LPS may affect the cardiovascular function and contribute to increased CVD risk: (1) Plasma LPS was elevated in at-risk subjects and may predict future CVD; (2) Vascular cells incubated with LPS showed a response similar to that found during atherogenesis, including inflammation, oxidative stress, cell apoptosis, smooth muscle cell proliferation, macrophage activation, and monocyte migration and adhesion; (3) Studies using animal models showed that the administration of low doses of LPS, similar to those verified in metabolic endotoxemia, leads to vascular inflammation and atherosclerosis (Battson et al. 2018). Nevertheless, further investigations are necessary for understanding how microbial LPS can influence CVD risk.

Other bacterial structural component that could be involved in the development of CVD is the peptidoglycan (PG), the main cell wall component of Gram-positive bacteria (Laman et al. 2002, Battson et al. 2018). PG promotes inflammation via TLR, particularly TLR2 (Laman et al. 2002, Battson et al. 2018). As PG is detected in human atherosclerotic lesions, mainly in macrophage-rich atheromatous regions, findings have shown that this bacterial component could be an additional pro-inflammatory factor in the lesion (Lee et al. 2011). Additionally, studies suggest that peptidoglycan recognition protein-1 (PGLYRP-1), a circulating protein that binds to PG and promotes inflammation through the activation of innate immune mechanism, might act as an atherosclerosis biomarker, since PGLYRP-1 levels have been associated to prevalent subclinical and acute atherosclerosis (Rohatgi et al. 2009, Brownell et al. 2016). However, additional studies are required to support the clinical relevance of PGLYRP-1 and its use as biomarker for cardiovascular diseases.

Intestinal microbial metabolites

Intestinal microorganisms enzymatically produce several metabolites which can have local or systemic action and can influence health or disease of the host (Brown and Hazen 2018). Recent advances in metabolomics have discovered various human disease-associated intestinal microbial metabolites (Brown and Hazen 2018). Thus, targeting the intestinal microbiota and its metabolites is becoming a promising approach in the prevention and/or treatment of CVD (Peng et al. 2018).

The main metabolites generated from intestinal microbiota which can be related to CVD risk belong to three different classes: trimethylamines, short chain fatty acids (SCFAs), and bile acids (Brown and Hazen 2018).

Trimethylamine-N-oxide (TMAO). It is an intestinal microbial co-metabolite that has drawn a lot of attention as both biomarker for CVD risk and a promoter of atherothrombotic diseases, which have supported the link between the gut microbiota and CVD (Brown and Hazen 2018). In this sense, the control of TMAO pathway could be considered as a promising target for CVD drugs focused on intestinal microbiome (Brown and Hazen 2018).

The TMAO biosynthesis may occur in response to dietary intake of nutrients that contain a trimethylamine (TMA) $[N(CH_3)_3]$ moiety, such as phosphatidylcholine, choline, and L-carnitine (Battson et al. 2018, Tang et al. 2019). Examples of dietary sources rich in these nutrients are red meat, fish, dairy, and egg yolk (Drosos et al. 2015). In the intestines, the microbial catabolism of these nutrients occurs through the action of microbial TMA-lyase (CutC) and its activactor CutD (Tang et al. 2019, Schiatarella et al. 2019). TMA produced in the gut is absorbed and enters the portal circulation, being transferred to the liver where it is oxidized into TMAO by hepatic flavin monooxygenases (FMO), particularly FMO3, and afterwards TMAO enters the systemic circulation (Tang et al. 2019). It is highlighted that the specific receptor or chemical sensor for TMAO that would promote the atherosclerotic process remains unknown (Tang et al. 2019).

Little is known about which microbial species or communities are the main TMA-producing in mouse or human gut (Brown and Hazen 2018). An *in vitro* study showed that specific bacteria such as *Anaerococcus hydrogenalis*, *Clostridium asparagiforme*, *Clostridium hathewayi*, *Clostridium sporogenes*, *Escherichia fergusonii*, *Proteus penneri*, *Providencia rettgeri*, and *Edwardsiellatarda* may contribute to the production of TMAO from TMA (Romano et al. 2015, Peng et al. 2018). Using gnotobiotic mice, the authors verified that TMAO can accumulate in the serum of animals colonized with TMA-producing species and not in the serum of animals colonized with intestinal microorganisms that do not produce TMA from choline *in vitro* (Romano et al. 2015). Some studies with mouse models displayed that the groups of microorganisms associated with circulating TMAO levels include the order of RF39; the families Erysipelotrichaceae, Lachnospiraceae, and Porphyromonadaceae; the genera *Prevotella* and *Anaeroplasma;* and the species *Akkermansia muciniphila* (Brown and Hazen 2018).

The *CutC/CutD* genes can be found in several taxa belonging to Firmicutes, Actinobacteria, and Proteobacteria phyla (Schiatarella et al. 2019). Although a study has shown the TMA-producing potential in human fecal samples through a catalog of TMA-producing microorganisms, further studies are required to identify the key microorganisms and the therapeutic potential of selective inhibition of TMA producers (Rath et al. 2017, Schiatarella et al. 2019).

Several studies have supported the relationship between TMAO and CVD risk (Battson et al. 2018). The first study that showed this link was conducted by Wang et al. (2011). Using a metabolomics approach, the authors identified three plasma metabolites of phosphatidylcholine (choline, TMAO, and betaine) which allowed predicting CVD risk in an independent large clinical cohort. Additionally, this work displayed that the dietary supplementation of mice with choline and TMAO led to the upregulation of multiple macrophage scavenger receptors related to atherosclerosis and promoted this pathology. This study also supported the role of dietary choline and intestinal microbiota in TMAO production, increased macrophage cholesterol accumulation, and foam cell formation. It is noteworthy that the suppression of intestinal microbiota through the use of broad-spectrum antibiotics in atherosclerosis prone-mice inhibited dietary choline induced-atherosclerosis, showing the importance of the gut microbiota in this process (Wang et al. 2011).

In an interesting study conducted by Gregory et al. (2015), the authors showed that the transplantation of cecal contents from mice with differing atherosclerosis potential and TMAO production capacity into recipient mice on choline diet and with intestinal microbiota suppressed by an oral poorly absorbed antibiotic mixture may lead to altered TMA and TMAO production capacity and atherosclerosis disease susceptibility in recipients, suggesting that atherosclerosis susceptibility is transmissible and influenced by the gut microbiota. In a large cohort study, Wang et al. (2014) assessed the relationship between fasting plasma choline and betaine levels and the risk of major adverse cardiovascular events (MACE: myocardial infarction, stroke, and death) over 3 years of follow-up in 3903 subjects who underwent elective diagnostic coronary angiography. The authors verified that high choline and betaine levels were associated with higher risk of MACE only with concomitant increased TMAO level, in other words, in individuals producing high levels of TMAO (TMAO producers). A meta-analysis of 19 prospective studies showed that elevated levels of

TMAO and its precursors were associated with increased risks of MACE, regardless of conventional risk factors (Helanza et al. 2017). Nevertheless, other studies have not shown a clear relationship between dietary intake of TMAO source nutrients and CVD risk (Brown and Hazen 2018). Conflicting results of studies might be explained by differences in the characteristics and disease status of subjects (Helanza et al. 2017). In this sense, research is needed to determine the correlation between TMAO plasma levels and increased cardiovascular risk (Al-Rubaye et al. 2019).

In general, some mechanisms of action have been proposed to explain the pro-atherosclerotic role of TMAO: (1) Inhibition of reverse cholesterol transport through reducing cholesterol 7-hydroxylase, a rate-limiting enzyme for cholesterol catabolism and bile acid synthesis; (2) Upregulation of macrophage scavenger receptors SRA and CD36 expression, leading to deposition of cholesterol in macrophages and development of foam cells; (3) Induction of endothelial dysfunction through activation of nuclear factor-κBNF-κB, protein kinase C, and nucleotide-binding oligomerization domain-like receptor family pyrin domain-containing 3 inflammasome, activating monocyte adhesion and increasing expression of vascular endothelial inflammation factors, such as vascular cell adhesion molecule-1; (4) Increasing of platelet responsiveness to several agonists including ADP, thrombin, and collagen, through increased Ca^{2+} release (Peng et al. 2018).

Short chain fatty acids (SCFAs). They are carboxylic acids produced by the intestinal microbiota from fermentation of dietary fibers (DF) which escape the host's enzymatic digestive system in the small intestine (Richards et al. 2016). The major SCFAs generated in the intestinal lumen are acetate (C2), propionate (C3), and butyrate (C4) (Battson et al. 2018). DF fermentation helps in the energy delivery for microbial growth and preservation of bacteria, and facilitates the excretion of oxidized fatty acids (Richards et al. 2016). In addition, these oxidized fatty acids may be absorbed by enterocytes and oxidized again to provide energy (Richards et al. 2016).

In general, SCFAs are found in the intestines at concentrations between 10 and 100 mM and present several physiological effects, including stimulation of ileal motility and mucus production, upregulation of the expression of tight junction proteins which helps in epithelial health, and as a main energy source for colonic epithelial cells (Battson et al. 2018).

Although SCFAs are produced and metabolized in the gastrointestinal tract, a small portion is absorbed and may enter the bloodstream via the hepatic portal vein (Peng et al. 2018). In the systemic circulation, SCFAs can modulate the cardiovascular functions (Battson et al. 2018). Overall, the effects of SCFAs on healthy cardiovascular tissue and CVD are mainly due to modulation of the metabolism of lipids and glucose (Richards et al. 2016). It is important to highlight that hyperglycemia, insulin resistance, and dyslipidemia may lead to endothelial dysfunction and low-grade inflammation in this tissue (Richards et al. 2016). SCFAs can bind to G protein-coupled receptors (GPCRs) including GPR41 (renamed to free fatty acid receptor – FFAR3) and GPR43 (FFAR2), and modulate the energy expenditure (Miele et al. 2015). Studies have shown that the decrease in blood glucose concentration can be mediated by SCFA receptors (FFAR2 and FFAR3). The activation of these receptors may lead to release of incretins glucagon-like peptide (GLP-1) and peptide YY (PYY) by L cells that are part of the enteroendocrine system in the intestinal epithelium (Richards et al. 2016). It is believed that GLP-1 is able to stimulate glucose-dependent insulin secretion from β-cells and to inhibit the glucagon secretion by α-cells in the pancreas (Richards et al. 2016). The release of PYY is related to the regulation of appetite and the decrease of gut motility (Richards et al. 2016).

Moreover, studies using experimental models suggest that SCFAs can regulate blood pressure and endothelial function (Battson et al. 2018). In general, the receptors GRP41 and Olfr78 expressed in the kidneys have been shown to modulate renin secretion in response to SCFAs from intestinal microbiota (Schiatarella et al. 2018). Studies have shown that GRP41 knockout mice are hypertensive, while Olfr78 knockout mice are hypotensive, suggesting that these pathways are significant regarding the effect of SCFAs on blood pressure control (Tang et al. 2019). Additionally, GRP41 and Olfr78 are expressed in smooth cells of small resistance vessels, where they differently mediate vascular tone (Tang et al. 2019).

Wang et al. (2017), investigating the role of sodium butyrate (NaBu) in angiotensin II (Ang II)-induced hypertension in an experimental model with Sprague-Dawley rats, verified that intramedullary infusion of NaBu exerts an antihypertensive action, probably by suppressing the (pro)renin receptor mediated intrarenal renin-angiotensin system. A clinical trial conducted by Roshanravan et al. (2017) in adults with type 2 diabetes mellitus showed that the supplementation with sodium butyrate capsules, inulin powder and inulin + butyrate capsules during 45 days significantly reduced diastolic blood pressure compared to the placebo group.

Although there are several studies linking microorganism-derived SCFAs to CVD risk factors, there is scarce literature data showing the direct effects of SCFAs on atherogenesis. In this line, Aguilar et al. (2014) verified that ApoE knockout mice fed on 1% butyrate-supplemented diet for 10 weeks showed slow progression of atherosclerosis by reducing adhesion and migration of macrophages and increasing plaque stability. The same research group found, in another study with similar experimental conditions, that one mechanism that could explain the influence of butyrate supplementation on atherosclerotic development is the decrease of oxidative stress in the lesion area, attenuating endothelium dysfunction and macrophage migration and activation in the atherosclerotic lesion (Aguilar et al. 2016).

In fact, several studies using animal models have pointed to the link between SCFAs and CVD risk factors; however, few clinical trials have supported this link, suggesting that research needs to be deepened in this field (Brown and Hazen 2018).

Bile acids (BAs). They are amphipathic molecules involved in the solubilization, absorption, and metabolism of lipid and fat-soluble vitamins (Fiorucci and Distrutti 2015, Battson et al. 2018). BAs are synthesized from cholesterol in the liver and conjugated by the amino acids glycine or taurine (Jones et al. 2014, Fiorucci and Distrutti 2015). The main human BAs are the primary BAs cholic acid (CA) and chenodeoxycholic acid (CDCA) and their respective secondary BAs deoxycholic acid (DCA) and lithocholic acid (LCA) (Jones et al. 2014).

It is believed that BAs might have a strong selective pressure on the intestinal microbiota (Islam et al. 2011). A study conducted by Islam et al. (2011) showed that the administration of cholic acid may regulate the composition of the intestinal microbiota in a rat model, leading to similar alterations to those induced by high-fat diets (with predominance of phylum Firmicutes in relation to phylum Bacteroidetes).

Microorganisms resistant to bile salts in the gut may lead to chemical diverse biologically active metabolites detected in the host. Examples of this biotransformation are the enzymatic hydrolysis of glycine-conjugated or taurine conjugated bile salts into free acids by bile salt hydrolase (BSH) and 7-dehydroxylase and 7-dehydroxylase activity producing secondary bile acids (deoxycholic, lithocholic, and ursodeoxycholic) (Brown and Hazen 2018). Besides deconjugation, 7α-dehydroxylase and 7β-dehydroxylase activity, reactions of oxidation, epimerization, esterification, and desulfation are involved in the transformation of the BA pool by intestinal microorganisms (Brown and Hazen 2018). The generated secondary bile acids are absorbed, enter the portal circulation, and can also function as endocrine-like signaling molecules with strong influence on the host physiology and disease (Brown and Hazen 2018). As the capacity to convert primary BAs to secondary BAs is present in only a limited number of bacterial species, any distortion in these species may lead to physiological consequences to the host (Long et al. 2017). In this sense, antibiotic use and dysbiosis of the intestinal microbiota have been associated with changes in BA metabolism (Jones et al. 2014).

The intestinal microbial metabolism of bile acids is linked to an important metabolic regulatory pathway of the host which may present a potential contribution to CVD (Hylemon et al. 2009, Brown and Hazen 2018). Thus, besides the involvement of BAs in intestinal fat absorption and their influence on the gut microbiota composition, they can act as ligands for nuclear receptors with strong effect on host metabolic regulation related to lipid, glucose, drug, and energy metabolism (Hylemon et al. 2009, Sayin et al. 2013, Schiatarella et al. 2018). In the heart and the systemic circulation, BAs interact with plasma membrane G-protein-coupled receptors (TGR5 and muscarinic receptors) and nuclear receptors (farnesoid X receptor [FXR] and pregnane xenobiotic receptor [PXR]) (Khurana et al. 2011). For example, CDCA and CA (DCA and LCA to a lesser extent) are the main FXR agonists and the secondary bile acids LCA and DCA are the most potent TGR5 agonists (Molinaro et al. 2018). It is noteworthy that intestinal microbiota can regulate FXR and TGR5 via conversion of primary BA into secondary BA, as well as by regulating BA synthesis (Sayin et al. 2013, Battson et al. 2018).

The interaction of BAs with FXR and PXR helps in regulating the cardiovascular function and may change vascular tone (Jones et al. 2014). Studies using experimental models have shown that FXR activation mitigates atherosclerosis and TGR5 activation may lead to atheroprotective effects (Battson et al. 2018). In general, scientific evidence, based mainly on animal studies, suggests that BAs cause cardioprotective effects. Additionally, the capacity of the intestinal microbiome to change the BA pool and receptor signaling might be a mechanism by which the intestinal microbiome would influence cardiovascular function (Battson et al. 2018). Nevertheless, further studies, particularly clinical trials, are necessary to elucidate the effects and the action mechanisms of BAs on CVD in humans (Battson et al. 2018).

Probiotics as Potential Modulators of Cardiovascular Disease

Several studies have explored the effects of probiotic consumption on the control of cardiovascular risk factors and the possibility of manipulating the composition and metabolism of the intestinal microbiota through the use of probiotics is an exciting aspect to be considered (Org et al. 2015). In this regard, the modulation of the intestinal microbiota composition through the ingestion of probiotic microorganisms to cause a decrease in TMA production and TMAO levels in the host has attracted the attention of the scientific community (Tang and Hazen 2015).

Clinical trials have shown the beneficial effects of probiotic consumption on blood pressure (BP) control (Khalesi et al. 2014, Hendijani and Akbari 2018). A meta-analysis of 9 randomized controlled trials showed that probiotic consumption ($\geq 10^{11}$ CFU daily) for at least 8 weeks may improve BP, with a greater decrease verified when multiple probiotic species were used (Khalesi et al. 2014).

Studies suggested that potential effects of probiotic strains on regulation of high pressure, lipid metabolism, diabetes, and immunomodulation may be due to the effectiveness of the bioactive peptides produced during fermentation (Ahtesh et al. 2018). Scientific evidence has shown that bioactive peptides released during a fermentative process by probiotic strains, including valyl-prolyl-proline (Val-Pro-Pro), and isoleucyl-prolyl-proline (Ile-Pro-Pro), may exhibit anti-hypertensive activity by inhibiting angiotensin-converting enzyme (ACE) (Singh et al. 2014, Miremadi et al. 2016, Ahtesh et al. 2018). ACE inhibitory peptides can be found in a variety of fermented milk products such as yoghurt, fermented milk beverages, and cheese (Miremadi et al. 2016). On the other hand, soybean fermented products with probiotics microorganisms have also been considered as a promising source of antihypertensive peptides (Singh et al. 2014). The ACE inhibitory peptides are produced by proteolytic degradation of soybean protein fractions (glycinin and β-conglycinin) (Singh et al. 2014).

Nevertheless, Xu et al. (2015) showed that the daily consumption of *Lactobacillus* (*L.*) *plantarum* DSM 15313 or bilberries fermented by the same bacterial strain during 3 months did not reduce the blood pressure in adults with hypertension. The authors also reported that the diversity and composition of the oral and fecal microbiota were not affected by these probiotic products. Ivey et al. (2015) also did not find any effect of *L. acidophilus* La-5 and *Bifidobacterium* (*B.*) *animalis* Bb-12, provided in either yoghurt or a capsule form, on blood pressure and serum lipid profile in overweight individuals. The researchers stressed that the lack of effects may be related to the strain specificity of probiotic actions and due to the relatively good baseline cholesterol levels of subjects.

A meta-analysis of 11 randomized, controlled clinical trials comparing probiotic supplementation with placebo or no treatment (control) showed that probiotic interventions (including fermented milk products and probiotics) may decrease total and low-density lipoprotein (LDL) cholesterol with higher effect verified in mildly hypercholesterolemic than in normocholesterolemic individuals (Shimizu et al. 2015). In a study with normocholesterolemic individuals, Bedani et al. (2015) found a significant decrease in LDL cholesterol and LDL cholesterol to HDL (high-density lipoprotein) cholesterol ratio in men who consumed a synbiotic soy product fermented with *L. acidophilus* La-5, *B. animalis* Bb-12, and *Streptococcus thermophilus* (starter culture) during 8 weeks. However, the changes (from baseline) of these biomarkers were not significantly different compared to the placebo group, suggesting a limited lipid-lowering effect of this synbiotic product in normocholesterolemic individuals.

A systematic review conducted by Hendijani and Akbari (2018) found that probiotic supplementation was successful in significantly reducing total cholesterol, LDL cholesterol, triglycerides, systolic blood pressure and diastolic blood pressure in type 2 diabetic patients, suggesting that probiotics might control dyslipidemia and hypertension in these individuals. However, the authors emphasized that more clinical trials with a large sample size and long follow-up time would be required to develop clinical practice guidelines for management of cardiovascular risk factors in type 2 diabetic subjects.

On the other hand, Xavier-Santos et al. (2018) reported that daily consumption of either a synbiotic mousse containing *L. acidophilus* La-5 and the prebiotics inulin and fructooligosaccharides or the placebo product led to the reduction of total cholesterol and HDL cholesterol in volunteers with metabolic syndrome (MetS), suggesting that the presence of probiotic microorganisms and prebiotic ingredients in the diet did not influence significantly the lipid profile of subjects with MetS.

Studies *in vitro* and *in vivo* have suggested that several lactic acid bacteria (LAB) strains, particularly *Lactobacillus* spp., have cholesterol-lowering effects (Ooi and Liong 2010, Choi and Chang 2015, Ishimwe et al. 2015). In this line, Choi and Chang (2015) showed that *L. plantarum* EM, isolated from kimchi, presented high cholesterol removal by growing, resting and dead cells. According to the authors, these results might be attributed to enzymatic assimilation including BSH activity and cell surface binding. Cholesterol bound to probiotic strains might represent lower intestinal absorption of cholesterol and reduction of serum cholesterol levels in humans (Ooi and Liong 2010). Nevertheless, other studies did not verify any significant effects regarding cholesterol decrease with probiotic consumption (de Roos et al. 1999, Simons et al. 2006, Ivey et al. 2015).

The cholesterol-lowering properties of probiotic microorganisms are strain- and species-specific (Ettinger et al. 2014). Although the precise pathways of cholesterol reduction by probiotic strains are still unclear, several mechanisms of probiotic effects have been proposed (most of them are based on *in vitro* studies), including cholesterol assimilation (Pereira and Gibson 2002), cholesterol binding to probiotics cell walls (Liong and Shah 2005), cholesterol incorporation into cellular membranes during probiotic growth (Lye et al. 2010a), enzymatic deconjugation of bile acids by probiotics bile salt hydrolase (BSH) (Begley et al. 2006), cholesterol co-precipitation with deconjugated bile (Liong and Shah 2006), conversion of cholesterol into coprostanol (Lye et al. 2010b), and SCFAs' production after fermentation of prebiotic fibres by probiotic microorganisms (De Preter et al. 2007).

It is noteworthy that several studies in animals and humans have not shown any probiotic beneficial effects on risk factors for CVD (Battson et al. 2018). However, the discrepancies between the intestinal microbiota of different human populations could explain why a generalized therapeutic approach is not effective for all humans (Singh et al. 2016). According to Battson et al. (2018), the probiotic efficacy would be greater if the probiotic was customized to specific bacterial deficiencies or imbalances of potential patients. In other words, a personalized approach that takes into account individual intestinal microbiota disposition (Singh et al. 2016). For example, there are consistent findings regarding reduced abundance of *Akkermansia* in individuals with metabolic disorder in both preclinical and clinical studies (Derrien et al. 2017). Everard et al. (2013) showed that the abundance of *Akkermansia* (*A.*) *muciniphila*, a mucin-degrading bacterium belonging to the phylum Verrucomicrobia with positive effects on metabolism, is reduced in obese and type 2 diabetic mice. In this line, Li et al. (2016) also reported that mice genetically susceptible to atherosclerosis (apolipoprotein E-deficient—ApoE$^{-/-}$) showed decreased levels of *A. muciniphila*.

The authors demonstrated that the replenishment with *A. muciniphila* by daily oral gavage reversed Western diet-induced exacerbation of atherosclerotic lesion formation without affecting hypercholesterolemia in ApoE$^{-/-}$ mice. They suggested that *A. muciniphila* attenuates atherosclerotic lesions by ameliorating metabolic endotoxemia-induced inflammation through restoration of the intestinal barrier (Li et al. 2016). Dietary intervention studies suggested that increased *A. muciniphila* abundance in the gut could be obtained by the use of viable *A. muciniphila* and prebiotics (fructooligosaccharides) (Zhou 2017). Moreover, supplementation with *B. animalis* could also increase *A. muciniphila* abundance by producing SCFAs and facilitating mucin growth to feed the microorganisms (Zhou 2017). Despite interesting results of this emerging probiotic species, larger studies in patients with atherosclerosis are required to prove any potential beneficial effects of *A. muciniphila* on cardiovascular risk factors (Jonsson and Backhed 2016, Zhou 2017).

Other probiotics such as VSL#3 (Mencarelli et al. 2012, Chan et al. 2016a), *L. rhamnosus* GG (Chan et al. 2016b), *L. acidophilus* ATCC 4358 (Huang et al. 2014) were able to attenuate atherosclerosis in ApoE$^{-/-}$ mice.

Even though evidence from an increasing number of animal and clinical studies has shown that probiotics may decrease several cardiovascular risk factors, the exact strains and the ideal probiotic dose that would be responsible for the potential effects on CVD are still uncertain. Therefore, further studies are necessary to prove the efficacy of probiotics on the cardiometabolic function as well as the probiotic mechanisms of action against CVD.

Conclusion

In recent years there has been a great deal of interest from the scientific community regarding the modulation of the human intestinal microbiota composition and the metabolic activity towards the prevention and/or treatment of CVD. However, whether CVD can be treated by targeting the gut microbiota is still not clear. In this sense, probiotics appear as a promising alternative to control cardiovascular risk factors through, for example, beneficial modulation of the intestinal microbiota and its metabolites. Therefore, future studies focusing on the identification of microbial metabolites involved in cardiovascular pathologies are extremely promising. In fact, more studies are required to understand the role of the gut microbiota (and its metabolites) and probiotics in cardiometabolic diseases, as well as to lead to novel therapeutic strategies to prevent and manage these illnesses.

References

Aguilar, E.C., A.J. Leonel, L.G. Teixeira, A.R. Silva, J.F. Silva, J.M. Pelaez, L.S. Capettini, V.S. Lemos, R.A. Santos and J.I. Alvarez-Leite. 2014. Butyrate impairs atherogenesis by reducing plaque inflammation and vulnerability and decreasing NFkappaB activation. Nutr. Metab. Cardiovasc. Dis. 24: 606–613.

Aguilar, E.C., LC. Santos, A.J. Leonel, J.S. de Oliveira, E.A. Santos, J.M. Navia-Pelaez, J.F. da Silva, B.P. Mendes, L.S. Capettini, L.G. Teixeira, V.S. Lemos and J.I. Alvarez-Leite. 2016. Oral butyrate reduces oxidative stress in atherosclerotic lesion sites by a mechanism involving NADPH oxidase down-regulation in endothelial cells. J. Nutr. Biochem. 34: 99–105.

Ahtesh, F.B., L. Stojanovska and V. Apostolopoulos. 2018. Anti-hypertensive peptides released from milk proteins by probiotics. Maturitas 115: 103–109.

Al-Rubaye, H., G. Perfetti and J.C. Kaski. 2019. The role of microbiota in cardiovascular risk: focus on trimethylamine oxide. Curr. Probl. Cardiol. 44: 182–196.

Ao, M., M. Miyauchi, T. Inubushi, M. Kitagawa, H. Furusho, T. Ando, N.F. Ayuningtyas, A. Nagasaki, K. Ishihara, H. Tahara, K. Kozai and T. Takat. 2014. Infection with *Porphyromonasgingivalis* exacerbates endothelial injury in obese mice. PLoS One 9: e110519.

Armingohar, Z., J. Jorgensen, A.K. Kristoffersen, E. Abesha-Brlay and I. Olsen. 2014. Bacteria and bacterial DNA in atherosclerotic plaque and aneurysmal wall biopsies from patients with and without periodontitis. J. Oral Microbiol. 6: 23408.

Backhed, F., C.M. Fraser, Y. Ringel, M.E. Sanders, R.B. Sartor, P.M. Sherman, J. Versalovic, V. Young and B.B. Finlay. 2012. Defining a healthy human gut microbiome: current concepts, future directions, and clinical applications. Cell Host Microbe 12: 611–22.

Bahekar, A.A., S. Singh, S. Saha, J. Molnar and R. Arora. 2007. The prevalence and incidence of coronary heart disease is significantly increased in periodontitis: a meta-analysis. Am. Heart J. 154: 830–837.

Battson, M.L., D.M. Lee, T.L. Weir and C.L. Gentile. 2018. The gut microbiota as a novel regulator of cardiovascular function and disease. J. Nutr. Biochem. 56: 1–15.

Bedani, R., E.A. Rossi, D.C.U. Cavallini, R.A. Pinto, R.C. Vendramini, E.M. Augusto, D.S.P. Abdalla and S.M.I. Saad. 2015. Influence of daily consumption of synbiotic soy-based product supplemented with okara soybean by-product on risk factors for cardiovascular diseases. Food Res. Int. 73: 142–148.

Begley, M., C. Hill and C.G. Gahan. 2006. Bile salt hydrolase activity in probiotics. Appl. Environ. Microbiol. 72: 1729–1738.

Boutagy, N.E., R.P. McMillan, M. Frisard and M.W. Hulver. 2016. Metabolic endotoxemia with obesity: is it real and is it relevant. Biochimie 124: 11–20.

Brownell, N., A. Khera, J.A. Lemos, C.R. Ayers and A. Rohatgi. 2016. Association between peptidoglycan recognition protein-1 and incident atherosclerotic cardiovascular disease events. J. Am. Coll. Cardiol. 67: 2310–2312.

Brown, J.M. and S.L. Hazen. 2018. Microbial modulation of cardiovascular disease. Nat. Rev. Microbiol. 16: 171–181.

Bui, F.Q., C.L. Almeida-da-Silva, B. Huynh, A. Trinh, J. Liu, J. Woodward, H. Asadi and D.M. Ojcius. 2019. Association between periodontal pathogens and systemic disease. Biomed. J. 42: 27–35.

Caesar, R., F. Fak and F. Bäckhed. 2010. Effects of gut microbiota on obesity and atherosclerosis via moculation of inflammation and lipid metabolism. J. Intern. Med. 268: 320–328.

Cani, P.D., J. Amar, M.A. Iglesias, M. Poggi, C. Knauf, D. Bastelica, A.M. Neyrinck, F. Fava, K.M. Tuohy, C. Chabo, A. Waget, E. Delmée, B. Cousin, T. Sulpice, B. Chamontin, J. Ferrières, J.F. Tanti, G.R. Gibson, L. Casteilla, N.M. Delzenne, M.C. Alessi and R. Burcelin. 2007. Metabolic endotoxemia initiates obesity and insulin resistance. Diabetes 56: 1761–1772.

Cani, P.D., R. Bibiloni, C. Knauf, A. Waget, A.M. Neyrinck, N.M. Delzenne and R. Burcelin. 2008. Changes in gut microbiota control metabolic endotoxemia induced inflammation in high-fat diet-induced obesity and diabetes in mice. Diabetes 57: 1470–1481.

Cani, P.D., S. Possemiers, T. van de Wiele, Y. Guiot, A. Everard, O. Rottier, L. Geurts, D. Naslain, A. Neyrinck, D.M. Lambert, G.G. Muccioli and N.M. Delzenne. 2009. Changes in gut microbiota control inflammation in obese mice through a mechanism involving GLP-2-driven improvement of gut permeability. Gut 58: 1091–1103.

Chan, Y.K., H. El-Nezami, Y. Chen, K. Kinnunen and P.V. Kirjavainen. 2016a. Probiotic mixture VSL#3 reduce high fat diet induced vascular inflammation and atherosclerosis in ApoE(–/–) mice. AMB Express 6: 61.

Chan, Y.K., M.S. Brar, P.V. Kirjavainen, Y. Chen, J. Peng, D. Li, F.C. Leung and H. El-Nezami. 2016b. High fat diet induced atherosclerosis is accompanied with low colonic bacterial diversity and altered abundances that correlates with plaque size, plasma A-FABP and cholesterol: a pilot study of high fat diet and its intervention with *Lactobacillus rhamnosus* GG (LGG) or telmisartan in ApoE–/– mice. BMC Microbiol. 16: 264.

Choi, E.A. and H.C. Chang. 2015. Cholesterol-lowering effects of a putative probiotic strain *Lactobacillus plantarum* EM isolated from kimchi. LWT - Food Sci. Technol. 62: 210–217.

De Preter, V., T. Vanhoutte, G. Huys, J. Swing, L. de Vuyst, P. Rutgeerts and K. Verbeke. 2007. Effects of *Lactobacillus casei* Shirota, *Bifidobacterium breve*, and oligofructose enriched inulin on colonic nitrogen-protein metabolism in healthy humans. Am. J. Physiol. Gastrointest. Liver Physiol. 292: 358–368.

De Roos, N.M., G. Schouten and M.B. Katan. 1999. Yogurt enriched with *Lactobacillus acidophilus* does not lower blood lipids in healthy men and women with normal to borderline high serum cholesterol levels. Eur. J. Clin. Nutr. 53: 277–280.

Derrien, M., C. Belzer and W.M. Vos. 2017. *Akkermansia muciniphila* and its role in regulating host functions. Microb. Pathog. 106: 171–181.

Drosos, I., A. Tavridou and G. Kolios. 2015. New aspects on the metabolic role of intestinal microbiota in the development of atherosclerosis. Metabolism 64: 476–481.

Ettinger, G., K. MacDonald, G. Reid and J.P. Burton. 2014. The influence of the human microbiome and probiotics on cardiovascular health. Gut Microbes 5: 719–728.

Everard, A., C. Belzer, L. Geurts, J.P. Ouwerkerk, C. Druart, L.B. Bindels, Y. Guiot, M. Derrien, G.G. Muccioli, N.M. Delzenne, W.M. de Vos and P.D. Cani. 2013. Cross-talk between *Akkermansia muciniphila* and intestinal epithelium controls diet-induced obesity. Proc. Natl. Acad. Sci. USA 110: 9066–9071.

Fiorucci, S. and E. Distrutti. 2015. Bile acid-activated receptors, intestinal microbiota, and the treatment of metabolic disorders. Trends Mol. Med. 21: 702–714.

Frankenfeld, C.L. 2017. Cardiometabolic risk and gut microbial phytoestrogen metabolite phenotypes. Mol. Nutr. Food Res. 61: 1.

Heianza, Y., W. Ma, J.E. Manson, K.M. Rexrode and L. Qi. 2017. Gut microbiota metabolites and risk of major adverse cardiovascular disease events and death: a systematic review and meta-analysis of prospective studies. J. Am. Heart Assoc. 6: e004947.

Hendijani, F. and V. Akbari. 2018. Probiotic supplementation for management of cardiovascular risk factors in adults with type II diabetes: a systematic review and meta-analysis. Clin. Nutr. 37: 532–541.

Hylemon, P.B., H. Zhou, W.M. Pandak, S. Ren, G. Gil and P. Dent. 2009. Bile acids as regulatory molecules. J. Lipid Res. 50: 1509–1520.

Ishimwe, N., E.B. Daliri, B.H. Lee, F. Fang and G. Du. 2015. The perspective on cholesterol-lowering mechanisms of probiotics. Mol. Nutr. Food Res. 59: 94–105.

Ivey, K.L., J.M. Hodgson, D.A. Kerr, P.L. Thompson, B. Stojceski and R.L. Prince. 2015. The effect of yoghurt and its probiotics on blood pressure and serumlipid profile; a randomised controlled trial. Nutr. Metab. Cardiovasc. Dis. 25: 46–51.

Gregory, J.C., J.A. Buffa, E. Org, Z. Wang, B.S. Levison, W. Zhu, M.A. Wagner, B.J. Bennett, L. Li, J.A. DiDonato, A.J. Lusis and S.L. Hazen. 2015. Transmission of atherosclerosis susceptibility with gut microbial transplantation. J. Biol. Chem. 290: 5647–5660.

Huang, Y., J. Wang, G. Quan, X. Wang, L. Yang and L. Zhong. 2014. *Lactobacillus acidophilus* ATCC 4356 prevents atherosclerosis via inhibition of intestinal cholesterol absorption in apolipoprotein E-knockout mice. Appl. Environ. Microbiol. 80: 7496–504.

Jones, M.L., C.J. Martoni, J.G. Ganopolsky, A. Labbe and S. Prakash. 2014. The human microbiome and bile acid metabolism: dysbiosis, dysmetabolism, disease and intervention. Expert Opin. Biol. Ther. 14: 467–482.

Jonsson, A.L. and F. Bäckhed. 2017. Role of gut microbiota in atherosclerosis. Nat. Rev. Cardiol. 14: 79–87.

Kasselman, L.J., N.A. Vernice, J. DeLeon and A.B. Reiss. 2018. The gut microbiome and elevated cardiovascular risk in obesity and autoimmunity. Atherosclerosis 271: 203–213.

Khalesi, S., J. Sun, N. Buys and R. Jayasinghe. 2014. Effect of probiotics on blood pressure: a systematic review and meta-analysis of randomized, controlled trials. Hypertension 64: 897–903.

Kholy, K.E., R.J. Genco and T.E.V. Dyke. 2015. Oral infections and cardiovascular disease. Trends Endocrinol. Metab. 26: 315–321.

Khurana, S., J.P. Raufman and T.L. Pallone. 2011. Bile acids regulate cardiovascular function. Clin. Transl. Sci. 4: 210–218.

Koren, O., A. Spor, J. Felin, F. Fak, J. Stombaugh, V. Tremaroli, C.J. Behre, R. Knight, B. Fagerberg, R.E. Ley and F. Bäckhed. 2011. Human oral, gut, and plaque microbiota in patients with atherosclerosis. Proc. Natl Acad. Sci. USA 108: 4592–4598.

Kuczynski, J., C.L. Lauber, W.A. Walters, L.W. Parfrey, J.C. Clemente, D. Gevers and R. Knight. 2011. Experimental and analytical tools for studying the human microbiome. Nat. Rev. Genet. 13: 47–58.

Laman, J.D., A.H. Schoneveld, F.L. Moll, M. van Meurs and G. Pasterkamp. 2002. Significance of peptidoglycan, a proinflammatory bacterial antigen in atherosclerotic arteries and its association with vulnerable plaques. Am. J. Cardiol. 90: 119–123.

Lee, S.A., S.M. Kim, Y.H. Son, C.W. Lee, S.W. Chung, S.K. Eo, B.Y. Rhim and K. Kim. 2011. Peptidoglycan enhances secretion of monocyte chemoattractans via multiple signaling pathways. Biochem. Biophys. Res. Commun. 408: 132–138.

Li, J., S. Lin, P.M. Vanhoutte, C.W. Woo and A. Xu. 2016. *Akkermansia muciniphila* protects against atherosclerosis by preventing metabolic endotoxemia-induced inflammation in Apoe–/– mice. Circulation 133: 2434–2446.

Liong, M.T. and N.P. Shah. 2005. Acid and bile tolerance and cholesterol removal ability of lactobacilli strains. J. Dairy Sci. 88: 55–66.

Long, S.L., C.G.M. Gahan and S.A. Joyce. 2017. Interactions between gut bacteria and bile in health and disease. Mol. Aspects Med. 56: 54–65.

Lye, H.S., G. Rusul and M.T. Liong. 2010a. Mechanisms of cholesterol removal by lactobacilli under conditions that mimic the human gastrointestinal tract. Int. Dairy J. 20: 169–175.

Lye, H.S., G. Rusul and M.T. Liong. 2010b. Removal of cholesterol by lactobacilli via incorporation and conversion to coprostanol. J. Dairy Sci. 93: 1383–1392.

Marchesi, J.R., D.H. Adams, F. Fava, G.D.A. Hermes, G.M. Hirschfield, G. Hold, M.N. Quraishi, J. Kinross, H. Smidt, K.M. Tuohy, L.V. Thomas, E.G. Zoetendal and A. Hart. 2016. The gut microbiota and host health: a new clinical frontier. Gut 65: 330–339.

Mencarelli, A., S. Cipriani, B. Renga, A. Bruno, C. D'Amore, E. Distrutti and S. Fiorucci. 2012. VSL#3 resets insulin signaling and protects against NASH and atherosclerosis in a model of genetic dyslipidemia and intestinal inflammation. PloS One 7: e45425.

Miele, L., V. Giorgio, M.A. Alberelli, E. Candia, A. Gasbarrini and A. Grieco. 2015. Impact of gut microbiota on obesity, diabetes, and cardiovascular disease risk. Curr. Cardiol. Rep. 17: 120.

Miremadi, F., F. Sherkat and L. Stojanovska. 2016. Hypocholesterolaemic effect and anti-hypertensive properties of probiotics and prebiotics: a review. J. Funct. Foods 25: 497–510.

Molinaro, A., A. Wahlström and H.U. Marschall. 2018. Role of bile acids in metabolic control. Trends Endocrinol. Metab. 29: 31–41.

Ooi, L.G. and M.T. Liong. 2010. Cholesterol-lowering effects of probiotics and prebiotics: a review of *in vivo* and *in vitro* findings. Int. J. Mol. Sci. 11: 2499–2522.

Org, E., M. Mehrabian and A. Lusis. 2015. Unraveling the environmental and genetic interactions in atherosclerosis: central role of the gut microbiota. Atherosclerosis 241: 387–399.

Peng, J., X. Xiao, M. Hu and X. Zhang. 2018. Interaction between gut microbiome and cardiovascular disease. Life Sci. 214: 153–157.

Pereira, D.I.A. and G.R. Gibson. 2002. Cholesterol assimilation by lactic acid bacteria and bifidobacteria isolated from the human gut. Appl. Environ. Microbiol. 68: 4689–4693.

Petersen, C. and J.L. Round. 2014. Defining dysbiosis and its influence on host immunity and disease. Cell. Microbiol. 16: 1024–1033.

Rath, S., B. Heidrich, D.H. Pieper and M. Vital. 2017. Uncovering the trimethylamine-producing bacteria of the human gut microbiota. Microbiome 5: 54.

Richards, L.B., M. Li, B.C.A.M. van Esch, J. Garssen and G. Folkerts. 2016. The effects of short-chain fatty acids on the cardiovascular system. PharmaNutrition 4: 68–111.

Rohatgi, A., C.R. Ayers, A. Khera, D.K. McGuire, S.R. Das, S. Matulevicius, C.H. Timaran, E.B. Rosero and J.A. Lemos. 2009. The association between peptidoglycan recognition protein-1 and coronary and peripheral atherosclerosis: observations from the Dallas Heart Study. Atherosclerosis 203: 569–575.

Romano, K.A., E.I. Vivas, D. Amador-Noguez and F.E. Rey. 2015. Intestinal microbiota composition modulates choline bioavailability from diet and accumulation of the proatherogenic metabolite trimethylamine-N-oxide. MBio 6: e02481.

Roshanravan, N., R. Mahdavi, E. Alizadeh, M.A. Jafarabadi, M. Hedayati, A. Ghavami, S. Alipour, N.M. Alamdari, M. Barati and A. Ostadrahimi. 2017. Effect of butyrate and inulin supplementation on glycemic status, lipid profile and glucagon-like peptide 1 level in patients with type 2 diabetes: a randomized double-blind placebo-controlled trial. Horm. Metab. Res. 49: 886–891.

Sanjukta, S. and A.K. Rai. 2016. Production of bioactive peptides during soybean fermentation and their potential health benefits. Trends Food Sci. Technol. 50: 1–10.

Sayin, S.I., A. Wahlström, J. Felin, S. Jäntti, H.U. Marschall, K. Bamberg, B. Angelin, T. Hyötyläinen, M. Oresic and F. Bäckhed. 2013. Gut microbiota regulates bile acid metabolism by reducing the levels of tauro-beta-muricholic acid, a naturally occurring FXR antagonist. Cell Metab. 17: 225–235.

Schiattarella, G.G., A. Sannino, G. Esposito and C. Perrino. 2019. Diagnostics and therapeutic implications of gut microbiota alterations in cardiometabolic diseases. Trends Cardiovasc. Med. 29: 141–147.

Sharma, A., E.K. Novak, H.T. Sojar, R.T. Swank, H.K. Kuramitsu and R.J. Genco. 2000. *Porphyromonas gingivalis* platelet aggregation activity: outer membrane vesicles are potent activators of murine platelets. Oral Microbiol. Immunol. 15: 393–396.

Shimizu, M., M. Hashiguchi, T. Shiga, H.O. Tamura and M. Mochizuki. 2015. Supplementation on lipid profiles in normal to mildly hypercholesterolemic individuals. Plos One 10: e0139795.

Simons, L.A., S.G. Armansec and P. Conway. 2006. Effect of *Lactobacillus fermentum* on serum lipids in subjects with elevated serum. Nutr. Metab. Cardiovasc. Dis. 16: 531–535.

Singh, B.P., S. Vij and S. Hati. 2014. Functional significance of bioactive peptides derived from soybean. Peptides 54: 171–179.

Singh, V., B.S. Yeoh and M. Vijay-Kumar. 2016. Gut microbiome as a novel cardiovascular therapeutic target. Curr. Opin. Pharmacol. 27: 8–12.

Tang, W.H.W. and S.L. Hazen. 2015. The contributory role of gut microbiota in cardiovascular disease. J. Clin. Invest. 124: 4204–4211.

Tang, W.H.W., F. Bäckhed, U. Landmesser and S. Hazen. 2019. Intestinal microbiota in cardiovascular health and disease. J. Am. Coll. Cardiol. 73: 2089–2105.

Tap, J., S. Mondot, F. Levenez, E. Pelletier, C. Caron, J.P. Furet, E. Ugarte, R. Muñoz-Tamayo, D.L. Paslier, R. Nalin, J. Dore and M. Leclerc. 2009. Towards the human intestinal microbiota phylogenetic core. Environ. Microbiol. 11: 2574–2584.

Thushara, R.M., S. Gangadaran, Z. Solati, H. Mohammed and M.H. Moghadasian. 2016. Cardiovascular benefits of probiotics: a review of experimental and clinical studies. Food Funct. 7: 632–642.

Turnbaugh, P.J., M. Hamady, T. Yatsunenko, B.L. Cantarel, A. Duncan, R.E. Ley, M.L Sogin, W.J. Jones, B.A. Roe, J.P. Affourtit, M. Egholm, B. Henrissat, A.C. Heath, R. Knight and J.L. Gordon. 2009. A core gut microbiome in obese and lean twins. Nature 457: 480–484.

Wang, H., W. Zhang, L. Zuo, J. Dong, W. Zhu, Y. Li, L. Gu, J. Gong, Q. Li, N. Li and J. Li. 2014. Intestinal dysbacteriosis contributes to decreased intestinal mucosal barrier function and increased bacterial translocation. Lett. Appl. Microbiol. 58: 384–392.

Wang, L., Q. Zhu, A. Lu, X. Liu, L. Zhang, C. Xu, X. Liu, H. Li and T. Yang. 2017. Sodium butyrate suppresses angiotensin II-induced hypertension by inhibition of renal (pro)renin receptor and intrarenal renin-angiotensin system. J. Hypertens. 35: 1899–1908.

Wang, Z., E. Klipfell, B.J. Bennett, R. Koeth, B.S. Levison, B. DuGar, A.E. Feldstein, E.B. Britt, X. Fu, Y.M. Chung, Y. Wu, P. Schauer, J.D. Smith, H. Allayee, W.H.W. Tang, J.A. DiDonato, A.J. Lusis and S.L. Hazen. 2011. Gut flora metabolism of phosphatidylcholine promotes cardiovascular disease. Nature 472: 57–63.

Wang, Z., W.H.W. Tang, J.A. Buffa, X. Fu, E.B. Britt, R.A. Koeth, S. Bruce, B.S. Levison, Y. Fan, Y. Wu and S.L. Hazen. 2014. Prognostic value of choline and betaine depends on intestinal microbiota-generated metabolite trimethylamine-*N*-oxide. Eur. Heart J. 35: 904–910.

WHO. Cardiovascular diseases (CVDs); Fact sheet, Geneva, Switzerland, May, 2017. Available at https://www.who.int/en/news-room/fact-sheets/detail/cardiovascular-diseases-(cvds). Accessed on August, 2019.

Xavier-Santos, D., E.D. Lima, A.N.C. Simão, R. Bedani and S.M.I. Saad. 2018. Effect of the consumption of a synbiotic diet mousse containing *Lactobacillus acidophilus* La-5 by individuals with metabolic syndrome: a randomized controlled trial. J. Funct. Foods 41: 55–61.

Xu, J., I.L. Ahrén, C. Olsson, B. Jeppsson, S. Ahrné and G. Molin. 2015. Oral and faecal microbiota in volunteers with hypertension in a double blind, randomized placebo controlled trial with probiotics and fermented bilberries. J. Funct. Foods 18: 275–288.

Zhou, K. 2017. Strategies to promote abundance of *Akkermansia muciniphila*, an emerging probiotics in the gut, evidence from dietary intervention studies. J. Funct. Foods 33: 194–201.

Zoetendal, E.G., M. Rajilic-Stojanovic and W.M. de Vos. 2008. High throughput diversity and functionality analysis of the gastrointestinal tract microbiota. Gut 57: 1605–1615.

13

Metabolites of Polyphenols Produced by Probiotic Microorganisms and Their Beneficial Effects on Human Health and Intestinal Microbiota

Carolina Beres,[1,*] *Ignacio Cabezudo*[2] and *Nurmahani M Maidin*[3]

Introduction

Polyphenols are secondary metabolites present in vegetables which can influence their nutritional and sensory quality. They are responsible for many antioxidant activities and beneficial effects (Middleton et al. 2000, Ky et al. 2014). As a consequence, much data has been collected on the polyphenol components in a variety of foods which related total antioxidant activities based on different extraction methods (Spigno et al. 2007, Gironi and Piemonte 2011). Numerous epidemiological investigations have shown that the consumption of diets rich in fruits and vegetables can diminish the risk of suffering chronic diseases such as cardiovascular diseases, specific cancers, or neurodegenerative illnesses (Rodríguez et al. 2009).

Structurally, phenolic compounds contain an aromatic ring, which has one or more hydroxyl substituents, and range from simple phenolic molecules to highly polymerized compounds (Yu and Ahmedna 2013).

[1] EMBRAPA, Avenida das Américas 29501, Rio de Janeiro, RJ, Brazil.
[2] Universidad Nacional de Rosario, Suipacha 531, Rosario, Santa Fe, S2002LRK, Argentina.
Email: nachocabe@hotmail.com
[3] Universiti MalaysiaTerengganu, Kuala Terengganu, Malaysia.
Email: Nurmahani@gmail.com
* Corresponding author: carolberes@gmail.com

Phenolic acids have a carboxylic acid functional group and are present in plants in free forms or linked to other compounds. They can be classified into two subgroups: hydroxybenzoic acids and hydroxycinnamic acids, according to their core structure (Figure 1a and 1b) (Sirotkin and Harrath 2014). Flavonoids are found in various foods; one of the most representative are catechins found in tea extracts and the cause of its antioxidant activity (Namal Senanayake 2013). Their structure consists of two aromatic rings linked through three carbon atoms, forming a bridge—usually in the form of a third heterocyclic ring (Figure 1c). Flavonoids include flavanones, isoflavones, and anthocyanins. This group of compounds are usually bound to different sugars as is shown in Figure 1d. Anthocyanins are responsible for providing colour to some vegetables and their use as a natural dye has increased due to their beneficial health properties (Khoo et al. 2017). Stilbenes are formed by two aromatic rings joined by an ethylene molecule, they are found in very few amounts in the human diet. Trans-resveratrol is one of the more ubiquitous compounds and has the highest biological activity (Figure 1e). It is related to the prevention of heart diseases and with beneficial effects to counteract inflammatory processes (Tang et al. 2014). Tannins are phenolics of relatively high molecular weight and can be classified into hydrolyzable and condensed. The first are derived from the family of gallic acid and can be found as polymers of glucose with ellagic, gallic acid or others (Middleton et al. 2000). The latter are polymers of the catechin, also called procyanidins (Figure 1f).

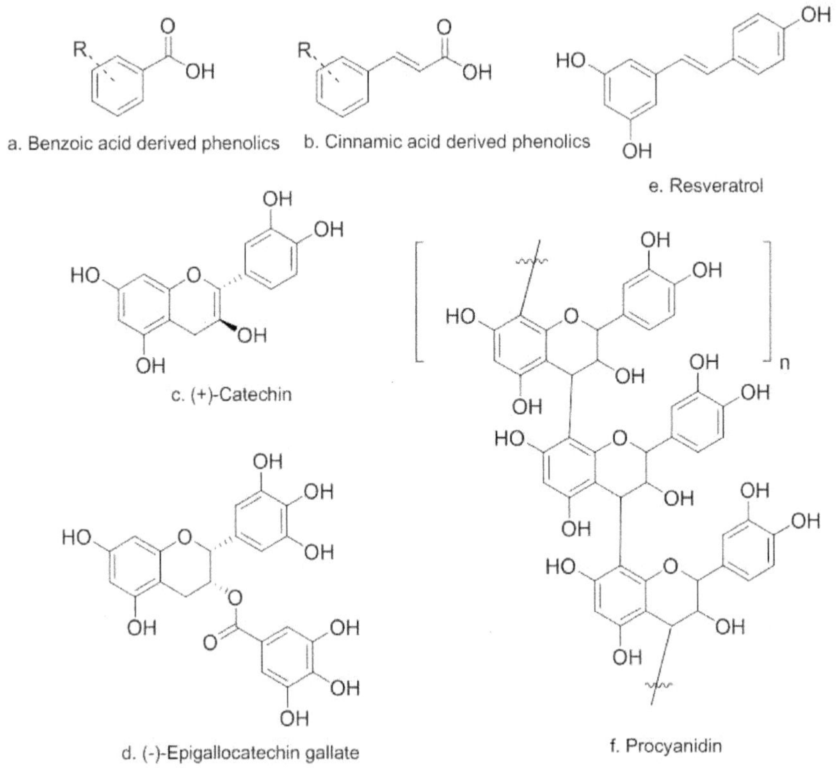

Figure 1: Structure of different phenolic compounds.

Given that many beneficial effects on human health are attributed to polyphenols, their study has become an increasingly important area of nutrition research, particularly their metabolism by the lactic acid bacteria present in the human gut or in dietary vegetables. In the colon, bacteria make up most of the flora where around 300–500 different species coexist. The prevailing bacteria are *Bacteroides*, which embody around 30% of the microorganisms in the gut, followed by *Clostridium*, *Faecalibacterium*, *Prevotella*, *Eubacterium*, *Ruminococcus*, *Peptococcus*, and *Bifidobacterium*. To lesser extent *Escherichia* and *Lactobacillus* are also present (Beaugerie and Petit 2004). The latter, lactic acid bacteria (LAB), are also a small part of the autochthonous microbiota of raw vegetables and constitute the most common source of probiotics (Souza et al. 2018).

Polyphenols undergo a complex metabolism and interact with bacterial enzymes, leading to the production of a large number of circulating and excreted polyphenol metabolites and catabolic products (Hervert-Hernandéz and Goñi 2011). These compounds can influence and induce a modulation of lactic acid bacteria composition by several interactions (Hervert-Hernandéz and Goñi 2011). Studies confirm that diet has a major influence on gut microbiota and is able to adjust their impact on health, with either beneficial or harmful consequences (Flint 2012). This chapter focuses on the mutual reciprocal relations between lactic acid bacteria and polyphenols.

Influence of Phenolics on Lactic Acid Bacteria Growth and Viability

Recent studies have shown that phenolic compounds may mostly modulate the composition of gut microbial communities through the inhibition of harmful flora and stimulation of beneficial flora. Phenolic compounds may exert a prebiotic function and increase the population of beneficial bacteria, including lactic acid bacteria (Souza et al. 2018). Phenolics exhibit an antimicrobial ability against various pathogens; nevertheless, the research of their effects on probiotic bacteria has been less extended.

The effect of several phenolics from diet on the growth of bacteria from human gut and their adhesion to enterocytes was studied. The effects on the growth of *Lactobacillus* (*L.*) *rhamnosus*, a probiotic, *Escherichia* (*E.*) *coli*, a commensal, and two pathogenic bacteria, *Staphylococcus* (*S.*) *aureus* and *Salmonella enterica* serovar Typhimurium, were determined. Naringenin and quercetin were the most effective compounds for the inhibition of all the four bacteria; the rest of the polyphenols had the most marked effect on the Gram-positive *S. aureus*. Also, the effects on adhesion of pathogenic and probiotic bacteria to Caco-2 cells were determined. Naringenin and phloridzin were the most effective inhibitors of *S. typhimurium* adherence to Caco-2 enterocytes while phloridzin and rutin increased the adherence of the probiotic *L. rhamnosus* (Parkar et al. 2008).

In order to ascertain whether resveratrol could exert anti-inflammatory activity *in vivo* at an attainable dietary dose, rats were induced colitis and were nourished with food containing resveratrol. The polyphenol increased the growth of lactobacilli and bifidobacteria, along with a reduction of enterobacterial growth upon treatment. Resveratrol better preserved the colonic mucosa and reduced body weight loss, among other beneficial properties (Larrosa et al. 2009).

Tea extracts have also been studied against *L. acidophilus* and other bacteria strains showing exceptional resistance to all extracts studied (white and green tea). *E. coli* was only inhibited very weakly for some of the extracts studied. In contrast, *Bacillus cereus* was very sensitive, followed by *Micrococcus luteus* and *Pseudomonas, Aeruginosa* (Almajano et al. 2008). Theobald et al. found that green tea could cause a growth stimulation of *Oenococcus oeni* as a result of the phenolic compounds present in green tea, especially epigallocatechin gallate. However, depending on its concentration, this compound could also inhibit *Oenococcus oeni* growth. They also described that individual catechins have a minor influence on the growth of oenococci (Theobald et al. 2008).

The effect of phenolic extracts from different dietary spices and herbs on different bacteria was evaluated. Extracts from Padang cassia, Chinese cassia, oregano, Japanese knotweed, pomegranate peel and clove were used on five common food-borne pathogenic bacteria, such as *S. aureus*, and five lactic acid bacteria, including *L. acidophilus*. From the results of minimum inhibitory concentration, Japanese knotweed, pomegranate peel, and clove extracts showed stronger inhibitory effects on *Salmonella flexneri* and *S. aureus*, about 313–2500 μg/mL, while the values of phenolic extracts on lactic acid bacteria were usually more than 2500 μg/mL. Overall, the phenolic extracts showed evident antibacterial ability against food-borne pathogenic bacteria, but not as much as lactic acid bacteria (Chan et al. 2018).

Fruit extracts from different tissues have exhibited antimicrobial and prebiotic activities attributed to their phenolic profile. The effect of five phenolic compounds (catechin, gallic, vanillic, ferulic, and protocatechuic acids) identified in different fruits, particularly in mango, was evaluated on the culture of two probiotic (*L. rhamnosus* and *L. acidophilus*) and two pathogenic (*E. coli* and S. typhimurium) bacteria. The minimal inhibitory and bactericidal concentrations of phenolic acids were relatively low against pathogens. Catechin also inhibited *E. coli* and *S. typhimurium*. On the other hand, in general, phenolics did not affect the viability of probiotics. Therefore, phenolic compounds could selectively inhibit the growth of pathogenic bacteria without affecting the viability of probiotics (Pacheco-Ordaz et al. 2018).

The fermentation of grapes to produce wine promotes the release or degradation of polyphenols making wine a beverage very rich in phenolics. Some studies focused on wine species such as *O. oeni*, involved in malolactic fermentation during winemaking. *O. oeni* growth was stimulated under wine conditions by 50 mg/L of phenolic compounds (Rozès et al. 2003). Moreover, the effects of flavonoids on growth and inactivation of *O. oeni* were studied, and quercetin and kaempferol exerted an inhibitory effect on *O. oeni*. Myricetin, catequin, and epicatequin did not affect considerably the growth of *O. oeni*. Condensed tannins were found to strongly affect *O. oeni* viability (Figueiredo et al. 2008).

Not only can polyphenols be found in dietary foods, but also in the residues of the agri-food industry including the production of processed vegetables, juices, and fermented beverages (Lai et al. 2017). In addition to the reports that use of isolated phenolic compounds, phenolic extracts also have effects on the growth of certain bacteria. After vinification, grape pomace retains a significant amount of antioxidant substances (González-Centeno et al. 2010). The influence of its polyphenols on *L. acidophilus* growth was studied using agar diffusion assays and cultivation in liquid media. None of the individual phenolic compounds exhibited an inhibitory effect on

Table 1: Some exemplary effects of dietary polyphenols on lactic acid bacteria.

Microorganism	Bioactive compound	Effect	Level	Reference
Lactobacillus	Resveratrol	Increased lactobacilli and bifidobacteria as well diminished the increase of enterobacteria upon induced colitis	*in vivo*	Larrosa et al. 2009
	Extracts from different dietary spices and medicinal herbs	No inhibition	*in vitro*	Chan et al. 2018
Lactobacillus acidophilus	Phenolic extracts from tea	No inhibition of *Lactobacillus acidophilus* - great inhibition of pathogen bacteria	*in vitro*	Almajano et al. 2008
	Catechin and gallic, vanillic, ferulic and protocatechuic acids	No inhibition	*in vitro*	Pacheco-Ordaz et al. 2018
	Extracts from grape pomace and grape seed, quercetin, gallic acid, catechin, caffeic acid, epicatechin, tannic acid	No inhibition - prebiotic effect	*in vitro*	Hervert-rnández et al. 2009
Lactobacillus rhamnosus	Caffeic acid, chlorogenic acid, o-coumaric acid, p-coumaric acid, catechin, epicatechin, phloridzin, rutin, daidzein, genistein, naringenin, quercetin	Polyphenols alter the total number of beneficial microflora in the gut, potentially conferring positive gut health benefits	*in vitro*	Parkar et al. 2008
	Catechin and gallic, vanillic, ferulic and protocatechuic acids	No inhibition	*in vitro*	Pacheco-Ordaz et al. 2018
L. rhamnosus LGG	By-product extracts from passion fruit, orange, acerola, and mango	Tropical fruit by-products can stimulate folate production by lactic acid bacteria	*in vitro*	Albuquerque et al. 2019
Oenococcus oeni	Green tea extracts, catechin, epigallocatechingallate	Growth stimulation	*in vitro*	Theobald et al. 2008
	Phenol carboxilic acids, malvidinglucoside, catechin	Growth stimulation under wine conditions	*in vitro*	Rozès et al. 2003
O. oeni and *L. hilgardii*	Phenolic aldehydes, flavonoids and condensed tannins	Phenolic aldehydes influenced the growth of *O. oeni. L. hilgardii* was negatively influenced by cinnamaldehydes. Kaempferol and quercetin inhibited the growth of *O. oeni* but only slightly to *L. hilgardii*. Some tannin fractions affected *O. oeni,* much more than *L. hilgardii*	*in vitro*	Figueiredo et al. 2008

L. acidophilus growth in agar. In addition, grape pomace phenolic extract (1 mg/mL) induced an important prebiotic effect increasing growth of *L. acidophilus* in liquid culture media (Hervert-Hernández et al. 2009).

Recently, the impact of tropical fruit by-product extracts (passion fruit, orange, acerola, and mango) on microbial growth, folate production, and adhesion ability of *S. thermophilus* TH-4 and *L. rhamnosus* LGG strains was studied. Mango extract presented the highest phenolic content and antioxidant activity. Orange and mango extracts showed the best anti-inflammatory potential by decreasing the highest nitric oxide levels when macrophages were stimulated with lipopolysaccharides from *S. typhimurium*. Folate production was only stimulated by mango. Passion fruit and orange increased the streptococci adhesion whereas acerola and mango extracts improved *L. rhamnosus* adhesion when strains were used individually (Albuquerque et al. 2019).

Digestion and Absorption of Bioactive Compounds

The nutritional value of a product is based on the nutrients offered by its matrix; however, part of these nutrients is not available for intestinal absorption. In this situation, chemical and enzymatic changes are required in order to increase their absorption. This scenario is called bioavailability which is the amount of nutrients that is absorbed and metabolized by enzymatic human digestion (Xie et al. 2013, Barba et al. 2015, Barba et al. 2017).

The *in vitro* and *in vivo* determination of bioavailability is important to avoid overestimated quantification of nutrients in food products. In addition, the amount of bioactive compounds that reaches the intestinal lumen is of vital importance since they play an important *in loco* role protecting the gastrointestinal tract from oxidative damage and cancer (Yang et al. 2001). The industrial appeal for functional properties may lead to deceive consumers' choices. Even with massively scientifically proved information about the beneficial health impact of bioactive compounds, the determination of bioavailability is essential to confirm their impact (Lima et al. 2014, Barba et al. 2015, Barba et al. 2017, Wu et al. 2017).

Different factors influence the bioavailability of bioactive compounds. Food matrix has a great influence on the release of phenolic compounds into the gastrointestinal tract, since these compounds can be chemically bound to food components producing stable compounds, in addition to the presence of suppressors or cofactors that reduce the digestion enzymatic process (Dueik and Bouchon 2016). A comparison between liquid juice and fiber products showed that the phenolic compound bioavailability in juices were higher due to the influence on pH caused by ascorbic acid content which promoted better phenolic extraction from the product matrix. Fiber-based products presented bioactive compounds linked to macromolecules and the formation of mineral complex which decrease their solubility (Bouyed et al. 2011). For example, when a juice was compared with a pure phenolic extract, the extract presented higher bioavailability due to a larger reduction in the food matrix influence (Wu et al. 2017). However, anthocyanins concentration was reduced during the small intestine digestion. This could be related to an increase in the pH value and, in this case, the food matrix could act as a protective effect due to the buffering capacity that protects the anthocyanins from degradation (Wu et al. 2017). The reaction suggests that each compound presents a specific behavior during the passage through the gastrointestinal tract.

The chemical characteristics of bioactive compounds and food matrix including solubility, stability, sugar content, moiety, hydrophobicity, molecular weight, and isomer configuration, also affect their absorption (Erlund et al. 2000, Appeldoorn et al. 2009, Rein et al. 2013, Wu et al. 2017). For instance, Kanakis et al. 2011 suggested that larger polyphenols with a higher number of aromatic rings and hydroxyl groups are more efficient to form complexes with proline-rich peptides. However some proteins and peptides decrease intestinal permeability due to tight junction closing, which reduce the paracellular compound transport. In opposite case, a combination with linoleic acid and medium chain fatty acids such as capric acid and lauric acid promote tight junction opening, increasing compounds' permeability (Shimizu 1999, Ulluwishewa et al. 2011). Anthocyanins have also shown to have an influence in the tight junction reorganization, mainly in the upper gastrointestinal tract, influenced by a lower pH, increasing the influx/efflux transport (Wu et al. 2017).

Usually, the bioavailability is determined using Caco-2 cell monolayer experiments to evaluate the bioactive compounds' transport. The location in the gastrointestinal tract and the phenolic type can interfere. For instance, it has been pointed out that the bioavailability of catechins is low, being one of the reasons for the instability of these compounds in neutral or slightly alkaline conditions during intestinal digestion (Xie et al. 2013). Proanthocyanindins, which are compounds of high molecular weight, have to be degraded into smaller molecules during digestion before transport into intestinal cells. β-glycosylated flavonoids are only absorbed in minimum amounts in the small intestine and if an additional sugar molecule is bound to the flavonoid, such as quercetin, the compound must reach the large intestine where the sugar moieties are hydrolyzed in order to be transported into the intestinal epithelial cells and possibly absorbed (Erlund et al. 2000, Appeldoorn et al. 2009, Rein et al. 2013). When not absorbed, bioactive compounds descend to the colon matrix and can be metabolized by the bacterial microbiota in the large intestine forming the so-called metabolites (Dueik and Bouchon 2016, Wu et al. 2017).

The microbiota activity interference in this phenomenon is not well elucidated; however, several reports suggest that alterations in membrane fluidity and permeability increase the solubility of phenolic compounds. The membrane alterations are related to production of short chain fatty acids such as propionate and butyrate which influence cell transport, decrease the viscosity of the mucus layer, and increase the interactions among the phenolics; the production of phenolic metabolites are all result of LAB fermentation (Boyer et al. 2005, Wiczkoski et al. 2010, González-Sarrías et al. 2013, Zhao and Shah 2016, Van Rymenant et al. 2017).

Metabolites from LAB Phenolic Digestion

Human intestine is populated by a microbial ecosystem called gut microbiota. Concentrated in the colon, this bacterial group has an important role in the host organism by regulating metabolic pathways through interactive and symbiotic signalling systems, and also by producing molecules that can induce modifications in critical cell processes (Lynch and Pedersen 2016, Vaiserman et al. 2017). The bacterial communities vary among individuals according to genetics, lifestyle, and environmental factors such as diet, changing not only its composition but also its functionality, which can modulate the susceptibility to disease (Turnbaugh et al. 2010, Sonnenburg and Bäckhed 2016). The equilibrium in this ecosystem is critical for maintaining the host's homeostasis related to metabolism, inflammation, immunity, cell proliferation, and also the regulation of the microbiota activity itself. Disturbance of this microbial ecosystem (gut dysbiosis) is involved in different chronic conditions such as inflammatory bowel disease, irritable bowel syndrome, colorectal cancer, obesity, diabetes, metabolic syndrome, cardiovascular diseases, allergy, asthma, neurological diseases, depression, and anxiety (Vaiserman et al. 2017).

Colonic fermentation by gut microbiota is responsible for the formation of metabolites that are effectively absorbed, and can be related with the beneficial health effects of the ingested polyphenols (Rein et al. 2013, Barba et al. 2017). Compounds with antioxidant capacity may be important in the gastrointestinal tract, maintaining the redox balance and consequently preventing the development of gastrointestinal diseases related to reactive oxygen species (ROS) generation during the digestion process, even when not absorbed (Barba et al. 2017). The health effects associated with phenolics is largely dependent on their bioavailability. The phenolics absorption occurs mainly in the small intestine following the gastric and intestinal digestion phases in the gastrointestinal tract. Once incorporated into the intestinal enterocytes, cellular conjugative enzymes, including UDP-glucuronosyltransferase, sulfotransferase and catechol-O-methyl transferase take part in the metabolism of flavonoids before they enter the circulatory system (Galijatovic et al. 2001, Van Duynhoven et al. 2011).

However, only a limited amount of phenolics can be absorbed by the intestinal epithelial cells. Large part of these phenolic compounds, particularly those that are oligomeric and polymeric continues to the colon where they are exposed to degradation or oxidation due to the slightly alkaline environment of the gut, and finally are metabolised by microorganisms producing metabolites that have an effect on modulating the colon microbial composition and function (Larrosa et al. 2010, Chen and Sang 2014). With the help of these bacteria, polyphenols can be biotransformed into compounds with enhanced bioavailability and bioactivity (Chiou et al. 2014, Zhao and Shan 2014). These gut microbiota transformations are grouped into three major catabolic processes: hydrolysis (O-deglycosylations and ester hydrolysis), cleavage (C-ring cleavage; delactonization, demethylation), and reductions (dehydroxylation and double bond reduction). The production of metabolites originated from phenolic compounds and their related absorption and bioactivity are mainly performed by LAB. Studies have shown that if dietary phenolic extracts are fermented by LAB before ingestion they can be more efficiently absorbed in the intestine and enter the circulatory system (Zhao and Shah 2016). The interaction with the gut microbiota leads to biochemical transformations of the native phytochemicals into more bioavailable metabolites.

Diversity of colonic microbiota can lead to variations of polyphenol metabolism (Van Duynhoven et al. 2011, Wu et al. 2017). In particular, *Lactobacillus* species have received attention due to their population in the gastrointestinal tract and their versatility. Many LAB have demonstrated their ability to deglycosylate, de-esterify, de-carboxylate and de-methylate dietary phenolic compounds (Hervert-Hernandez and Goni 2011). Studies have shown that *L. plantarum* 292 and *L. brevis* 145 were able to metabolise some of the phenolic compounds, such as gallic acid, after

48 h fermentation, increasing free quercetin, pyrogallol and kaempferol. In addition, part of the gallic acid freed from catechin gallates or gallic acid derivates was also further converted to pyrogallol (Espín et al. 2017). During fermentation, bacteria removed sugar moieties and hydrolysed galloyl moieties of a variety of phenolic compounds, resulting in different phenolic profiles. *L. brevis* 145 presented the ability to deglycosylate most types of monosaccharide conjugates but not rutinoside, and hydrolysed the ester bond linked with galloyl groups (Zhao 2016). As previously reported by Parkar et al. 2013, *L. plantarum* 292 presented the ability to hydrolyse quinic acid esters and to produce caffeic acid. Beta-glucosidase is one of the major enzymes responsible for the hydrolysis of flavonoid conjugates during bacterial fermentation (Rodríguez et al. 2009), and it has been shown to exhibit higher affinity towards glucosides than other glycosides. One of the special effects exerted by LAB that can benefit phenolic compounds stability is the acid production during fermentation creating a buffered environment by providing hydrogen ions and attenuating the adverse impact of pH on phenolic compounds; and it is also possible that the change in viscosity during bacterial fermentation should also be taken into consideration (Zhao and Shah 2014).

Gut microbiota metabolites from tannins presented 'prebiotic-like' activity, such as the urolithins produced from ellagitannins, which positively modulated lactobacilli, bifidobacteria, and enterobacteria growth (Larrosa et al. 2010). The classical definition of 'prebiotic' was related to dietary carbohydrates that induce the growth of lactobacilli and bifidobacteria. This concept has been re-evaluated, and a novel definition states that prebiotic is 'a substrate that is selectively utilized by host microorganisms conferring a health benefit' (Gibson et al. 2017). Therefore, there are other dietary constituents, such as phenolic compounds, that also promote the growth of lactobacilli and bifidobacteria, and of other bacteria included in probiotics group presenting beneficial effects to the host health. Previous studies included *Akkermansia muciniphila*, *A. muciniphila*, *Faecalibacterium prausnitzii*, and *Roseburia* spp. as species that provide health benefits and are stimulated by the action of phenolic compounds on gut microbiota. In addition the *Firmicutes/Bacteroidetes* ratio which is higher in obese host and those with metabolic syndrome can be influenced by phenolic compounds consumption, like proanthocyanidins (Anhê et al. 2015, Roopchand et al. 2015). A beneficial microbiota modulation effect was also observed when flavonol -quercetin (Etxeberria et al. 2015), caffeic acid (Zhang et al. 2016), anthraquinone (Neyrinck et al. 2017), anthocyanins, flavonols, and ellagitannins (Heyman-Lindén et al. 2016) were analyzed as bioactive compounds sources. All the studies showed that induced decrease in the *Firmicutes/Bacteroidetes* ratio, changing gut microbiota to a healthier composition.

Among the major phenolics, the non-gallated epicatechin and epigallocatechin have relatively higher cell absorption *in vivo*, while others are poorly absorbed (Henning et al. 2008). The presence of sugar moieties may protect aglycones from degradation when being exposed to digestive enzymes and alkaline environment during digestion (Häkkinen et al. 2000). Phenolic acid is relatively stable in the small intestine, but is largely metabolized in the large intestine, especially in the ascending colon (Loo et al. 2016, Xie et al. 2016). In this scenario the production of metabolites are necessary to achieve beneficial health effects. Metabolites are produced according to enzymatic metabolism of microbiota, for instance *Lactobacillus*, *Bifidobacterium,*

Bacteroides, *Enterococcus* and *Enterobacter* strains have been reported to have rhamnosidase activity that leads to flavonoid-O-rhamnosyl deconjugation (Braune et al. 2016). Other metabolites from polyphenols compounds are shown in Table 2.

Investigations with faecal fermentations have demonstrated the critical role of the gut microbiota in the complex transformations of polyphenols, and therefore, on their biological activities. However, the evidence of the biological activity and the mechanisms underlying the effects of these metabolites *in vivo* is still unclear (Espín et al. 2017).

Table 2: Metabolites from polyphenols compounds produced by microorganisms enzymatic metabolism.

Polyphenols	Metabolites	Microorganism	Reference
Chlorogenic acid	Caffeic acid, quinic acid, caffeoyl glycerol, sulphate, glucuronides of caffeic, ferrulic, dihydrocaffeic, hydroferulic acid and dihydrogenic acid	Probiotics and gut microbiota	Stalmach 2014, Tomás-Barerán et al. 2014, Raimondi et al. 2015, Xie et al. 2016, Braune et al. 2016, Zhao and Shah 2016, Wu et al. 2017
Ellagitannins	Hexahydroxydiphenic acid, glucose and ellagic acid	gut microbiota	Aguilera-Carbo et al. 2008
Ellagic acid	Urolithin	gut microbiota	González-Sarrías et al. 2013, Zhao and Shah 2016
Procyanidin	Epicatechins, phenolic acid	gut microbiota	Appeldoorn et al. 2009, Kahle et al. 2011

Food Applications of Polyphenols

Earlier studies have highlighted the interactions between polyphenols and probiotics. More importantly, studies have shown that polyphenols do not inhibit the growth of probiotics. Polyphenols occurred extensively in plant and fruits which are being consumed by human. However, in the event of extracting the polyphenols out of its natural form, the term natural might not be completely accurate. For many years, synthetic antioxidants have been added to food in order to add value as enriched with antioxidants. Nevertheless, most synthetic antioxidants added to processed food and beverages do not deliver the desired health benefits to consumers, because antioxidants are excreted instead of assimilated. Table 3 shows the applications of some polyphenols and their tested food applications.

Beverages

The beverage industry can also utilize natural antioxidants such as the anthocyanins that are present in fruits and vegetables or their by-products to provide natural colorants, and at the same time the vitamin C could serve as natural antioxidants and/or preservative to the beverages. Grapes is one of the widest fruit crop cultivated in the world (Amendola et al. 2010), and is rich in anthocyanins. Wine-making industry is seen as the biggest industry that utilizes grapes and at the same time, leaving a huge amount of grape pomace that is still considerably high in polyphenols,

Table 3: The applications of some polyphenols and their tested food application/matrices.

Polyphenols origins	Extraction method	Compounds	Application	Impact	Reference
Grape pomace	70% acetone	–	Yoghurt and salad dressing	–	(Tseng and Zhao 2013)
	water	–	Fermented milk	–	(Aliakbarian et al. 2015)
	–	–	Minced fish, and chicken breast	–	(Sánchez-Alonso et al. 2008)
	95% ethanol	–	Beef patties	–	(Sagdic et al. 2011)
	–	–	Flour (for Bars, pancakes, noodles)	–	(Rosales Soto et al. 2012)
	water	–	Pasta	–	(Marinelli et al. 2015)
Green tea	Water infusion	–	Yoghurt	No effect on lactic acid level	(Jaziri et al. 2009)
		Catechin Epictechin gallate	Low fat hard cheese	Successfully retained in cheese	(Rashidinejad et al. 2014)
	Green tea powder	–	Cookies	Enhance visco-elastic and stability of wheat dough Increase overall sensory acceptability	(Ahmad et al. 2015)
Moringa leaves	–	Phenolic compounds, Glucosinolates	Ready to eat snack	–	(Devisetti et al. 2016)
Orange peel	Soxhlet Ultrasound Assisted Extraction	Hesperidin, gallic acid, Naringin	–	–	
–	–	Tannic acid	Crosslinking agent in protein-based food packaging	Improved physicochemical properties of casein film	Picchio et al. 2018
Spent coffee grouds	Ohmic-heating	Ascorbic acid, gallic acid, chlorogenic acid, catechin, caffeic acid	Biscuits	α-glucosidade inhibition	Vázquez-Sánchez et al. 2018

particularly anthocyanins. Anthocyanins which are responsible for colours such as blue, red and purple, just to name a few, had been extensively extracted out and studied for their stability (Cardona et al. 2009, Clemente and Galli 2011, MohdMaidin et al. 2019), and also their incorporation in food system such as in yoghurts (Tseng and Zhao 2013). The major group of anthocyanins that has been found in grapes and/or in grape pomace include the malvidins, cyanidins, and delphinidins which are mainly responsible for the dark purple, red and blue colours (Schieber et al. 2001, Clemente and Galli 2011, MohdMaidin et al. 2019). Anthocyanins from grapes have also been well established as one of the natural food colouring permitted by the European Union (EU), typically referred to as E163. In fact, one of the leading beverage companies have been using grape anthocyanins as the colouring ingredient to their beverages. High anthocyanins and antioxidant content of grapes could potentially be used in the beverage industry as a natural colour enhancer, and in the same time could act as a functional ingredient in the beverage. Moreover, many studies have been conducted to change these anthocyanins into powder for easy applications. Encapsulation is seen as one of the popular methods, using carriers such as maltodextrin and arabic gum, and undergone a high temperature processing in the spray drier (Souza et al. 2014, De and Secado 2015). A recent study on the microwave assisted extraction of grape anthocyanins undergone maltodextrin-skimmed milk encapsulation, revealed that the degradation of anthocyanins was significantly slower in those which are encapsulated as compared to non-encapsulated anthocyanins although the antioxidant capacity in both samples remained the same (Tsali and Goula 2018). Interestingly, the application of this encapsulated anthocyanins' powder has been also applied in food, such as in apple puree, whereby 72% of the anthocyanins retained after storage with higher antioxidant capacity observed in the encapsulated anthocyanins compared to the non-encapsulated control (Lavelli et al. 2017).

Another source of food anthocyanins that is widely studied is roselle extracts. The roselle, known as *Hibiscus sabdariffa* flower in the Africa had gained enormous attention for the past 20 years, mainly due to the attractive colours it offers. The major anthocyanins in roselle flower extracts is delphinidins (318 mg/L) and its derivatives followed by cyanidins (112 mg/L) (Rasheed et al. 2018). Anthocyanins from roselle that were encapsulated could have a half-life from 7 days when stored at room temperature and up to 180 days when stored at 5°C (de Moura et al. 2018). Moreover, a study comparing the freeze-dried roselle powder in a beverage system with beverages containing colourants such as SAN RED RC® and synthetic carmoesin was carried out by Duangmal and Sueeprasan 2004. The results obtained showed that roselle drink was more stabilised with the addition of maltodextrin and that the colour changes during storage were more correlated to the increase in lightness and decrease in chroma, with overall acceptance of 56 days.

Flours

Flours are one of the important sources in dietary foods. A wide variety of flours that have added value, such as high concentration of dietary fibre and natural antioxidants, have been produced. Many studies have recently emerged on the production of flour-based food, such as cookies and pasta. Grapes are a good example to describe this application due to the amount of wine-making by-products left. For instance, Beres et al. (2016) produced a high dietary fibre flour from grape pomace with a high antioxidant content. In addition to that, Acun and Gül (2014) found that the use of grape pomace flour at less than 10% did not significantly affect the properties of the cookies made out of it. A grape marc flour, initially extracted using ultrasonic method was also found to produce acceptable pasta with high phenolics and antioxidant content (Marinelli et al. 2015). Interestingly, the pasta fortified with grape pomace flour was acceptable in terms of sensory with low cooking loss. Roselle seed flour has also been made into powder for the production of cookies and the results suggested that 15% inclusion had been the best result in nutritional improvement, nutritional composition and reduction of total carbohydrate due protein enhancement (Rimamcwe 2018). Jerusalem artichoke has been studied for wheat flour substitution in biscuits production as artichoke is highly rich in inulin and also phenolic compounds. The results indicated that the substitution at 15% was highly accepted by the panels, and thus promised a biscuit that is high in antioxidant dietary fibre and low in calorie (Díaz et al. 2019).

As an overall view, these phenolics can be obtained from cheap sources, for example, food by-products or food waste. Therefore the utilization or incorporation of these phenolics in food applications is welcome and helps in reducing the amount of food waste generated during processing.

Conclusion

The protective effect of compounds with antioxidant activity is well known. Nevertheless, the mechanisms underlying this effect have not been elucidated completely. Studies have proved that polyphenols from food sources are responsible for health benefits on different assays at cell level; however, not all bioactive compounds reach cell level in the human body. Recent advances imply that the bioavailability of these compounds depends directly on the microbiota of human gut. Part of the phenolic compounds cross the brush border of the intestinal tract; however a large amount of these compounds reaches the colon. Microbiota mainly formed by LAB is responsible for enzymatic and fermentative reactions that produce metabolites which are able to go through the cell wall and stimulate cell changes reducing human health concerns, such as inflammatory disease. Several products are supplemented with bioactive compounds in order to attend the desired antioxidant capacity; however, the absorption mechanism of these compounds and their metabolic activity and the LAB role are still under discussion.

References

Acun, S. and H. Gül. 2014. Effects of grape pomace and grape seed flours on cookie quality. Qual. Assur. Saf. Crop. 6(1): 81–88.

Aguilera-Carbo, A., C. Augur, L.A. Prado-Barragan, E. Favela-Torres and C.N. Aguilar. 2008. Microbial production of ellagic acid and biodegradation of ellagitannins. Appl. Microbiol. Biotechnol. 78: 189–199.

Ahmad, M., W.N. Baba, T.A. Wani, A. Gani, A. Gani, U. Shah and F.A. Masoodi. 2015. Effect of green tea powder on thermal, rheological & functional properties of wheat flour and physical, nutraceutical & sensory analysis of cookies. Food Sci. Technol J. 52(9): 5799–5807.

Albuquerque, M.A.C. de, R. Levit, C. Beres, R. Bedani, A. de Moreno de LeBlanc, S.M.I. Saad and J.G. LeBlanc. 2019. Tropical fruit by-products water extracts as sources of soluble fibres and phenolic compounds with potential antioxidant, anti-inflammatory, and functional properties. J. Funct. Foods. 52: 724–733.

Aliakbarian, B., M. Casale, M. Palini, A.A. Casazza, S. Lanteri and P. Perego. 2015. Production of a novel fermented milk fortified with natural antioxidants and its analysis by NIT spectroscopy. LWT - Food Sci. Technol. 62(1): 376–383.

Almajano, M.P., R. Carbó, J.A.L. Jiménez and M.H. Gordon. 2008. Antioxidant and antimicrobial activities of tea infusions. Food Chem. 108(1): 55–63.

Amendola, D., D.M. De Faveri and G. Spigno. 2010. Grape marc phenolics: Extraction kinetics, quality and stability of extracts. J. Food Eng. 97: 384–392.

Anhê, F.F., D. Roy, G. Pilon, S. Dudonné, S. Matamoros, T.V. Varin, C. Garofalo, Q. Moine, Y. Desjardins, E. Levy and A. Marette. 2015. A polyphenol-rich cranberry extract protects from diet-induced obesity, insulin resistance and intestinal inflammation in association with increased *Akkermansia* spp. population in the gut microbiota of mice. Gut. 64: 872–883.

Appeldoorn, M.M., J.P. Vincken, A.M. Aura, P.C.H. Hollman and H. Gruppen. 2009. Procyanidin dimmers are metabolized by human microbiota with 2–(3,4-dihydroxyphenyl)acetic acid and 5-(3,4-dihydroxyphenyl)-γ-valerolactone as the major metabolites. J. Agric. Food Chem. 57(3): 1084–1092.

Barba, F.J., N.S. Terefe, R. Buckow, D. Knorr and V. Orlien. 2015. New opportunities and perspectives of high pressure treatment to improve health and safety attributes of foods. A review. Food Res. Int. 77(4): 725–742.

Barba, F.J., L.R.B. Mariutti, N. Bragagnolo, A.Z. Mercadante, G.V. Barbosa-Cánovas and V. Orlien. 2017. Bioaccessibility of bioactive compounds from fruits and vegetables after thermal and nonthermal processing. Trends Food Sci. Technol. 67: 195–206.

Beaugerie, L. and J.-C. Petit. 2004. Antibiotic-associated diarrhoea. Best Pract. Res. Clin. Gastroenterol. 18(2): 337–352.

Beres, C., F.F. Simas-Tosin, I. Cabezudo, S.P. Freitas, M. Iacomini, C. Mellinger-Silva and L.M.C. Cabral. 2016. Antioxidant dietary fibre recovery from Brazilian *Pinot noir* grape pomace. Food Chem. 201: 145–152.

Bouayed, J., L. Hoffmann and T. Bohn. 2011. Total phenolics, flavonoids, anthocyanins and antioxidant activity following simulated gastrointestinal digestion and dialysis of apple varieties: Bioacessibility and potential uptake. Food Chem. 128: 14–21.

Boyer, J., D. Brown and R.H. Liu. 2005. *In vitro* digestion and lactase treatment influence uptake of quercitin and quercitin glucoside by the Caco-2 cell monolayer. Open Nutr. J. 4(1): 1.

Bindels, L.B., N.M. Delzenne, P.D. Cani and J. Walter. 2015. Towards a more comprehensive concept for prebiotics. Nat. Rev. Gastroenterol. Hepatol. 12: 303–310.

Braune, A. and M. Blaut. 2016. Bacterial species involved in the conversion of dietary flavonoids in the human gut. Gut Microbes. 7: 216–234.

Cardona, J.A., J.H. Lee and S.T. Talcott. 2009. Color and polyphenols stability in extracts produced from muscadine grape (*vitis rotundifolia*) pomace. J. Agric. Food Chem. 57(18): 8421–8425.

Chan, C.-L., R.-Y. Gan, N.P. Shah and H. Corke. 2018. Polyphenols from selected dietary spices and medicinal herbs differentially affect common food-borne pathogenic bacteria and lactic acid bacteria. Food Control. 92: 437–443.

Chen, H. and S. Sang. 2014. Biotransformation of tea polyphenols by gut microbiota. J. Func. Foods. 7: 26–42.

Chiou, Y.S., J.C. Wu, Q. Huang, F. Shahidi and Y.J. Wang. 2014. Metabolic and colonic microbiota transformation may enhance the bioactivities of dietary polyphenols. J. Func. Foods. 7: 3–25.

Clemente, E. and D. Galli. 2011. Stability of the anthocyanins extracted from residues of the wine industry. Ciênc. Tecnol. 31(3): 765–768.

de Moura, S.C.S.R., C.L. Berling, S.P.M. Germer, I.D. Alvim and M.D. Hubinger. 2018. Encapsulating anthocyanins from *Hibiscus sabdariffa* L. calyces by ionic gelation: Pigment stability during storage of microparticles. Food Chem. 241: 317–327.

De, I. and P. Secado. 2015. Influence of maltodextrin and spray drying process conditions on sugarcane juice powder quality. Rev. Fac. Nac. Agron. Medellín. 68(1): 7509–7520.

Devisetti, R., Y.N. Sreerama and S. Bhattacharya. 2016. Processing effects on bioactive components and functional properties of moringa leaves: development of a snack and quality evaluation. J. Food Sci. Technol. 53(1): 649–657.

Díaz, A., R. Bomben, C. Dini, S.Z. Viña, M.A. García, M. Ponzi and N. Comelli. 2019. Jerusalem artichoke tuber flour as a wheat flour substitute for biscuit elaboration. LWT - Food Sci. Technol. 108: 361–369.

Duangmal, K. and S. Sueeprasan. 2004. Roselle anthocyanins as a natural food colorant and improvement of its colour stability. In AIC 2004 Color and Paints, Interim Meeting of the International Color Association, Proceedings 155–158.

Dueik, V. and P. Bouchon. 2016. Development of polyphenol-enriched vaccum and atmospheric fried matrices: Evaluation of quality parameters and *in vitro* bioavailability of polyphenols. Food Res. Int. 88: 166–172.

Espín, J.C., A. González-Sarrías and F.A. Tomás-Barberán. 2017. The gut microbiota: A key factor in the therapeutic effects of (poly)phenols. Biochem. Pharmacol. 139: 2–93.

Erlund, I., T. Kosonen, G. Alfthan, J. Mäenpää, K. Perttunen, J. Kenraali, J. Parantainen and A. Aro. 2000. Pharmacokinetics of quercitin from quercetin aglycone and rutin in healthy volunteers. Eur. J. Clin. Pharmacol. 56(8): 545–553.

Etxeberria, U., N. Arias, N. Boqué, M.T. Macarulla, M.P. Portillo, J.A. Martínez and F.L. Milagro. 2015. Reshaping faecal gut microbiota composition by the intake of transresveratrol and quercetin in high-fat sucrose diet-fed rats. J. Nutr. Biochem. 26: 651–660.

Figueiredo, A.R., F. Campos, V. de Freitas, T. Hogg and J.A. Couto. 2008. Effect of phenolic aldehydes and flavonoids on growth and inactivation of *Oenococcus oeni* and *Lactobacillus hilgardii.* Food Microbiol. 25(1): 105–112.

Flint, H.J. 2012. The impact of nutrition on the human microbiome. Nutrit. Rev. 70: S10–S13.

Galijatovic, A., Y. Otake, U.K. Walle and T. Walle. 2001. Induction of UPD-glucuronosyltransferase UGT1A1 by the flavonoid chrysin in CaCo-2 cells—Potential role in carcinogen bioactivation. Pharm. Res. 18: 374–279.

Gironi, F. and V. Piemonte. 2011. Temperature and solvent effects on polyphenol extraction process from chestnut tree wood. Chem. Eng. Res. Des. 89(7): 857–862.

González-Centeno, M.R., C. Rosselló, S. Simal, M.C. Garau, F. López and A. Femenia. 2010. Physico-chemical properties of cell wall materials obtained from ten grape varieties and their byproducts: grape pomaces and stems. LWT - Food Sci. Technol. 43(10): 1580–1586.

González-Sarrías, A., V. Miguel, G. Merino, R. Lucas, J.C. Morales, F. Tomás-Barberán and J.C. Espín. 2013. The gut microbiota ellagic acid-derived metabolite urolithin A and its sulfate conjugate are substrates for the drug efflux transporter breast cancer resistance protein (ABCG2/BCRP). J. Agri. Food Chem. 61: 4352–4359.

Häkkinen, S.H., S.O. Kärenlampi, H.M. Mykkänen and A.R. Törrönen. 2000. Influence of domestic processing and storage on flavonol contents in berries. J. Agri. Food Chem. 48: 2960–2965.

Henning, S.M., J.J. Choo and D. Heber. 2008. Nongallated compared with gallated flavan-3-ols in green and black tea are more bioavailable. J. Nutr. 138: 1529S–1534S.

Hervert-Hernández, D., C. Pintado, R. Rotger and I. Goñi. 2009. Stimulatory role of grape pomace polyphenols on *Lactobacillus acidophilus* growth. Int. J. Food Microbiol. 136(1): 119–122.

Hervert-Hernández, D. and I. Goñi. 2011. Dietary polyphenols and human gut microbiota: a review. Food Rev. Int. 27(2): 154–169.

Heyman-Lindén, L., D. Kotowska, E. Sand, M. Bjursell, M. Plaza, C. Turner, C. Holm, F. Fak and K. Berger. 2016. Lingonberries alter the gut microbiota and prevent low-grade inflammation in high-fat diet fed mice. Food Nutr. Res. 60: 29993.

Jaziri, I., M. Ben Slama, H. Mhadhbi, M.C. Urdaci and M. Hamdi. 2009. Effect of green and black teas (*Camellia sinensis* L.) on the characteristic microflora of yogurt during fermentation and refrigerated storage. Food Chem. 112(3): 614–620.

Kahle, K., M. Kempf, P. Schreier, W. Scheppach, D. Schrenk, T. Kautenburguer and E. Richling. 2011. Intestinal transit and systemic metabolism of apple polyphenols. Eur. J. Nutr. 50: 507–522.

Kanakis, C.D., I. Hasni, P. Bourassa, P.A. Tarantilis, M.G. Polissiou and H.A. Tajmir-Riahi. 2011. Milk β-lactoglobulin complexes with tea polyphenols. Food Chem. 127: 1046–1055.

Khoo, H.E., A. Azlan, S.T. Tang and S.M. Lim. 2017. Anthocyanidins and anthocyanins: colored pigments as food, pharmaceutical ingredients, and potential health benefits. Food Nutr. Res. 61: 1–21.

Ky, I., B. Lorrain, N. Kolbas, A. Crozier and P.-L. Teissedre. 2014. Wine by-products: phenolic characterization and antioxidant activity evaluation of grapes and grape pomaces from six different French grape varieties. Molecules. 19(1): 482–506.

Lai, W.T., N.M.H. Khong, S.S. Lim, Y.Y. Hee, B.I. Sim, K.Y. Lau and O.M. Lai. 2017. A review: Modified agricultural by-products for the development and fortification of food products and nutraceuticals. Trends Food Sci. Technol. 59: 148–160.

Larrosa, M., M.J. Yañéz-Gascón, M.V. Selma, A. González-Sarrías, S. Toti, J.J. Cerón and J.C. Espín. 2009. Effect of a low dose of dietary resveratrol on colon microbiota, inflammation and tissue damage in a DSS-induced colitis rat model. J. Agric. Food Chem. 57(6): 2211–2220.

Larrosa, M., A. González-Sarrías, M.J. Yáñez-Gascón, M.V. Selma, M. Azorín-Ortuño, S. Toti, F. Tomás-Barberán, P. Dolara and J.C. Espín. 2010. Antiinflammatory properties of a pomegranate extract and its metabolite urolithin A in a colitis rat model and the effect of colon inflammation on the phenolic metabolism. J. Nutr. Biochem. 21: 717–725.

Lavelli, V., P.S.C.S. Harsha, M. Laureati and E. Pagliarini. 2017. Degradation kinetics of encapsulated grape skin phenolics and micronized grape skins in various water activity environments and criteria to develop wide-ranging and tailor-made food applications. Innov. Food Sci. Emerg. Technol. 39: 156–164.

Lima, A.C.S., D.J. Soares, L.M.R. Silva, R.W. Figueiredo, P.H.M. Sousa and E.A. Menezes. 2014. *In vitro* bioaccessibility of copper, iron, zinc and antioxidant compounds of whole cashew apple juice and cashew apple fibre (*Anacardium occidentale* L.) following simulated gastro-intestinal digestion. Food Chesm. 161: 142–147.

Loo, B.-M., I. Erlund, R. Koli, P. Puukka, J. Hellström, K.Wähälä and A. Jula. 2016. Consumption of chokeberry (*Aronia mitschurinii*) products modestly lowered blood pressure and reduced low-grade inflammation in patients with mildly elevated blood pressure. Nutr. Res. 36: 1222–1230.

Lynch, S.V. and O. Pedersen. 2016. The human intestinal microbiome in health and disease. N. Engl. J. Med. 375: 2369–2379.

Marinelli, V., L. Padalino, D. Nardiello, M.A. Del Nobile and A. Conte. 2015. New approach to enrich pasta with polyphenols from grape marc. J. Chem. 2015: 1–8.

Middleton, E., C. Kandaswami and T.C. Theoharides. 2000. The effects of plant flavonoids on mammalian cells: Implications for inflammation, heart disease, and cancer. Pharmacol. Rev. 52(4): 673–751.

MohdMaidin, N., M.J. Oruna-Concha and P. Jauregi. 2019. Surfactant TWEEN20 provides stabilisation effect on anthocyanins extracted from red grape pomace. Food Chem. 271: 224–231.

Namal Senanayake, S.P.J. 2013. Green tea extract: Chemistry, antioxidant properties and food applications—A review. J. Func. Foods. 5(4): 1529–1541.

Neyrinck, A.M., U. Etxeberria, B. Taminiau, G. Daube, M. Van Hul, A. Everard, P.D. Cani, L.B. Bindels and N.M. Delzenne. 2017. Rhubarb extract prevents hepatic inflammation induced by acute alcohol intake, an effect related to the modulation of the gut microbiota. Mol. Nutr. Food Res. 61(1): 1–12.

Pacheco-Ordaz, R., A. Wall-Medrano, M.G. Goñi, G. Ramos-Clamont-Montfort, J.F. Ayala-Zavala and G.A. González-Aguilar. 2018. Effect of phenolic compounds on the growth of selected probiotic and pathogenic bacteria. Lett. Appl. Microbiol. 66(1): 25–31.

Parkar, S.G., D.E. Stevenson and M.A. Skinner. 2008. The potential influence of fruit polyphenols on colonic microflora and human gut health. Int. J. Food Microbiol. 124(3): 295–298.

Parkar, S.G., T.M. Trower and D.E. Stevenson. 2013. Fecal microbial metabolism of polyphenols and its effects on human gut microbiota. Anaerobe. 23: 12–19.

Picchio, M.L., Y.G. Linck, G.A. Monti, L.M. Gugliotta, R.J. Minari and C.I. Alvarez Igarzabal. 2018. Casein films crosslinked by tannic acid for food packaging applications. Food Hydrocol. 84: 424–434.

Raimondi, S., A. Anighoro, A. Quartieri, A. Amaretti, F.A. Tomás-Barberán, G. Rastelli and M. Rossi. 2015. Role of bifidobacteria in the hydrolysis of chlorogenic acid. Microbiol. Open. 4: 41–52.

Rasheed, D., A. Porzel, A. Frolov, H.R. El Saedi, L.A. Wessjohann and M.A. Farag. 2018. Comparative analysis of Hibiscus sabdariffa (roselle) hot and cold extracts in respect to their potential for α-glucosidase inhibition. Food Chem. 1(250): 236–244.

Rashidinejad, A., E.J. Birch, D. Sun-Waterhouse and D.W. Everett. 2014. Delivery of green tea catechin and epigallocatechin gallate in liposomes incorporated into low-fat hard cheese. Food Chem. 156: 176–183.

Rein, M.J., M. Renouf, C. Cruz-Hernandez, L. Actis-Goretta, S.K. Thakkar and M. da Silva Pinto. 2013. Bioavailability of bioactive food compounds: A challenging journey to bioefficacy. Br. J. Clin. Pharmacol. 75(3): 588–602.

Rimamcwe, K.B., U.D. Chavan and R.S. Gaikwad. 2017. Nutritional quality of roselle seed flour cookies. Int. J. Curr. Res. 9(12): 63053–63058.

Rodríguez, H., J.A. Curiel, J.M. Landete, B. de las Rivas, F.L. de Felipe, C. Gómez-Cordovés and R. Muñoz. 2009. Food phenolics and lactic acid bacteria. Int. J. Food Microbiol. 132(2): 79–90.

Roopchand, D.E., R.N. Carmody, P. Kuhn, K. Moskal, P. Rojas-Silva, P.J. Turnbaugh and I. Raskin. 2015. Dietary polyphenols promote growth of the gut bacterium *Akkermansia muciniphila* and attenuate high-fat diet-induced metabolic syndrome. Diabetes. 64: 2847–2858.

Rosales Soto, M.U., K. Brown and C.F. Ross. 2012. Antioxidant activity and consumer acceptance of grape seed flour-containing food products. Int. J. Food Sci. Techn. 47(3): 592–602.

Rozès, N., L. Arola and A. Bordons. 2003. Effect of phenolic compounds on the co-metabolism of citric acid and sugars by *Oenococcus oeni* from wine. Lett. Appl. Microbiol. 36(5): 337–341.

Sagdic, O., I. Ozturk, M.T. Yilmaz and H. Yetim. 2011. Effect of grape pomace extracts obtained from different grape varieties on microbial quality of beef patty. J. Food Sci. 76(7): 515–521.

Sánchez-Alonso, I., A. Jiménez-Escrig, F. Saura-Calixto and A.J. Borderías. 2008. Antioxidant protection of white grape pomace on restructured fish products during frozen storage. LWT - Food Sci. Techn. 41(1): 42–50.

Schieber, A., F. Stintzing and R. Carle. 2001. By-products of plant food processing as a source of functional compounds—recent developments. Trends Food Sci. Technol. 12(2001): 401–413.

Shimizu, M. 1999. Modulation of intestinal functions by food substances. Nahrung-Food 43: 154–158.

Sirotkin, A.V. and A.H. Harrath. 2014. Phytoestrogens and their effects. Eur. J. Pharmacol. 741: 230–236.

Sonnenburg, J.L. and F. Bäckhed. 2016. Diet-microbiota interactions as moderators of human metabolism. Nature. 535: 56–64.

De Souza, V.B., A. Fujita, M. Thomazini, E.R. Da Silva, J.F. Lucon, M.I. Genovese and C.S. Favaro-Trindade. 2014. Functional properties and stability of spray-dried pigments from Bordo grape (*Vitis labrusca*) winemaking pomace. Food Chem. 164: 380–386.

Souza, E.L., T.M.R. de Albuquerque, A.S. dos Santos, N.M.L. Massa and J.L.B. Alves. 2018. Potential interactions among phenolic compounds and probiotics for mutual boosting of their health-promoting properties and food functionalities—A review. Crit. Rev. Food Nutr. 1–15.

Spigno, G., L. Tramelli and D.M. De Faveri. 2007. Effects of extraction time, temperature and solvent on concentration and antioxidant activity of grape marc phenolics. J. Food Eng. 81(1): 200–208.

Stalmach, A., G. Williamson and A. Crozier. 2014. Impact of dose on the bioavailability of coffe chlorogenic acids in humans. Food Funct. 5: 1727–1737.

Tang, P.C.T., Y.F. Ng, S. Ho, M. Gyda and S.W. Chan. 2014. Resveratrol and cardiovascular health–promising therapeutic or hopeless illusion? Pharmacol. Res. 90: 88–115.

Theobald, S., P. Pfeiffer, U. Zuber and H. König. 2008. Influence of epigallocatechin gallate and phenolic compounds from green tea on the growth of *Oenococcus oeni*. J. Appl. Microbiol. 104(2): 566–572.

Tomas-Barberan, F., R. García-Villalba, A. Quartieri, S. Raimondi, A. Amatti, A. Leonardi and M. Rossi. 2014. *In vitro* transformation of chlorogenic acid by human gut microbiota. Mol. Nutr. Food Res. 58: 1122–1131.

Tsali, A. and A.M. Goula. 2018. Valorization of grape pomace: Encapsulation and storage stability of its phenolic extract. Powder Technol. 340: 194–207.

Tseng, A. and Y. Zhao. 2013. Wine grape pomace as antioxidant dietary fibre for enhancing nutritional value and improving storability of yogurt and salad dressing. Food Chem. 138(1): 356–365.

Turnbaugh, P.J., C. Quince, J.J. Faith, A.C. McHardy, T. Yatsunenko, F. Niazi, J. Affourtit, M. Egholm, B. Henrissat, R. Knight and J.I. Gordon. 2010. Organismal, genetic, and transcriptional variation in the deeply sequenced gut microbiomes of identical twins. Proc. Natl. Acad. Sci. USA 107: 7503–7508.

Ulluwishewa, D., R.C. Anderson, W.C. McNabb, P.J. Moughan, J.M. Wells and N.C. Roy. 2011. Regulation of tight junction permeability by intestinal bacteria and dietary components. J. Nutr. 141: 769–776.

Van Duynhoven, J., E.E. Vaughan, D.M. Jacobs, R. Kemperman, E.J. Van Velzen, G. Gross, L.C. Roger, S. Possemiers, A.K. Smilde, J. Dore, J.A Westerhuis and T. Van de Wiele. 2011. Microbes. Health Sackler Colloquium. Metabolic fate of polyphenols in the human superorganism. Proc. Natl. Acad. Sci. USA 108(Suppl.1): 4531–4538.

Van Rymenant, E., L. Abrankó, S. Tumova, C. Grootat, J. Van Camp, G. Williamson and A. Kerimi. 2017. Chronic exposure to short-chain fatty acids modulates transport and metabolism of microbiome-derived phenolics in human intestinal cells. J. Nutr. Biochem. 39: 156–168.

Vaiserman, A.M., A.K. Koliada and F. Marotta. 2017. Gut microbiota: a player in aging and a target for anti-aging intervention. Ageing Res. Rev. 35: 36–45.

Vázquez-Sánchez, K., N. Martinez-Saez, M. Rebollo-Hernanz, M.D. Del Castillo, M. Gaytán-Martínez and R. Campos-Vega. 2018. *In vitro* health promoting properties of antioxidant dietary fiber extracted from spent coffe (*Coffee arabica* L.) grounds. Food Chem. 261: 253–259.

Wu, T., C. Grootaert, S. Voorspoels, G. Jacobs, J. Pitart, S. Kamiloglu, S. Possemiers, M. Heinonen, N. Kardum, M. Glibetic, G. Smagghe, K. Raes and J. Van Camp. 2017. Aronia (*Aronia melanocarpa*) phenolics bioavailability in a combined *in vitro* digestion/Caco-2 cell model is structure and colon region dependent. J. Funct. Foods. 38: 128–139.

Wiczkowski, W., E. Romaszko and M.K. Piskula. 2010. Bioavailability of cyaniding glycosides from natural chokeberry (*Aronia melanocarpa*) juice with dietary-relevant dose of anthocyanins in humans. J. Agric. Food Chem. 58(23): 12130–12136.

Xie, Y., A. Kosińska, H. Xu and W. Andlauer. 2013. Milk enhances intestinal absorption of green tea catechins in *in vitro* digestion/Caco-2 cells model. Food Res. Int. 53: 793–800.

Xie, L., S.G. Lee, T.M. Vance, Y. Wang, B. Kim, J.-Y. Lee and B.W. Bolling. 2016. Bioavailability of anthocynins and colonic polyphenol metabolites following consumption of aronia berry extract. Food Chem. 211: 860–868.

Yang, F., H.S. Oz, S. Barve, W.J.S. de Villiers, C.J. McClain and G.W. Varilek. 2001. The green tea polyphenol (–)-epigallocatechin-3-gallate blocks nuclear factor–κB activation by inhiniting IκB kinase activity in the intestinal epithelial cell line IEC-6. Mol. Pharmacol. 60: 528–533.

Yu, J. and M. Ahmedna. 2013. Functional components of grape pomace: their composition, biological properties and potential applications. Int. J. Food Sci. Technol. 48(2): 221–237.

Zhang, Z., X. Wu, S. Cao, L. Wang, D. Wang, H. Yang, Y. Feng, S. Wang and L. Li. 2016. Caffeic acid ameliorates colitis in association with increased Akkermansia population in the gut microbiota of mice. Oncotarget. 7: 31790–31799.

Zhao, D. and N.P. Shah. 2014. Antiradical and tea polyphenol-stabilizing ability of functional fermented soymilk-tea beverage. Food Chem. 158: 262–269.

Zhao, D. and N.P. Shan. 2016. Lactic acid bacterial fermentation modified phenolic composition in tea extracts and enhanced their antioxidant activity and cellular uptake of phenolic compounds following *in vitro* digestion. J. Func. Foods. 20: 182–194.

14

Application of Lactic Acid Bacteria in Time-Temperature Integrators

A Tool to Monitor Quality and Safety of Perishable Foods

Amélie Girardeau,[1,2] *Vanessa Biscola,*[2] *Sophie Keravec,*[2] *Georges Corrieu*[3] and *Fernanda Fonseca*[1,*]

Introduction

Fermented foods containing lactic acid bacteria (LAB) have been around for centuries. The ability of these microorganisms to acidify their environment by producing lactic acid gives them a strong preservative effect. In recent years, this technological property of LAB is also being used as a tool to measure the shelf-life of perishable foods.

Most refrigerated perishable foods are not sterile. The microorganisms they contain can multiply, degrade their environment and often produce unsavory metabolites. Their growth affects the microbiological quality of food and can lead to spoilage. Food quality thus evolves over time, more or less rapidly, as a function of a number of parameters such as storage temperature, pH, water activity, gas composition, etc.

[1] UMR GMPA, AgroParisTech, INRA, Université Paris-Saclay, 78850, Thiverval-Grignon, France. Email: amelie.girardeau@inra.fr

[2] CRYOLOG, R&D Department, 44261 Nantes, France. Emails: v_biscola@yahoo.com.br; sophie.keravec@cryolog.com

[3] Bioval Process, La Llagonne, France. Email: georges.corrieu@grignon.inra.fr

* Corresponding author: fernanda.fonseca@inra.fr

Among these factors, temperature has the most critical effect on spoilage rates due to the rapid temperature changes that can take place during transport and storage. Ensuring cold chain integrity from production to consumption is therefore crucial to maintain and control the microbiological quality of refrigerated food. In practice, cold chain integrity is not always kept throughout production, distribution and domestic storage. Breaking the cold chain can affect expected shelf-life, making it difficult to predict the microbiological quality and safety of food at the time of consumption. Food industrials are responsible for the safety of products they have imported, produced, processed, manufactured or distributed (EFSA Regulation No. 764/2008). Consequently, there are frequent conservative estimations of shelf-life by actors of the food industry who add substantial security margins that lead to considerable waste of perfectly good products. Among the 28 countries of the European Union, annual food waste attributed to date marking issues was estimated to be between 6.9 and 8.9 million tonnes, across the manufacturing, processing, retailing, and household sectors (European Commission 2018). The emergence of innovative technologies has therefore been encouraged by the European Union to help reduce food waste, while still ensuring consumer safety.

Time-temperature integrators (TTI) can be considered as part of the said innovative technologies. They are simple tools that provide information on the microbiological quality of a food product throughout the cold chain, based on its true time-temperature history. These smart devices are reliable shelf-life indicators that can contribute to reduce food waste, without compromising food safety. TTI can be used to track all kinds of products (food, medical and pharmaceutical products, cut flowers, etc.), for which storage and transportation at low temperatures are essential to ensure proper preservation (Kerry et al. 2006, Sahin et al. 2007, McMeekin et al. 2008). Despite having undergone significant technical evolutions with over 100 patents delivered since the 1960s, few prototypes were able to reach marketing stage (Sahin et al. 2007, Wang et al. 2015).

The present chapter describes the design and working principles of LAB-based time-temperature integrators and focuses on how they are configured to cover a wide range of shelf-lives. Current applications and future prospects of this innovative way of using lactic acid bacteria are also discussed.

Time-Temperature Integrators

Time-temperature integrators are relatively small and cost-effective devices that relay through an irreversible and easily detectable sign, the cumulative effect of time and temperature on the food they are tracking. To be able to integrate all cold chain irregularities during the food's entire life-cycle, and adjust shelf-life information accordingly, TTIs generally come in the form of labels attached directly on the food's packaging. These labels can be divided into three main categories, according to their working principles: chemical, physical or biological TTI (Table 1). The first group is comprised of polymerization-based, photochromic-based, and oxidation reaction-based systems (Galagan and Su 2008, Salman et al. 2009, Gou et al. 2010). Physical TTIs encompass devices that are diffusion-based (Monitor Mark®, USA), nanoparticle-based, electronic, and so forth (Zweig 2005, Zeng et al. 2010). Biological TTI are based on enzymatic reactions and/or the use of microorganisms such as yeasts and lactic acid bacteria (Wang et al. 2015). Whatever its working principle is, a TTI's response is always strongly correlated to changes occurring within the tracked product's matrix, at a given time-temperature profile. Most of the commercialized TTIs today are of the chemical or physical types (Table 1), but biological TTIs, and more specifically LAB-based TTIs, are the ones that actually reproduce the microbiological reactions that take place in food, leading to spoilage.

Table 1: Examples of currently commercialized TTI belonging to different categories and their basic functioning principles.

Commercial name/company	Category	Functioning principle
Fresh-Check® and HEATmarker™ (TempTime, New Jersey, USA)	Chemical	Polymerization-based
OnVu® (BASF, Ludwigshafen, DE)	Chemical	Photochromic-based
Freshpoint® (Freshpoint Quality Assurance, Nesher, IS)	Chemical	Photochromic-based
Monitor Mark® (3M, St. Paul, MN, USA)	Physical	Diffusion-based
Tempix® (TEMPIX, Gävle, SE)	Physical	Diffusion-based
Tsenso® (Tsenso, Stuttgart, DE)	Physical	Electronic
Koovea® (Koovea, Montpellier, FR)	Physical	Electronic
CheckPoint® (Vitsab, Limhamn, SE)	Biological	Enzymatic
TopCryo® (Cryolog, Nantes, FR)	Biological	LAB growth

Design principle of lactic acid bacteria-based TTIs

The active core of biological time-temperature integrators based on lactic acid bacteria, contains three main components: a growth medium, lactic acid bacteria and a (or several) chromatic indicator(s). With these labels, alteration is generally determined by a progressive color shift brought about by a pH drop of the medium contained within the label, due to the growth and lactic acid production of LAB (Figure 1A). Perishable food quality loss kinetics greatly depend on the growth and metabolic activity of the microorganisms they contain which in turn, are heavily influenced by temperature. The influence of temperature on the color shift kinetics of a biological TTI label is illustrated in Figure 1B: the endpoint of a TTI label tracking a food product stored, for example, on the door of a household refrigerator (typically 8°C) will be reached twice as fast as one tracking a similar food product placed at the back of the refrigerator (typically 4°C).

Microorganisms contained within TTI labels must have similar growth and/or metabolic activity responses to temperature variations, as selected microorganisms contained within the food being tracked (e.g., spoilage LAB, *Listeria monocytogenes, Staphylococcus aureus,* etc.). The underlying mechanisms of these labels, therefore rely on predictive microbiology: the determination and validation of models for growth and metabolic activity of microorganisms as a function of various environmental conditions (temperature, pH, water activity, etc.). Examples of studies reported on the development of LAB-based TTI systems to monitor the quality and safety of perishable foods using different strains and varying designs, are presented in Table 2. Although many examples of LAB-based TTI prototypes can be found in literature, only three have been brought to the market to date: eO®, TRACEO®, and

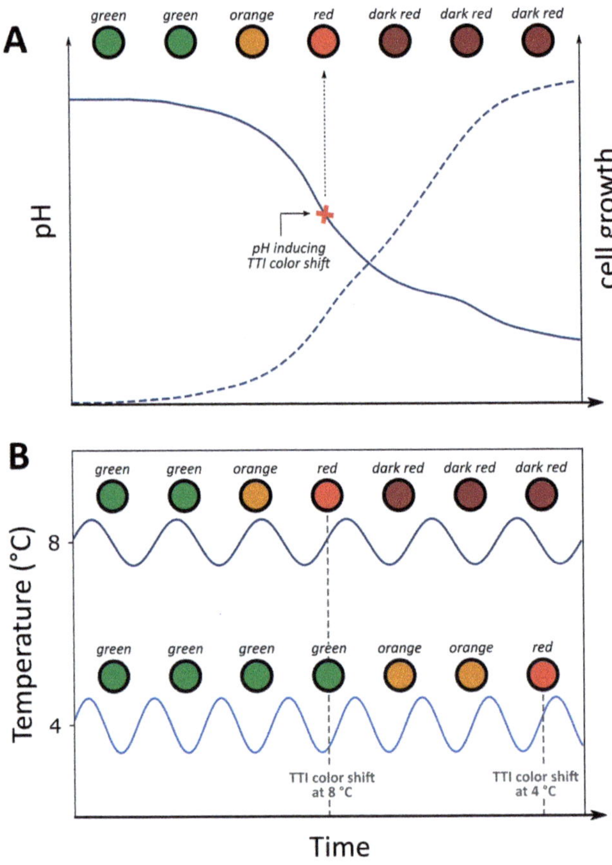

Figure 1: Schematic representation of the growth (dotted line) and acidification activity (solid line) of LAB in a TopCryo® TTI label stored at 8°C, inducing a color shift when a specific pH is reached **(A)** and the color shift's kinetic dependence on the average temperature profile: at 4°C the color shift takes place later than at 8°C **(B)**. The color shift response is also dependent on the TTI's label composition.

TopCryo®. All were developed and commercialized by Cryolog Clock-T° (Nantes, France). The present chapter will use Cryolog's latest self-adhesive label, TopCryo®, as a concrete example to explain LAB-based TTI parametrization.

TopCryo® was patented in 2005 (Louvet et al. 2005) as a device intended to monitor the quality of thermosensitive products. Their active core contains chromatic pH indicators and a gel-like growth medium inoculated with *Carnobacterium maltaromaticum* CNCM I-3298 cells, a psychrotrophic lactic acid bacteria. TopCryo® labels are meant to operate within a temperature range of 2°C to 12°C, whilst integrating breaks of the cold chain at room temperature. *C. maltaromaticum* cells contained within the label then start to multiply and acidify the growth medium. The acidification results in a color shift from bright green (pH > 8) to deep red (pH < 6.5), more or less rapidly, depending on the label's time-temperature history (Figure 1). TopCryo® labels are designed to cover different shelf-lives, ranging from 30 hours to 8 days (192 h) at 4°C (Table 3), depending on the type of food that needs to be

Table 2: A few studies reported on the development of various LAB-based TTI systems to monitor the quality and safety of perishable foods using different LAB strains.

LAB used	Design of TTI system	Applicability and performance evaluation when available	References
Lactobacillus sakei	- Laboratory screw-cap bottles containing nutrient-rich growth medium, chlorophenol red and microorganisms; - Activation at inoculation of TTI medium; - Color shift from red to yellow, via orange.	- Modelling of TTI evolution isothermally between 0°C and 16°C; - Shelf-life studies correlating TTI endpoint with end of minced beef shelf-life.	Vaikousi et al. 2008, 2009
Carnobacterium maltaromaticum	- Thin, self-adhesive label containing a gel like growth medium, chromatic indicators and microorganisms; - Activation by rapid freeze-thaw at room temperature; - Color shift from green to red.	- Modelling of the growth and acidification activity of the strain used within the TTI label medium, as a function of temperature, assuring accurate TTI response time; - Shelf-life studies and challenge tests done on ground beef, minced chicken and smoked salmon ensuring that TTI endpoint corresponds to the tracked food's shelf-life.	Louvet et al. 2005 Ellouze et al. 2008, 2010, 2011
Weissella cibaria	- Laboratory cell culture-ware containing a nutrient rich growth medium, bromothymol purple, methyl red and microorganisms; - Activation at inoculation of TTI medium; - Color shift from purple to yellow.	- Strong correlation between Ea of the system's pH and color evolution and the quality degradation of ground beef and chicken breast.	Kim et al. 2012 Park et al. 2013
Weissella koreensis	- Laboratory test tubes containing MRS-based growth medium, bromothymol blue, methyl red and microorganisms; - Color shift from green to red, via yellow.	- Strong correlation between Ea of TTI system and the Ea of Korean fermented food (kimchi) evolution; - TTI system thus proposed to indicate ripeness of kimchi.	Lim et al. 2014

Table 2 contd. ...

... Table 2 contd.

LAB used	**Design of TTI system**	**Applicability and performance evaluation when available**	**References**
Lactobacillus ssp.	- Two-dimensional code made of nanobeads containing microorganism immobilized on a nutrient-rich base; - Time-temperature effect translated into both a color shift and an easily scannable 2D code; - Must be stored below 5°C, gradually activated between 5 and 15°C.	- Limited temperature range: 5 to 15°C.	Lee and Jung 2013
Lactobacillus rhamnosus	- Transparent disk-shaped container including a nutrient-rich growth medium, bromothymol green, methyl red and microorganisms.	- Proposed for commercial, chilled products.	Lu et al. 2013

Ea: The Arrhenius activation energy parameter, quantifying a reaction's dependency to temperature.

Table 3: Shelf-life (in hours) of the TopCryo® TTI label range offered by Cryolog, as a function of storage temperatures.*

TopCryo® label range	Storage temperature				
	2°C	3°C	4°C	8°C	12°C
TopCryo® B	50	36	30	18	12
TopCryo® C	76	57	48	27	18
TopCryo® D	112	85	72	40	25
TopCryo® E	142	113	96	54	32
TopCryo® F	184	141	120	57	40
TopCryo® G	216	168	144	80	46
TopCryo® H	254	196	168	93	55
TopCryo® I	292	224	192	106	64

* Information obtained from the Cryolog website (http://cryolog.com/topcryo), January 2019.

tracked. A given label (e.g., TopCryo® E) tracking perishable food products stored at a temperature below 4°C, such as 2°C, will reach its shelf-life endpoint slower than it would at 4°C. Inversely, this same label will reach its shelf-life endpoint faster than at 4°C if it is used for tracking foods stored at 12°C.

Producing a LAB-based TTI such as TopCryo® involves two main processes: the production of frozen LAB concentrates followed by the production of the TTI labels themselves (Figure 2).

The production of frozen bacterial concentrates, such as *C. maltaromaticum*, includes the following key steps:

1. *Fermentation*: It is done in pure culture and batch mode. pH, temperature, and harvest time are controlled and set to values that result in the growth of the

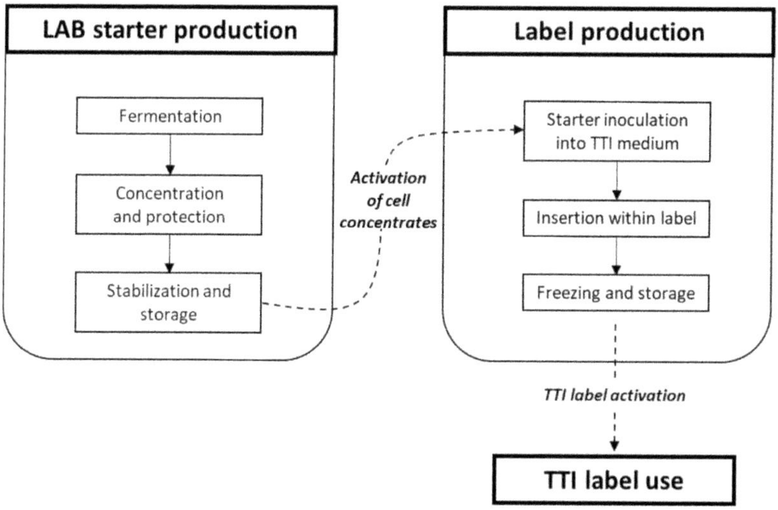

Figure 2: Diagram of LAB-based TTI production steps.

highest bacterial population (biomass) exhibiting the expected acidification activity for a given TTI.

2. *Concentration and protection*: Harvested cells are separated from the culture medium by centrifugation and resuspended in solutions containing protective molecules to reduce cell damage potentially caused by stabilization.
3. *Stabilization and storage*: The stabilization of bacterial concentrates by freezing or freeze-drying, enables the storage of LAB concentrates for long periods of time (6 to 24 months), without losing cell viability or any other technological function. Storage is generally done between –60°C and –80°C for frozen concentrates or between –20°C and 4°C for freeze-dried concentrates.

The production of the LAB-based TTI label itself includes the following key steps:

1. *Starter inoculation into the TTI medium*: In order to produce labels, the stored bacterial concentrates are freeze-thawed (or rehydrated) and a precise starter dosage is mixed into the label medium, depending on the desired shelf-life of the TTI being produced.
2. *Insertion of medium within the label*: This is achieved using a specially designed and patented machine that, in the case of TopCryo® TTI, seals the medium into self-adhesive labels (Vaillant and Peteuil 2012).
3. *Freezing and storage*: The inoculated labels are then inactivated by blast freezing and stored at –40°C until use. Activation is achieved by rapid freeze-thaw at room temperature upon placing the label on the food to be tracked.

Parametrization

TTI label parametrization consists in identifying the production conditions to achieve a target shelf-life. It can be carried out according to various parameter modulation strategies involved in the TTI production process. So far, much of TTI shelf-life modulation—such as the range covered by TopCryo® labels—are achieved by adjusting the initial cell concentration inoculated into the label. There is indeed a negative linear relationship between initial cell concentration and the label's shelf-life response time. Topcryo®'s 8-day (192 h) shelf-life at 4°C (TopCryo® I) is reached by inoculating the lowest cell concentration allowing reproducible acidification rates. The 30-hour shelf-life is analogously achieved by inoculating the highest possible cell concentration into the label, beyond which the TTI's initial color would no longer be green. The color shift response times of 30 hours and 8 days at 4°C correspond thus respectively to the shortest and the longest shelf-life that can be achieved by these labels. However, time-temperature integrators are ideally meant to track the microbiological quality of different food types displaying varying lengths of shelf-lives outside of the range offered by TopCryo®, such as 3 to 21 days for various meats and 0.5 to 3 days for shellfish (Dalgaard 1995). To be used at a much larger scale, LAB-based TTI are thus expected to cover a wider shelf-life range than they currently do. Three different TTI parametrization strategies can be pursued to further increase response time ranges, beyond simply adjusting initial cell concentration: (1) modulating the physiological state of the selected strain; (2) tweaking the label medium's composition; (3) using a different bacterial strain or species, alone or altogether.

Modulating the physiological state of a selected strain

LAB industrial starter production generally involves a succession of steps from fermentation to stabilization processes (Figure 2), and it is generally optimized to achieve the highest possible yield of concentrated cells exhibiting fast metabolic activities. However, in productions intended for TTI labels, other objectives are pursued: production parameters are chosen according to the response time needed to be covered by the TTI label. For long shelf-lives, starters inoculated into the label must indeed harbor slow acidification activities, while high acidification rates lead to labels exhibiting short shelf-lives. It is known that physiological state modulation of LAB is highly dependent on the fermentation parameters applied during their production (temperature, pH, fermentation duration, medium composition, etc.) (van de Guchte et al. 2002). The effect of culture conditions on the physiological state of some LAB species proposed for LAB-based TTI are presented in Table 4.

Viable cell concentration and acidification activity are the technological properties that characterize the physiological state of cells for use in LAB-based TTI labels. The range of acidification activity and viability that can be covered by a strain of *C. maltaromaticum*, simply by changing three fermentation parameters (temperature, pH and harvest time), has recently been mapped (Girardeau et al. 2019). Figure 3 shows the effect of fermentation pH and harvest time, the two most influential parameters on the biological activity of *C. maltaromaticum* concentrates characterized by their specific acidification activity ($dtpH1.5_{spe}$, in min/log/[{CFU/mL}]). Specific acidification activity is defined as the ratio of dtpH1.5 (the time necessary for the produced concentrates to acidify a TTI-like medium by 1.5 upH) to the corresponding log of cell concentration (colony forming unit [CFU]/mL). Consequently, the $dtpH1.5_{spe}$ descriptor gives an accurate and meaningful measurement of *C. maltaromaticum*'s physiological state, including its acidification activity and viability.

C. maltaromaticum cells produced at low pH (e.g., pH 6) and harvested at the beginning of the stationary phase exhibited faster acidification activities (solid arrow, Figure 3) than cells produced in standard conditions (pH 7, 30°C) (dashed arrow, Figure 3). These cells could potentially reduce the current lower limit of TopCryo®'s response time by two hours. Conversely, cells produced during fermentations carried out at high pH (e.g., pH 9.5) and harvested up to 10 hours into stationary phase (empty solid arrow, Figure 3), exhibited slower acidification activities, thus potentially increasing TopCryo®'s upper limit response time by 68 hours. Although the effect of temperature was less pronounced than pH and harvest time, it was still significant: low temperatures (20°C instead of the 30°C reference condition) led to increased acidification activities, whatever the harvest time up to 10 hours of stationary phase. This study thus showed that identifying the range of acidification activities that can be achieved by a certain strain, reveals the possible shelf-life range extensions that can be reached by TTI labels based on this strain. Following this approach for a given strain, once the temperature, pH, and harvest time have been studied for identifying upper and lower limits of the range of acidification activities, further extension of the TTI's response time range can be realized by tweaking the label's medium composition.

It is important to note, however, that the study on *C. maltaromaticum* CNCM I-3298 carried out by Girardeau et al. (2019) investigated the effect of fermentation parameters on the acidification activity of freeze-thawed cells, after confirming the

Table 4: Effect of culture conditions on the physiological state of LAB proposed for LAB-based TTI.

LAB strains	Investigated culture conditions	Effect on physiological state	References
Lactobacillus sakei 23 K	Fermentation at optimal growth temperature (26°C) and upon reaching exponential phase: → Heat shock (HS): 20 min at 42°C → Cold shock (CS): 20 min at 4°C	- No difference in viability; - Both HS and CS induced better acidification activity; - Best acidification activity after CS.	Hüfner and Hertel 2008
Carnobacterium divergens V41	→ Temperature: 15 to 30°C → pH: 6.5 to 8 → Carbohydrates (2 to 20 g L^{-1})	- Positive effect of low temperature and high carbohydrate concentration on viability; - No effect of tested pH range on viability; - Low pH increased bacteriocin production.	Brillet-Viel et al. 2016
Carnobacterium maltaromaticum CNCM I-3298	→ Temperature: 20 to 37°C → pH: 6 to 9.5 → Harvest times: 0 to 10 hours into stationary phase	- No significant effect of fermentation conditions on viability of harvested cells; - Low temperatures and pH increased acidification activity of harvested cell concentrates; - High pH and extended harvest times decreased acidification activity of harvest cell concentrates.	Girardeau et al. 2019
Lactobacillus rhanmnosus GG*	→ Temperature: 37°C → pH: unregulated → Glucose (10 to 34 gL^{-1}) → Peptone (8 to 12 g L^{-1})	- Growth favored by high peptone concentrations; - Lactic acid production yields favored by high glucose concentration.	Mel et al. 2008
Weisella cibaria DBPZ1006*	→ Temperature: 10 to 45°C → pH: 4 to 8 → a_w: 0.935 to 0.994	- Maximum growth rate achieved at 36.3°C, pH 6.6 and a_w of 0.994.	Ricciardi et al. 2009

* The results presented for these strains are the effect of culture conditions on the growth and metabolic activity of the cells during fermentation, not on the physiological state of the harvested cell concentrates.

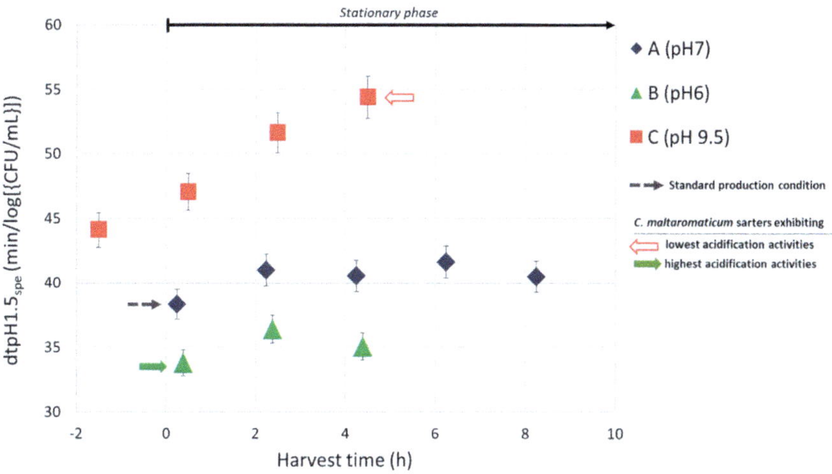

Figure 3: Specific acidification activity of *C. maltaromaticum* concentrates harvested at increasing times of fermentation, carried out at 30°C and different pH values (Adapted from Girardeau et al. 2019).

strain's high resistance to freezing, even when produced at extreme fermentation conditions (i.e., pH 6, 20°C or pH 9.5, 30°C). Such resistance to freezing is very much dependent on the considered LAB genus, specie or even strain. Freezing or freeze-drying exposes bacterial cells to big changes in environmental conditions like low temperatures, high solute concentration and extreme dehydration (Meneghel et al. 2017). Bacteria passively respond to stresses leading to cellular injuries and rather severe deterioration of biological functionalities after freeze-thaw or rehydration. It has been shown that LAB resistance to different stresses is also greatly dependent on their fermentation parameters (Palmfeldt and Hahn-Hägerdal 2000, Wang et al. 2005, Velly et al. 2014). During fermentation, stress proteins can be synthesized and membrane lipid compositions can be changed (Schoug et al. 2008). Promoting active cell response to moderate stressors during fermentation can help minimize the effect of subsequent stressors that cells face during stabilization steps. The use of protective solutions containing various substances such as polyols, polysaccharides and disaccharides has also been shown to help lessen cellular damage and improve storage stability (Linders et al. 1997, Conrad et al. 2000, Reddy et al. 2009). In LAB-based TTI production (Figure 2), cells undergo two stabilization/activation cycles: the first one after fermentation and the second one within the TTI label. It is therefore crucial to verify that the cells maintain their viability and acidification activity from the moment they are harvested after fermentation until the moment they are reactivated within the TTI label. The resistance to stabilization of potential strains for TTI use must thus be thoroughly investigated and optimized.

Tweaking the label medium's composition

pH and chromatic indicators. The choice of chromatic indicator contained within a label greatly influences the time it takes for the color shift to occur. In TopCryo® labels for example, the chromatic pH indicator shifts from bright green at pH 8 to dark red at pH 6.5. In Vaikousi's proposed prototype (Vaikousi et al. 2008), the pH indicator is chlorophenol red, shifting from red at pH 6.1 to yellow at pH 5.2. In both TTI, the response time is thus dependent on the time it takes for the microorganisms to acidify the label medium down to the pH value at which a distinct visual color can be observed. Color shift response time can therefore be modulated by either increasing or decreasing the difference between initial and final pH.

Nutrient availability. Nutrient availability within the TTI label medium could also be a response time modulation lever. Indeed, LAB are known to have many nutritional requirements and their production often demands growth in nutritionally rich environments. Most nutritional studies performed on LAB are focused on maximizing growth rates and/or metabolic activity. In the case of a TTI application, research on nutritional levers that either accelerate or slow down metabolic activity are both relevant when developing a large range of shelf-life response times. LAB growth requires first and foremost the presence of a carbon source through carbohydrates, their main energy source. The nature of the carbohydrates (monosaccharides, disaccharides, pentoses, hexitols or penitols) varies according to the considered species or strain and sometimes even between strains. It is frequent for a strain to be able to catabolize more than one type of carbohydrate, but its growth and metabolic kinetics will change depending on the available type. In a study aimed at investigating the growth and metabolic performances of *Lactobacillus* (*L.*) *sakei* strains on different carbon sources, it was shown that using ribose as a sole carbon source decreased growth and increased acetate production compared to glucose (McLeod et al. 2010). Changing the nature of the carbon source within the TTI label medium can thus have an impact on the growth and/or acidification activity rates of a selected strain, therefore resulting in the adjustment of the TTI's response time. Nutritional studies have also shown that LAB growth can be stimulated by the addition of ingredients rich in amino acids and small peptides, such as yeast extract or peptone. *C. maltaromaticum* for example exhibits substantially better growth in milk when it is enriched with 1 g/L of yeast extract (Edima et al. 2008). Similarly, certain vitamins and nitrogenous bases were shown to stimulate growth, although requirements vary greatly even between strains of the same species. This was clearly evidenced in a study aimed at screening for vitamin and amino acid requirements on 24 *L. plantarum* strains, where a very high diversity of requirements between strains was demonstrated (Ruiz-Barba and Jiménez-Díaz 1994). Therefore, although it can take a lot of time, great potential for TTI shelf life modulation can arise from modifying nutrient availability in the TTI medium.

Water activity. One of the oldest methods of preventing food spoilage is drying. Its direct consequence is to reduce water activity (a_w), thus limiting the available water required for the growth of microorganisms. Adjusting the a_w of an aqueous growth medium is often done by the addition of salts, saccharides or polyols. Growth rate and maximum viability reductions, as well as acidification activity modulations are observed in LAB when a_w levels are reduced from 0.99 to 0.95 or 0.93, depending on the species and compounds used to lower a_w (Ruiz-Barba and Jiménez-Díaz 1994). Interestingly, the effect of a_w on growth and viability can also be linked to temperature, pH and available nutrients. Such linkages were for example evidenced in a study of the effect of suboptimal temperatures and high salt concentrations on the growth and long-term survival of *L. sakei* (Marceau et al. 2002). Incubation at 4°C or in a 9% NaCl solution respectively resulted in *L. sakei* survival of 16 and 10 days, respectively. However, when cells were incubated at 4°C in a 9% NaCl solution, cell viability was stable between 24 and 28 weeks. The relationship between temperature and a_w, and its impact on the kinetics of cell development and acidification are thus particularly relevant in TTI research.

Tweaking a TTI's medium composition therefore offers many possibilities to modulate a biological TTI's shelf-life response time and therefore widen the shelf-life range that can be achieved by a single strain. There are however various technological and design constraints linked to label production at industrial scale that must be taken into account, such as viscosity, cost, consumer behavior, etc.

Using different bacterial species or strains

In addition to being dependent on initial cell concentrations, TTI response time evidently relies on the inherent growth and acidification kinetics of the selected LAB. Another strategy to widen the covered shelf-life range is to switch bacterial species entirely. Due to their diversity, LAB exhibit very different growth and acidification kinetics, depending on the considered genus, species or strain, on whether they are homo- or hetero-fermentative and on their culture conditions.

Homo-fermentative LAB, such as *L. bulgaricus* or *Lactococcus lactis*, are widely used in the dairy industry because they exclusively metabolize sugars into lactic acid (*pKa* = 3.86), resulting in high acidification activities. Hetero-fermentative LAB however, tend to have much slower acidification activities because their fermentation of sugars results in numerous other metabolites, including acids that might not have the acidification power of lactic acid, such as acetic acid (*pKa* = 4.76), butyric acid (*pKa* = 4.82) or propionic acid (*pKa* = 4.88), and ethanol. Proportions among different produced metabolites may also vary depending on the microorganism and applied fermentation parameters. A good example of a slow acidifying hetero-fermentative LAB is *C. maltaromaticum*, the strain contained within Cryolog labels.

LAB used in biological TTI must follow certain conditions. They need to grow within the temperature range in which perishable foods are stored at (typically between 0 and 30°C) and their growth and metabolic activity must be highly dependent on temperature, allowing an accurate monitoring and recording of all temperature variations throughout the product's shelf-life. Additionally, growth must not be inhibited by the pH-indicator used to indicate end of shelf-life. Furthermore, most LAB-based TTI developed to date use psychrotrophic strains that present growth rates similar (or slightly higher) to those of main pathogenic and spoilage bacteria contained within the food being tracked (e.g., *Lysteria monocytogenes*). This way, the TTI endpoint is attained before the growth of those microorganisms reaches unacceptable levels.

C. maltaromaticum follows all the above conditions needed to be an adequate microorganism for use in TTI: it is a psychrotrophic LAB commonly found in many highly perishable, high protein foods such as meat, fish and shellfish, and it has sometimes been regarded as a spoilage bacterium, although non-pathogenic. *C. maltaromaticum* was initially not considered to be a LAB species of industrial interest due to its slow acidification activity compared to other starter LAB such as *Lactococcus lactis* or *Streptococcus thermophilus* (Edima et al. 2008). However, since it has been regarded as a good candidate for biological TTI, its kinetic properties have been extensively studied and several growth and acidification models were developed to predict its behavior in different TTI labels offered by Cryolog (Ellouze et al. 2008, Girardeau et al. 2019). Other LAB strains that have been proposed in biological TTI systems are *L. sakei*, *L. rhamnosus* and *Weisella* (*W.*) *cibaria* (Table 2), all commonly used in the food industry. *Lactobacillus sakei* is a hetero-fermentative psychrotrophic lactic acid bacterium used in various fermented meats such as *saucisson* (French dry cured sausage). This strain presents similar kinetic behaviors as many spoilage bacteria of meat products (Vaikousi et al. 2009), and has in addition been shown to inhibit *Listeria monocytogenes* in chicken cold cuts (Katla et al. 2002). *L. rhamnosus* and *W. cibaria* are both hetero-fermentative, mesophilic bacteria. The first is known for its probiotic properties (Saad et al. 2013), and the former is commonly isolated from fermented foods and is increasingly being investigated to be used as a starter for sourdough and vegetable fermentation (Ricciardi et al. 2009).

An additional approach to further extend TTI response ranges would be to investigate the use of co-cultures. Bacterial cultures applied in food fermentations mostly consist of consortia involving multiple species and/or strains, where each member is affected by the growth and metabolism of the other. A perfect example of this is the yoghurt consortium, consisting of a protocooperation between *Streptococcus thermophilus* and *L. delbrueckii* subsp. *bulgaricus*. In milk fermentation, to produce yoghurt, the first microbial exponential growth phase is the one of *S. thermophilus*, a species that is more tolerant to neutral pH and more effective in metabolizing amino acids and trace elements than *L. bulgaricus* (Sieuwerts 2016). The growth of *S. thermophilus* results in medium acidification, reduction of oxygen and increase of CO_2 concentrations, thereby creating a favorable environment for the more acidophilic and less oxygen tolerant *L. bulgaricus* (Sasaki et al. 2014). A similar type of interaction could be imagined within a biological TTI, where a strain such as *C. maltaromaticum*, very tolerant to high pH values (up to pH 9.5), would begin to acidify the TTI medium down to a value that favors the growth of a second, more acid tolerant strain, such as *W. cibaria*. The combination potential between two or more microorganisms within a TTI is immense, as microbial interactions are governed by a multitude of molecular and physiological mechanisms. Still, systematic optimization and repeatability studies are required to choose and confirm the applicability of new bacterial species in a biological TTI system. Moreover, the processes of acquiring a take-to-market authorization for a single or a combination of species are both long and expensive. Mapping the range of shelf-lives that can be achieved with a selected strain as well as screening new strains and co-culture exploration must therefore be concurrent R&D activities.

Concluding summary on LAB-based TTI parametrization

LAB-based TTI parametrization can be done according to different strategies. For a (or several) given microorganism(s) they include: adjusting the initial starter concentration, modulating the physiological state of a selected strain or tweaking the TTI's core medium by adjusting its chemistry (initial pH, a_w, available nutrients, etc.), and switching chromatic pH indicators. The selection of different microorganisms (genus, species or strain), in pure cultures or in associations (co-cultures) are also relevant approaches.

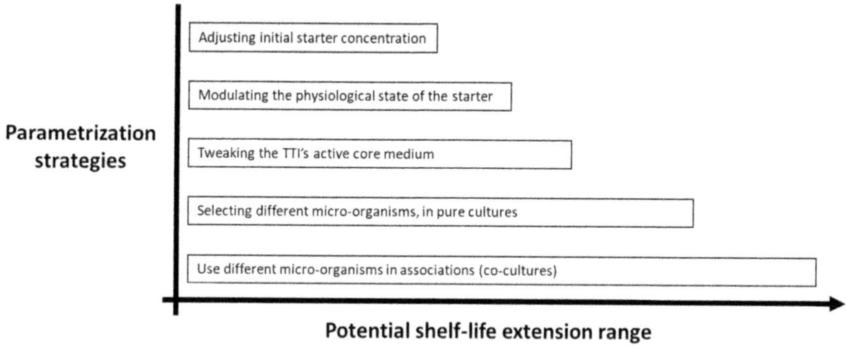

Figure 4: Schematic representation of biological TTI parametrization strategies and their expected potential shelf-life extension ranges.

Figure 4 summarizes the different parametrization strategies described in this chapter and their corresponding potential in extending the shelf-life ranges covered by a biological TTI device. All of these strategies can be pursued sequentially or in parallel to achieve response times covering the widest possible range of perishables food shelf-lives, from shellfish to sandwiches.

Performance evaluation

In many of the studies proposing biological TTI prototypes (Table 2), the Arrhenius activation energy (Ea), indicating temperature sensitivity of the TTI's response, is used as the main performance evaluation parameter (Vaikousi et al. 2009, Kim et al. 2012, Lim et al. 2014). According to this approach, a TTI is considered effective if it presents Ea response values similar to those of the spoilage reaction being tracked. However, relying solely on activation energy values is not sufficient to prove the applicability of a system intended to track the microbiological quality of specific perishable foods. In order to properly evaluate performance and be granted a take-to-market authorization, a LAB-based TTI must undergo rigorous testing, much like shelf-life date labels: durability studies and challenge tests.

First, food business operators generally start by consulting predictive microbiological models to estimate the microbiological quality evolution of a product. While these models provide good estimations of spoilage kinetics by taking into account composition and manufacturing processes, they are developed assuming consistent microbial responses in very controlled environments and can thus sometimes fail to accurately predict inconsistent microbial responses and variations in a complex food matrix. This first shelf-life estimation must therefore be followed by durability studies. Durability studies (or storage tests) are conducted by quantitatively determining growth of target spoilage and pathogenic microorganisms within their food product, under "reasonably foreseeable conditions of distribution, storage and use" (European Commission Regulation No. 2073/2005). A recommended temperature scenario, or reference profile, is to maintain the food product one third of its estimated shelf-life time at 4°C and two thirds at 8°C. For highly perishable foods that may contain pathogens such as *Listeria monocytogenes*, EU guidelines recommend additional challenge tests. Challenge tests are similar to the durability studies mentioned above, with the difference being that specific cold-tolerant pathogens are added to the food product before it is packaged. This is especially important when the microorganism's environment within the food product (chemical characteristics, temperature, a_w, etc.), is close to its growth/no growth limit.

The applicability of Cryolog's commercialized TTI was therefore evaluated with both durability studies and challenge tests using ground beef, cooked chicken and smoked salmon (Ellouze et al. 2010, 2011). These food products and their corresponding TTI labels were exposed to three dynamic temperature profiles (Figure 5):

A. A reference profile that was used to estimate shelf-life: one third of the time at 4°C and the rest of the time at 8°C.
B. A well-managed cold chain scenario with temperature fluctuations averaging below the reference condition (4°C).
C. A cold-chain failure scenario with temperature fluctuations averaging well above reference condition.

For each temperature profile, the growth of endogenous food microorganisms and added pathogens was evaluated at the end of the estimated shelf-life and at the TTI's endpoint (Ellouze et al. 2010, 2011). Results showed that for the first profile (reference profile, Figure 5A), the TTI endpoint coincided with the estimated shelf-life, confirming proper TTI parametrization according to current established guidelines within the food industry. For the second temperature profile (well managed cold chain, Figure 5B), the TTI endpoint came after the estimated shelf-life, but still before pathogenic or spoilage microorganism levels reached consumption limits. This scenario thus demonstrates how TTI can contribute to reducing food waste. In the third profile (cold-chain failure, Figure 5C), the TTI endpoint was reached before the estimated shelf-life, thus evidencing how the use of TTI helps avoid consumption of potentially spoiled food. Cryolog thus successfully proved their TTI system's ability to integrate the time-temperature history of highly perishable foods and adjust shelf-life information accordingly. Smart devices such as these can thus contribute to reduce food waste (Figure 5B), without compromising food safety (Figure 5C).

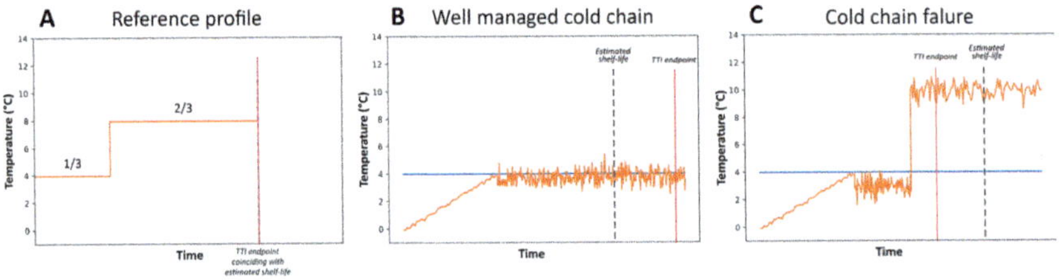

Figure 5: Simulation of different, realistic storage conditions used to evaluate LAB-based TTI performance.

Current applications and prospects

Research on various time-temperature integrator devices increased back in the early 1980s, in the hope of replacing shelf-life date labels entirely, with a label that was more representative of the real time-temperature history of a product and also more comprehensible by consumers (Blixt and Tiru 1976, Labuza 1982, Taoukis and Labuza 1989). Half a decade later, although food regulations in most countries still currently require a date on shelf-life labelling of pre-packaged perishable products, an increasing number of food business operators use TTI, along specific stages of their manufacturing process to control proper cold-chain adherence and raise staff awareness.

A good example of this kind of TTI usage is that of an industrial catering kitchen using TopCryo® labels to ensure cold-chain integrity of products that go through preliminary preparations before packaging. Following a preliminary hazard analysis, the kitchen's quality control manager determined that products undergoing a preliminary preparation phase, such as slicing, must then be stored at a maximum temperature of 3°C and used within three days. Before using TTI labels, if one of those intermediary products had been temporarily exposed to temperatures above 3°C the established precautionary principle at that company was to either use it within 24 hours or discard it. Since using TopCryo® label parametrized for a three-day/3°C shelf-life, rules have changed and produce is safely used within the three-day limit, as long as its TTI label remains green. Another example of current LAB-based TTI use is in the deli section of a supermarket, where the sanitary control plan established an internal shelf life of four days at 4°C for all sliced meats. During a test phase, corresponding TTI labels placed next to the meats met their endpoint under four days in more than half of the cases, highlighting poor cold-chain management (display temperature too high, long waiting time between cutting and refrigeration). The quality control manager was then able to bring awareness to the staff and implement corrective actions. LAB-based TTI are also currently making their way onto catering trays, take-away boxes and food delivery packages to inform customers of proper storage and transportation within the cold-chain.

As TTI research continues to grow, LAB-based TTI labels will certainly become an inherent part of production, storage, transportation and distribution of perishable food products. Supermarkets could add TTI tracking devices on their drive-through grocery pickup or delivery bags and postal services could paste TTI labels on temperature sensitive packages. Ultimately, TTI information such as color/pH recognition could be digitally integrated and used on mobile apps, allowing remaining shelf-life calculations, data collection, etc.

Conclusion

Time-temperature integrators are devices that do not simply aim to report a temperature limit breach, but translate all inevitable temperature variations experienced by perishable foods, from production to distribution. They thus integrate the time and temperature history of the product they are tracking and adjust shelf-life information accordingly. Among the different types of TTI systems developed to date, biological TTIs have the advantage of basing their response upon the same, temperature dependent, microbial growth and metabolic activity that causes food spoilage and end of shelf-life. To date, the available biological TTI systems still offer a limited range of shelf-life. A lot of work has yet to be carried out in biological TTI research and in overcoming regulatory obstacles that keep this technology from being fully deployed within the food industry. Nevertheless, it is a promising and a reliable technology that has great potential in reducing food poisoning risks and food waste.

Acknowledgments

This work was supported by Clock-°T (Cryolog). This project has also received funding from the European Union's Horizon 2020 research and innovation program under grant agreement No. 777657.

References

Blixt, K. and M. Tiru. 1976. An enzymatic time/temperature device for monitoring the handling of perishable commodities. Dev. Biol. Stand. 36: 237–241.

Brillet-Viel, A., M.-F. Pilet, P. Courcoux, H. Prévost and F. Leroi. 2016. Optimization of growth and bacteriocin activity of the food bioprotective *Carnobacterium divergens* V41 in an animal origin protein free medium. Front. Mar. Sci. 3: 168.

Conrad, P.B., D.P. nMiller, P.R. Cielenski and J.J. de Pablo. 2000. Stabilization and preservation of *Lactobacillus acidophilus* in saccharide matrices. Cryobiology 41: 17–24.

Dalgaard, P. 1995. Quality changes and shelf life of chilled fish. *In*: Huss, H.H. (ed.). Quality and Quality Changes in Fresh Fish. Food and Agriculture Organization of the United Nations, Rome, Italy. Available at: http://www.fao.org/3/V7180E/V7180E07.htm.

Edima, H.C., C. Cailliez-Grimal, A-M. Revol-Junelles, E. Rondags and J.-B. Millière. 2008. Short communication: impact of pH and temperature on the acidifying activity of *Carnobacterium maltaromaticum*. J. Dairy Sci. 91: 3806–3813.

[EFSA] European Food Safety Authority. 2015. Regulation No. 764/2008 of the European Parliament and of the Council. *In*: Core EU Legislation. Macmillan Education UK, London, pp. 183–186.

Ellouze, M., M. Pichaud, C. Bonaiti, L. Coroller, O. Couvert, D. Thuault and R. Vaillant. 2008. Modelling pH evolution and lactic acid production in the growth medium of a lactic acid bacterium: Application to set a biological TTI. Int. J. Food Microbiol. 128: 101–107.

Ellouze, M. and J.-C. Augustin. 2010. Applicability of biological time temperature integrators as quality and safety indicators for meat products. Int. J. Food Microbiol. 138: 119–129.

Ellouze, M., J.-P. Gauchi and J.-C. Augustin. 2011. Use of global sensitivity analysis in quantitative microbial risk assessment: Application to the evaluation of a biological time temperature integrator as a quality and safety indicator for cold smoked salmon. Food Microbiol. 28: 755–769.

European Commission. 2005. Regulation No. 2073/2005 on microbial criteria for foodstuffs. *In*: Official Journal of the European Union, L 338–26.

European Commission. 2018. Market study on date marking and other information provided on food labels and food waste prevention. Available at: https://circulareconomy.europa.eu/platform/sites/default/files/market_study_on_date_marking.pdf.

Galagan, Y. and W.-F. Su. 2008. Fadable ink for time–temperature control of food freshness: Novel new time–temperature indicator. Food Res. Int. 41: 653–657.

Girardeau, A., C. Puentes, S. Keravec, P. Peteuil, I.C. Trelea and F. Fonseca. 2019. Influence of culture conditions on the technological properties of *Carnobacterium maltaromaticum* CNCM I-3298 starters. J. Appl. Microbiol. 126(5): 1468–1479.

Gou, M., G. Guo, J. Zhang, K. Men, J. Song, F. Luo, X. Zhao, Z. Qian and Y. Wei. 2010. Time–temperature chromatic sensor based on polydiacetylene (PDA) vesicle and amphiphilic copolymer. Sensor. Actuat. B-Chem. 150: 406–411.

Hüfner, E. and C. Hertel. 2008. Improvement of raw sausage fermentation by stress-conditioning of the starter organism *Lactobacillus sakei*. Curr. Microbiol. 57: 490–496.

Katla, T., T. Møretrø, I. Sveen, I.M. Aasen, L. Axelsson, L.M. Rørvik and K. Naterstad. 2002. Inhibition of *Listeria monocytogenes* in chicken cold cuts by addition of sakacin P and sakacin P-producing *Lactobacillus sakei*. J. Appl. Microbiol. 93: 191–196.

Kerry, J.P., M.N. O'Grady and S.A. Hogan. 2006. Past, current and potential utilisation of active and intelligent packaging systems for meat and muscle-based products: A review. Meat Sci. 74: 113–130.

Kim, Y.-A., S.-W. Jung, H.-R. Park, K.-Y. Chung and S.-J. Lee. 2012. Application of a prototype of microbial time temperature indicator (TTI) to the prediction of ground beef qualities during storage. Food Sci. Anim. Resour. 32: 448–457.

Labuza, T.P. 1982. Scientific evaluation of shelf-life. pp. 41–87. *In*: Labuza, T.P. (ed.). Shelf-life Dating of Foods. Food & Nutrition Press, Westport, USA.

Lee, S. and S. Jung. 2013. USA Patent No. US 9476083B2. Time-temperature indicator, method for manufacturing the time-temperature indicator, quality guarantee system using the time-temperature indicator, and quality guarantee method using the quality guarantee system.

Lim, S.H., W.Y. Choe, B.H. Son and K. Hong. 2014. Development of a microbial time-temperature integrator system using lactic acid bacteria. Food Sci. Biotechnol. 23: 483–487.

Linders, L.J., W.F. Wolkers, F.A. Hoekstra and K. van't Riet. 1997. Effect of added carbohydrates on membrane phase behavior and survival of dried *Lactobacillus plantarum*. Cryobiology 35: 31–40.

Louvet, O., D. Thuault and R. Vaillant. 2005. World Intellectual Property Organization Patent No. WO 2005/026383A1. Method and device for determining if a product is in condition for use or consuption.

Lu, L., Z. Jia and Y. Cai. 2013. Chinese Patent Application 2013/102507576A. Microbial type time–temperature indicator for low-temperature circulation items.

Marceau, A., M. Zagorec and M.C. Champomier-Vergès. 2002. Positive effects of growth at suboptimal temperature and high salt concentration on long-term survival of *Lactobacillus sakei*. Res. Microbiol. 154: 37–42.

McLeod, A., M. Zagorec, M.-C. Champomier-Vergès, K. Naterstad and L. Axelsson. 2010. Primary metabolism in *Lactobacillus sakei* food isolates by proteomic analysis. BMC Microbiol. 10: 120.

McMeekin, T., J. Bowman, O. McQuestin, L. Mellefont, T. Ross and M. Tamplin. 2008. The future of predictive microbiology: Strategic research, innovative applications and great expectations. Int. J. Food Microbiol. 128: 2–9.

Mel, M., M. Karim, M. Salleh and N. Amin. 2008. Optimizinig media of *Lactobacillus rhamnosus* for lactic acid fermentation. J. Appl. Sci. 8: 3055–3059.

Meneghel, J., S. Passot, S. Dupont and F. Fonseca. 2017. Biophysical characterization of the *Lactobacillus delbrueckii* subsp. *bulgaricus* membrane during cold and osmotic stress and its relevance for cryopreservation. Appl. Microbiol. Biotechnol. 101: 1427–1441.

Palmfeldt, J. and B. Hahn-Hägerdal. 2000. Influence of culture pH on survival of *Lactobacillus reuteri* subjected to freeze-drying. Int. J. Food Microbiol. 55: 235–238.

Park, H.R., Y.A. Kim, S.W. Jung, H.C. Kim and S.J. Lee. 2013. Response of microbial time temperature indicator to quality indices of chicken breast meat during storage. Food Sci. Biotechnol. 22: 1145–1152.

Reddy, K.B.P.K., S.P. Awasthi, A.N. Madhu and S.G. Prapulla. 2009. Role of cryoprotectants on the viability and functional properties of probiotic lactic acid bacteria during freeze drying. Food Biotechnol. 23: 243–265.

Ricciardi, A., E. Parente and T. Zotta. 2009. Modelling the growth of *Weissella cibaria* as a function of fermentation conditions. J. Appl. Microbiol. 107: 1528–1535.

Ruiz-Barba, J.L. and R. Jiménez-Díaz. 1994. Vitamin and amino acid requirements of *Lactobacillus plantarum* strains isolated from green olive fermentations. J. Appl. Bacteriol. 76: 350–355.

Saad, N., C. Delattre, M. Urdaci, J.M. Schmitter and P. Bressollier. 2013. An overview of the last advances in probiotic and prebiotic field. LWT-Food Sci. Technol. 50: 1–16.

Sahin, E., M. Zied Babaï, Y. Dallery and R. Vaillant. 2007. Ensuring supply chain safety through time temperature integrators. Int. J. Log. Manag. 18: 102–124.

Salman, H., E. Tenetov, L. Feiler and T. Raimann. 2009. European Patent No. EP 2121870A1. Time-temperature indicator based on oligomeric spiroaromatics.

Sasaki, Y., H. Horiuchi, H. Kawashima, T. Mukai and Y. Yamamoto. 2014. NADH oxidase of *Streptococcus thermophilus* 1131 is required for the effective yogurt fermentation with *Lactobacillus delbrueckii* subsp. *bulgaricus* 2038. Biosci. Microbiota. Food Health 33: 31–40.

Schoug, Å., J. Fischer, H.J. Heipieper, J. Schnürer and S. Håkansson. 2008. Impact of fermentation pH and temperature on freeze-drying survival and membrane lipid composition of *Lactobacillus coryniformis* Si3. J. Ind. Microbiol. Biotechnol. 35: 175–181.

Sieuwerts, S. 2016. Microbial interactions in the yoghurt consortium: current status and product implications. SOJ Microbiol. Infect. Dis. 4: 1–5.

Taoukis, P.S. and T.P. Labuza. 1989. Applicability of time-temperature indicators as shelf life monitors of food products. J. Food Sci. 54: 783–788.

Troller, J.A. and J.V. Stinson. 1981. Moisture requirements for growth and metabolite production by lactic acid bacteria. Appl. Environ. Microbiol. 42(4): 682–687.

Vaikousi, H., C.G. Biliaderis and K.P. Koutsoumanis. 2008. Development of a microbial time/temperature indicator prototype for monitoring the microbiological quality of chilled foods. Appl. Environ. Microbiol. 74: 3242–3250.

Vaikousi, H., C.G. Biliaderis and K.P. Koutsoumanis. 2009. Applicability of a microbial Time Temperature Indicator (TTI) for monitoring spoilage of modified atmosphere packed minced meat. Int. J. Food Microbiol. 133: 272–278.

Vaillant, R. and P. Peteuil. 2012. USA Patent No. US 8215088B2. Method and means for the manufacturing of individual packaging for a liquid or solid product.

van de Guchte, M., P. Serror, C. Chervaux, T. Smokvina, S.D. Ehrlich and E. Maguin. 2002. Stress responses in lactic acid bacteria. pp. 187–216. *In*: Siezen, R.J., J. Kok, T. Abee and G. Schasfsma (eds.). Lactic Acid Bacteria: Genetics, Metabolism and Applications. Springer, Dordrecht, Netherlands.

Velly, H., F. Fonseca, S. Passot, A. Delacroix-Buchet and M. Bouix. 2014. Cell growth and resistance of *Lactococcus lactis* subsp. *lactis* TOMSC161 following freezing, drying and freeze-dried storage are differentially affected by fermentation conditions. J. Appl. Microbiol. 117: 729–740.

Wang, S., X. Liu, M. Yang, Y. Zhang, K. Xiang and R. Tang. 2015. Review of time temperature indicators as quality monitors in food packaging: review of time temperature indicators. Packag. Technol. Sci. 28: 839–867.

Wang, Y., G. Corrieu and C. Béal. 2005. Fermentation pH and temperature influence the cryotolerance of *Lactobacillus acidophilus* RD758. J. Dairy Sci. 88: 21–29.

Zeng, J., S. Roberts and Y. Xia. 2010. Nanocrystal-based time–temperature indicators. Chem. Eur. J. 16: 12559–12563.

Zweig, S.E. 2005. USA Patent No. US 6950028B2. Electronic time-temperature indicator.

15

Impact of Probiotics on Animal Health

Sabrina da Silva Sabo,[1] *Elías Figueroa Villalobos,*[2] *Anna Carolina Meireles Piazentin,*[1] *André Moreni Lopes*[3] and *Ricardo Pinheiro de Souza Oliveira*[1,*]

Introduction

The concept of probiotics is very well known and their effects on human's health have been extensively reported along the last years. Meanwhile, the application of probiotics in feed nutrition are far less explored and documented. Probiotics started to be described in 1974, when Parker stated that probiotics are "organisms and substances which contribute to intestinal microbial balance" thus including both living organisms and non-living substances. Later, Fuller (1989) defined probiotics as "a live microbial feed supplement which beneficially affects the host animal by improving its intestinal microbial balance". At the same moment, the Food and Drug Administration (FDA) adopted the term "direct-fed microbials". Except for the health aspect stated in the probiotic definition, direct-feed administration was described as feed supplements that consist of "viable naturally occurring microorganisms", which means that the manufacturers are instructed to use this term on their ingredient lists, but FDA does not allow them to make therapeutic claims (Zoumpopoulou et al. 2018). Over the years,

[1] Department of Biochemical and Pharmaceutical Technology, Faculty of Pharmaceutical Sciences, University of São Paulo, São Paulo, Brazil.
Emails: sabrinasabo@gmail.com; legecarol@gmail.com

[2] Núcleo de Investigación en Producción Alimentaria, Departamento de Ciencias Biológicas y Químicas, Facultad de Recursos Naturales, Universidad Católica de Temuco, Temuco, Chile.
Email: efigueroa@proyectos.uct.cl

[3] Faculty of Pharmaceutical Sciences, University of Campinas – UNICAMP, Campinas, São Paulo, Brazil.
Email: amorenilopes@gmail.com

* Corresponding author: rpsolive@usp.br

several researches have been demonstrating that such feed supplementation might improve the host health intestinal microbiota promoting a balance between beneficial commensal and potentially pathogenic bacteria (Gaggìa et al. 2010). Thus, nowadays, both terms, probiotics and direct-feed microbials, are accepted by regulatory agencies to designate animals' supplementation that consist of live microorganisms that, when administered in adequate amounts, confer a health benefit on the host (FAO 2016).

The most important objectives for using probiotics (or direct fed microbials) in animals feed are to maintain and improve the performance (productivity and growth) of the host, including the control of enteric pathogens. During the last decades, probiotics are becoming increasingly popular as one of the alternatives for antibiotics, commonly used in farming animals in subtherapeutic doses to achieve the same purposes aforementioned. They are also known as Antibiotic Growth Promoters (AGPs) (FAO 2016).

The use of antibiotics in animal production originated more than 70 years ago, when chlortetracycline was shown as an improvement on the growth and health of birds (Moore et al. 1946). Since then, antibiotics have been widely used in the treatment and prevention of diseases or even to improve growth and productivity (Broom et al. 2018). The reason for this occurrence can be associated with the intensification of livestock farming as a consequence of increased consumer demand and the need to achieve efficiency conversion of natural resources into food animal products (Ronquillo and Hernandez 2017). The different applications of antibiotics in food animals have been described as therapeutic, prophylactic, and subtherapeutic uses and can be applied to treat a single animal with clinical disease or a large group of animals (Landers et al. 2012).

The concerns regarding the use of AGPs are associated that these substances can accumulate significantly in edible tissues and products (Ronquillo and Hernandez 2017), and also in the environment through excreting via urine, faeces (Ostermanna et al. 2013), milk (Arikan et al. 2009), and eggs (Idowu et al. 2010). These facts may represent a real threat to the consumer, either through exposure to the residues; increased allergies resulted from antibiotic presence in foods, or the development of bacteria resistant to drugs (Dibner and Richards 2005). Bacteria possessing resistance genes are a potential risk for humans if they are transferred to persons (Castanon 2007). In fact, there is a substantial body of evidence suggesting an epidemiological link between the antibiotic use in food animals and antibiotic resistance in humans (Koike et al. 2017).

The first evidence of this correlation was reported in 1995, when after banning avoparcin, which is chemically related to the glycopeptide vancomycin commonly used as a growth promoter for broilers, there was a reduction in the prevalence of vancomycin-resistant enterococci (VRE) from 72% in 1995 to 2% in 2005 (DANMAP 2005). An equivalent decline in prevalence of *Enterococcus* (*Ec.*) *faecium* resistant to avilamycin was observed after the prohibition of this growth promoter (DANMAP 2005).

Also in Denmark, the prevalence of VRE in pigs persisted for three years after the ban of avoparcin, until growth promoter tylosin (a macrolide) was prohibited. The ban on tylosin was related to a reduction in the prevalence of VRE in pigs. The genetic characterization of VRE isolated from these animals revealed the presence of plasmids encoding resistance to glycopeptides and macrolides, suggesting that the resistance to vancomycin could have been a consequence of using tylosin after the avoparcin prohibition (Hasman and Aarestrup 2005).

Increasing evidences led the World Health Organization (WHO) to become aware of the importance of this subject, suggesting periodically over the last few years the phase out or even the complete banishment of using AGPs (WHO 1997, 2004, 2017). The European Commission (EC) banned the marketing and use of AGPs in feed nutrition since January 1st 2006 (EC Regulation No. 1831/2003), although ionophores continue to be administered (Economou and Gousia 2015). In the United States, the use of fluoroquinolones in poultry was banned in 2005. After that, the rise of fluoroquinolone-resistant *Campylobacter* infections in humans has significantly decreased (FDA 2018). However, other than this action, little has been effectively done by regulatory authorities to reduce inappropriate and unnecessary use of antibiotics on livestock. In South America, Brazil, known as the second largest poultry producer (Ferreira et al. 2018), following the recommendations of the international agencies such as Food and Agriculture Organization of the United States (FAO), OIE World Organization for Animal Health, and the EC, and based on scientific evidence for the analysis of risks associated with veterinary drug residues in foods, has restricted since 1998 the use of antibiotics avoparcin, amphiphilic, tetracyclines, penicillins, cephalosporins, quinolones, sulfonamides, erythromycin, spiramycin, and recently colistin (Bezerra et al. 2017).

Such phasing out of using AGPs represents an unreasonable pressure on the livestock industry (Casewell et al. 2003). Motived to reduce the massive use of AGPs, the seeking for viable alternatives that can improve the natural defense mechanisms of these animals has increased exponentially (Gaggìa et al. 2010). In this sense, using probiotic microorganisms can be an efficient solution for such circumstances.

As mentioned before, probiotics are defined as "living microorganisms which, when administered in adequate quantities, confer benefits to the host's health" (Hill et al. 2014). The benefits proposed by using them as a food additive are associated with the balance and activity of the intestinal microbiota, increasing the resistance against pathogenic bacteria, and improving the immunity of the intestinal mucosa providing health to the host (Choct 2009). Such benefits are supplied by probiotic strains capable of producing fatty acids and organic acids, as well as the synthesis of compounds such as vitamins and antimicrobial biomolecules (bacteriocins, for example), which act specifically against certain pathogens or in the activity of certain enzymes (Hill et al. 2014). It is important to highlight that the most significant effects of probiotics in feed nutrition have been reported when they are included in the diet of animals during particularly stressful periods, such as weaning, the lactation period, after sudden change in diet, and other situations that may threaten the host's health (Chaucheyras-Durand and Durand 2010).

Although several strains have already been described in the literature as probiotic, in 2010 the International Scientific Association for Probiotics and Prebiotics (ISAPP) published a consensus report on the use of the term "probiotic", in which it states that the beneficial effects generated by these microorganisms are strain-specific (Sanders 2010). Additionally, many studies have pointed out that the effects generated by probiotics are also host-specific (Nader-Macias and Tomás 2015). Thus, strains isolated from a particular host have a high capacity to colonize and/or survive in the gastrointestinal tract conditions, exerting physiological effects in individuals of the same species, which means that probiotics isolated in humans, probably will not have the same effect when administered in animals and vice versa (Gilliland 2011).

Another feature required is that the probiotic strains should not compete with the microorganisms of the normal microbiota, instead they should interact in symbiosis. Moreover, they must be resistant to bile and acids from the gastrointestinal tract and have the ability to adhere to the intestinal mucosa cells, rapidly multiplying in order to always be present in the intestinal lumen and not be eliminated with intestinal peristalsis. Another fundamental criterion is the safety for use in humans or animals (Musa et al. 2009).

The probiotic microorganisms usually used in the food, pharmaceutical, and veterinary industries belong to the group of lactic acid bacteria (LAB), since they are known as desirable members of the intestinal microbiota being historically applied over the years in the manufacture of dairy foods and as starter cultures receiving the status of "generally regarded as safe" (GRAS) (Wessels et al. 2004).

Based on the above context, this chapter will explore the main targets for probiotic LAB uses in the mean polygastric (cattle, goat, sheep and lamb) and monogastric (poultry, swine, horses and pets) animals as well aquatic animals. The current knowledge on the contribution of probiotics to the well-being of the host intestinal microbiota will be also discussed.

Application of Probiotics in Polygastric Animals

Animals classified as polygastric are also known as ruminants, comprising up to 150 domestic and wild species, while economic interest lies mainly in the breeding of cattle, sheep, goats, and lambs (Zoumpopoulou et al. 2018).

Ruminants' digestive tract has some peculiarities, it contains four-compartmentalized stomach chambers: the rumen, the reticulum, the omasum, and the abomasum, each one performing in different processes. Briefly, in the rumen the fibers and solid feeds are fermented by commensal microbiota. The liquids are transferred to the reticulum, also serving to the entrapment of large feed particles, which are regurgitated subsequently for a complete digestion. In the omasum the liquids are filtered and various nutrients are absorbed. Finally, the enzymatic digestion of the feed takes place in the abomasum (Hofmann 1989, Zoumpopoulou et al. 2018).

The rumen is the main chamber of food digestion, for this reason probiotics have been selected mainly to target this compartment. The rumen microbiota is predominantly composed of a wide diversity of strictly anaerobic bacteria, ciliate protozoa, anaerobic fungi, and archaea, playing an important role on degradation and fermentation of dietary substances (Chaucheyras-Durand and Durand 2010). The large amount of energy recovery from dietary polysaccharides has been attributed to the action of the rumen's microbial community; however, this process also depends on the composition of the microbiota inherent in this organ (Uyeno et al. 2015).

The balance of rumen microbiota is essential to ensure health and productivity of the ruminants, since several physiological parameters of such animals are strictly correlated with the microorganisms composing the rumen (Jami et al. 2014).

Probiotics have been widely administrated to ruminants to promote the microbiota balance and, consequently, improving their health and well-being through treatment of digestive disorders and reduction of gut pathogens (Chaucheyras-Durand and Durand 2010). Furthermore, rumen fermentation is beneficial, especially on feed digestibility, improving the cellulolysis and the synthesis of microbial protein during digestion, enhancing the nutrient absorption (Yoon and Stern 1995). Additionally, probiotics are capable of stabilizing the rumen pH and lactate levels, reduce ruminal acidosis, improve milk yield and composition, stimulate the immune system, enhance animal performance, increase feed conversion efficiency, and decrease stress (Chaucheyras-Durand and Durand 2010).

Studies on the application of probiotics in ruminants have been performed in the pre-ruminant's life and in adult ruminants, aiming both at the health status of the animals and economic parameters (Gaggìa et al. 2010). The next sections will explore the use of probiotic on health and welfare on the most important animals from the economic point of view: cattle, sheep, goats and lambs.

Cattle

In recent years, the global population growth, the greater demand for food, and the consequent reduction of agricultural areas are situations that require the intensification of animal production systems (Frizzo et al. 2018). Following this trend, the livestock producers have been adopting strategies to intensify and to increase the productivity. These strategies can be characterized as stressful conditions to the animals, unbalancing the intestinal microbiota and, as consequence, reducing animals' natural resistance against contamination or pathogen colonization, which can result in digestive disorders (Frizzo et al. 2018). In cattle production, the high temperatures in feedlot, transportation to long distances, changes in diet, management methods, early weaning and artificial habitat, are some examples of such stress factors (Lamy et al. 2012). It is important to highlight that the intestinal unbalance, besides promoting digestive disorders, can also cause consequences such as reducing the profitability of livestock production and facilitating disease transmission to humans by direct contact with ill animals or their feces or even through the food chain (Frizzo et al. 2018).

Before the ban or the restriction on using antibiotics in animal production, they were indiscriminately applied in order to prevent/control intestinal infections and to obtain better productivities of animals (Frizzo et al. 2018). Considering all facts previously mentioned regarding antibiotics in animal production, cattle producers have been alternatively using probiotics in different stages of cattle growth aiming to mitigate stressful factors, to control infectious diseases, to improve productive performances, to increase dry matter digestibility in cattle, to increase milk yield and quality, and to increase feed efficiency (Huber 1997).

Calves, before weaning, are susceptible to many pathogens affecting their performance during subsequent rearing, and it can be modified by diseases, especially gastrointestinal infections (Frizzo et al. 2011). A meta-analysis conducted by Signorini et al. (2012) demonstrated that probiotic LAB supplementation has shown a protective effect reducing the incidence of diarrhea in calves when they were fed with whole milk mixed with a multistrain inoculum. On the other hand, in the meta-analysis published by Frizzo et al. (2011), despite demonstrating positive effect in growth of calves, such an effect was not observed when probiotic LAB was added to whole milk, but instead it was observed when they were added to milk replacers.

For dairy cattle, the use of probiotics is related to the milk productivity and nutritional quality (Krehbiel et al. 2002, Xu et al. 2017). In addition to these targets, probiotics have been used to reduce the prevalence of metritis (Genís et al. 2018). Metritis is an inflammation of the uterus occurring due to bacterial infection during the first 21 days after calving (LeBlanc et al. 2008). Metritis affects up to 40% of dairy cows and it is usually treated with antibiotics (Genís et al. 2018). In spite of their effectiveness, there is an increasing concern about excessive use of antibiotics due to the aforementioned drawbacks. In this sense, researchers have been successfully using probiotics via intravaginal administration to prevent postpartum uterine infections and inflammation (Deng et al. 2015, 2016, Genís et al. 2016, 2017a,b, 2018). Another important and prevalent disease of dairy cattle is mastitis, defined as the inflammation of the mammary glands where the tissue of the udder is severely affected (Pellegrino et al. 2018). To avoid using antibiotics to control mastitis, researches have been evaluating the effect of probiotics and their bioproducts (bacteriocin, as example) to prevent and/or for treatment of this disease (Twomey et al. 2000, Bouchard et al. 2015, Pellegrino et al. 2018). For such approach, studies are purposing intramammary infusion of live probiotics cultures, associating the positive results with stimulation of the host intramammary immune system (Crispie et al. 2008, Klostermann et al. 2008, Pellegrino et al. 2018).

Regarding beef cattle, besides promoting weight gain and increasing feed efficiency, probiotic LAB has been used to reduce fecal shedding of human pathogens, as *Escherichia* (*E.*) *coli* serotype O157:H7, responsible for many hemorrhagic colitis outbreaks, directly or indirectly associated with beef products (Zhao et al. 1998, Brashears et al. 2003). This strategy has also been successfully used with dairy cattle, and plays an important role in controlling this pathogen, considering studies suggesting that O157 is ubiquitous in the environment of dairy farms, implicating dairy cattle as principle O157 reservoirs (Stenkamp-Strahm et al. 2017). Additionally, recent studies have been evaluating the potential of probiotic LAB as preventive and therapeutic agents in upper respiratory tract infections and otitis by competing with pathogens for nutrients, space and attachment to host cells (Popova et al. 2012, Amat et al. 2017). Considering that bovine respiratory disease is responsible for economic losses that exceed those incurred by all other diseases of beef cattle combined (Duff and Galyean 2007), controlling such infections using probiotic can be a promising approach.

Other than the aforementioned applications, probiotics have also been shown positive effect in reducing the negative environmental impact, such as methane emission associated with the ruminant production (Martin et al. 2002); and more recently, it has been proven their interaction with intestinal epithelial cells and their effect on antiviral immunity (Villena et al. 2018). Additionally, the use of probiotics has been shown evidences that they are capable stabilizing the ruminal pH and thus decreasing the risk of acidosis (Ghorbani et al. 2002, Chiquette et al. 2012, Qadis et al. 2014), one of the most important digestive disorder in high producing dairy or beef cattle, varying from acute forms due to lactic acid overload to subacute forms due to the accumulation of volatile fatty acids (Chaucheyras-Durand and Durand 2010).

Sheep, goat, and lambs

As mentioned before, sheep, goat, and lambs are ruminant animals. They are mostly recognized for the production of fleece, meat, and milk. Due to indiscriminate use of antibiotics to treat mastitis in lambs, sheep, and goats, for example, these substances can be liberated through the milk. If this kind of milk is not treated correctly, it can cause food allergy, antibiotic resistance, or prevent the fermentation of cheeses and drink yoghurts produced with such milks (Berruga et al. 2016).

With the recently changes in the legislation of antibiotic used in animals, various researchers have been trying to substitute such substances by probiotics, mainly aiming at animal weight gain and to reduce some risks of intestinal infections. In the case of small ruminants, generally, such benefits are achieved through manipulation of rumen microbiota using probiotics to accelerate the feed digestion and fermentation (Angelakis 2017).

Recently, some researchers have published studies intending to prove that the antibiotics subtherapeutic doses used to promote weight gain and prevent infections in goat, sheep, and lambs could be substituted by probiotics (Jayabal et al. 2008, Jinturkar et al. 2009). Nevertheless, Apás et al. (2010) reported that a mix of probiotic strains offered to goats did not increase the weight. Instead, they improved their milk quality by stimulating higher production of fatty acids, such as conjugated linoleic acid, oleic, and linoleic acids. Kritas et al. (2006) and Maragkoudakis et al. (2010) also observed that supplementing ewes and dairy goats feed with probiotics may have beneficial effect on subsequent milk yields, fat, and protein content. Besides the fat acids improvement, Utz et al. (2018) demonstrated that the use of probiotics in the feeding of the goats can reduce the mutagen compounds produced after an infection caused by *Salmonella enterica* serovar Typhimurium.

In addition to the mentioned applications, probiotics have been used to reduce faecal shedding of Shiga toxin-producing *E. coli* (STEC) in sheep. STEC are zoonotic, foodborne pathogens of humans. Ruminants, including sheep, are the primary reservoirs of STEC and it is extremely necessary to develop intervention strategies to reduce the entry of STEC into the food chain (Rigobelo et al. 2014).

Although studies applying probiotics in goats, sheep, and lambs feed are restricted, they are necessary principally because cheeses made of goats' milk are worldwide consumed. Besides, there are some regions that do not have enough space to raise cattle, so there is a high consumption of goat and sheep milk considering that these animals are cheaper and easier to maintain compared with cattle.

Application of Probiotic in Monogastric Animals

Monogastrics are classified as animals having a simple or single-chambered stomach with the main agricultural species being pigs, poultry, and horses (Zoumpopoulou et al. 2018).

The microbiota within the gastrointestinal tract of monogastric animals is predominantly composed of *Bacteroides, Clostridium, Bifidobacterium, Eubacterium, Lactobacillus, Enterobacteriaceae, Streptococcus, Fusobacterium, Peptostreptococcus,* and *Propionibacterium* (Gaggìa et al. 2010). Such commensal bacteria are essential to ensure health and well-being of the host with a nutritional function and a protective influence on the intestinal structure and homeostasis (O'Hara and Shanahan 2007). Particularly, the close proximity of individuals in livestock or poultry farms could facilitate the host–host transmission of microbiota. Therefore, in high population density, it is important to maintain a "healthy" microbiota as a barrier against pathogen infection (Gaggìa et al. 2010).

As mentioned before, physiological and psychological stressors caused by the intensification on agricultural systems can lead to a dysfunction of the intestinal barrier, increasing the intestinal permeability, impacting on gut microbial composition, and on susceptibility to enteric pathogens. Probiotics have been widely used in monogastric, as well in polygastric animals, aiming for a balance between beneficial microorganisms and potential pathogenic ones (Ahasan et al. 2015).

The most frequently used probiotic in monogastric animals are LAB (FAO 2016), targeting the colon and cecum colonization, which harbors an abundant and very diverse microbial population (Chaucheyras-Durand and Durand 2010). Similar to the benefits proposed for polygastric animals, probiotic LAB promise the increase of body weight, the reduction of the risk of diarrhea, the improvement of feed efficiency, and food digestibility (Ahasan et al. 2015).

Details of using probiotic LAB in monogastric animals will be discussed in the following sections. It is important to highlight that although pigs, poultry, and horses are monogastric animals of major agricultural importance, the use of probiotics in pets (cats and dogs) will be also reported, since the concern with the diet of such animals has been received increasing attention, not only because most of pet owners consider the animals to be family members (Cohen 2002), but also because recent studies demonstrated that there is a relationship between pets and owners microbiota; in the worst case, including pathogenic bacteria (Grześkowiak et al. 2015).

Poultry

Poultry meat and eggs are the cheapest source of animal proteins compared to all other livestock products, contributing significantly to supplying the growing demand for animal food products around the world (Farrell 2013). The consumption of poultry products has been increasing rapidly in the last few decades, which make it the first largest source of meat (FAO 2018). For that reason, the poultry industry has become an important economic activity in many countries (Mirza et al. 2018). On the other hand, in large-scale rearing facilities, poultry are exposed to stressful conditions, problems related to diseases and deterioration of environmental conditions, resulting in serious economic losses (Kabir 2009). These occurrences can affect the immune system functions, predisposing broilers to colonization of the gastrointestinal tract by pathogenic bacteria, which could be a real threat to birds' health and to food safety (Gaggìa et al. 2010).

Among poultry pathogens, *Salmonella* spp., *Campylobacter jejuni*, and *Clostridium* (*C.*) *perfringens* have been shown to infect chicken and hen increasing the risk of contamination through the food chain and resulting in a harmful condition both for poultry and human (Ahasan et al. 2015). Prevention and control of diseases caused by such pathogens have led to a substantial increase in the use of veterinary medicines (Kabir 2009). On the other hand, besides the problem of banishment and the increasing restrictions, the decline in the use of antibiotics seems to be inevitable in the future, and the practice of using antimicrobials may prove economically impractical due to market limitations and export restrictions (Dibner and Richards 2005, Gadde et al. 2017). To fill this gap, probiotics have been considered as successful and eco-friendly alternatives for antibiotics in poultry nutrition (Alagawany et al. 2018).

A variety of well-characterized probiotic strains have been selected to modulate the intestinal microbiota of birds and to promote the protection against a variety of pathogens (Gaggìa et al. 2010). In fact, some of them are commercially available as feed supplement. The most known are GalliPro® (*Bacillus licheniformis*), BioPlus®YC (*Bacillus subtilis* DSM5750 and *Bacillus licheniformis* DSM5749), PoultryStar® (multistrain probiotic) and Bactocell® (*Pediococcus acidilactici* MA 18/5M). These products were only possible to be released for sale after several studies proved their benefits in poultry.

Although the use of probiotics in animals health has been only received attention in recent years, the application of such beneficial bacteria for the digestive tract of poultry is not a new concept (Kabir 2009). A strikingly crucial occurrence to the development of probiotics to such animals was first reported in 1973 by Nurmi and Rantala, who described that newly hatched chickens could be protected against *Salmonella enterica* serovar Infantis colonization by feeding them with a suspension of gut contents derived from healthy adult chicken. This concept is called competitive exclusion (Kabir 2009). Since then, several experiments with undefined and defined probiotic cultures have been developed and successfully applied to control and reduce not only *Salmonella* colonization, but also *C. jejuni*, *Listeria monocytogenes*, pathogenic *E. coli*, and *C. perfrigens* (Gaggìa et al. 2010).

In order to elucidate the concept of competitive exclusion in the short life of chickens, if the microbial colonization of intestinal tract is postponed, the birds' intestine can be open to infections. In the natural environmental, the mother is responsible for feeding her hatching chicks with the feed stored in her crop. In the mother's crop, the feed is fermented and mixed with beneficial microorganisms, which are passed to the babies supporting their intestinal and crop microbiota colonization. Additionally, the hatching chicks' intestinal colonization is ensured by the habit of eating feces. However, with the increased demand for food, commercial chickens are reared in sterile incubators, which make the young chickens have no contact with the environment, resulting in prolonged intestinal tract colonization (Mirza et al. 2018). The purpose of using probiotics in poultry farms is to provide for broilers an intestinal microbiota colonization as similar as possible to that they would get if they were in a natural environment.

Probiotics have been applied in poultry production mainly in an attempt to balance the intestinal microbiota eliminating pathogenic bacteria; to enhance the quality and composition of meat and eggs; to improve the digestion, and absorption of nutrients, consequently, boosting the productivity and health of poultry; and to stimulate the immune system (Kabir 2009, Ahasan et al. 2015, Alagawany et al. 2018, Mirza et al. 2018).

Most of the studies developed so far investigated the effects of probiotics in reducing the numbers of pathogenic microorganisms in the gastrointestinal tract, especially the aforementioned pathogens. Carter et al. (2017) provided *in vitro* evidences that probiotics strains *Lactobacillus* (*L.*) *salivarius* 59 and *Ec. faecium* PXN-33 can significantly reduce *Salmonella* Enteritidis S1400 colonization in poultry, especially when the strains were mixed, providing a superior performance comparing to individual bacterial preparations. Using another multi-strain probiotic supplementation, Prado-Rebolledo et al. (2017) reported successful results after testing a commercial supplement named FloraMax-B11®, composed of probiotic LAB capable to reduce *Salmonella enterica* serovar Enteritidis colonization and intestinal permeability in broiler chickens. Besides *Salmonella* species, probiotics have demonstrated to be efficient in controlling *Campylobacter jejuni* (Baffoni et al. 2012, Saint-Cyr et al. 2017, Smialek et al. 2018), *E. coli* (Huff et al. 2015, Tarabees et al. 2018, Zhang et al. 2016), *C. perfringens* (Jayaraman et al. 2013, Wang et al. 2017), and coccidiosis, significant enteric diseases in poultry with the potential to inflict a considerable economic effect on farm profitability (Lee et al. 2007, Abdelrahman et al. 2014).

Despite the fact that the majority of studies are mainly focused in controlling pathogenic bacteria, a considerable amount of studies are evaluating the effects of probiotics on improving growth, performance and welfare of poultry without any apparent disease. Cavazzoni et al. (1998) reported that when chickens received a probiotic *Bacillus coagulans* strain in their diet, a growth-promoting and prophylactic effect compared to the AGP virginiamycin was observed. Similarly, Mountzouris et al. (2007) demonstrated that a multi-strain probiotic formulation, besides displayed a growth-promoting effect comparable to avilamycin treatment, have modulated the composition and the activities of the broilers cecal microbiota. Wei et al. (2017) shown that the dietary supplementation of *Bacillus amyloliquefaciens* LFB112 enhanced the average daily gain and average feed intake, besides improving the meat quality from broilers. Kalavathy et al. (2003) demonstrated that feed supplementation of a mixture of 12 *Lactobacillus* strains in the diet of broilers had a beneficial effect by improving growth performance and feed conversion ratio, reducing serum total cholesterol, LDL cholesterol, triglycerides and abdominal fat deposition. Khaksefidi and Ghoorchi (2006) demonstrated that broilers receiving a dietary supplementation with *Bacillus subtilis* had improved their immune system performance. Such occurrence was possible to be seen when the broilers were immunized with Newcastle disease virus vaccine and the antibody production post immunization was significantly higher in the probiotic supplemented group compared to the control group. Furthermore, probiotics have proven to positively affect poultry's intestinal histomorphology (FAO 2016). The structure of the intestinal mucosa is an important determinant of intestinal function (digestive and absorptive) affecting growth performance of poultry. Generally, an increase in villus height and villus height:crypt ratio increases the absorption of nutrients due to a larger surface area (Afsharmanesh and Sadaghi 2014, FAO 2016). Additionally, despite divergent effects (FAO 2016), probiotics have been also used to improve the laying performance, hatching of laying hens, and sperm quality of roosters (Sobczak and Koztowski 2015, Tang et al. 2015, Mazanko et al. 2018).

Besides many such applications, researchers have recently evaluated the *in ovo* probiotic supplementation to improve the performance and immunocompetence of broiler chicks when challenged with pathogenic bacteria (Yamawaki et al. 2013, Oliveira et al. 2014, Pruszynska-Oszmalek et al. 2015, Roto et al. 2016). *In ovo* probiotic feeding consist of administrating a liquid feed composed of probiotic strains into the amnion of an egg several days before hatch. To provide nutritional substance to the birds at an early age can stimulate the development of the intestinal tract sooner, allowing more efficient digestion upon hatch (Cox and Dalloul 2014). Considering the previously discussed problem regarding commercial broilers, which has delayed intestinal tract colonization, this technique may be an interesting strategy to overcome such complication in the poultry industry.

Swine

As mentioned before, environmental stresses can affect farm animal production causing discrepancy in the intestinal diversity and facilitating pathogen infections (Gaggìa et al. 2010). Similar to cattle, in swine production, the most stressful conditions are related to the weaning and post-weaning periods, which include separation from the sow; end of the lactation immunity; early and critical transition from milk to a diet based on plant polysaccharides; and transportation to long distances (Ahasan et al. 2015). These factors can reduce feed intake, promote the inhibition of growth performance, and negatively influence the immune function and the intestinal microbiota homeostasis leading to the susceptibility of gut disorders, infections and diarrhea in pigs (Modesto et al. 2009, Ahasan et al. 2015). Additionally, sows' body weight loss during lactation negatively impacts on the following steps in the reproductive cycle, directly influencing the number of piglets born alive, the number of stillborn piglets, and the body weight of piglets at birth and weaning, due to nutrient absorption during gestation and lactation (Kritas et al. 2015).

While the legislations prohibited the use of antibiotics as AGP, there is a high consumer's demand for safe pork, leading the researchers seeking for alternative feed additives (e.g., probiotics) instead of antibiotics in swine diets aiming not only to improve performance and preserve health in such animals, but also to ensure food safety and protect consumers' health (Yirga 2015, Zimmermann et al. 2016).

Due to the banning of antibiotics, there have been an increased interest in using probiotics to decrease pathogen load and to attenuate gastrointestinal disease symptoms in pigs. For such purpose, besides *in vitro* tests to identify the best potential probiotics, several researchers are carrying out *in vivo* tests utilizing different probiotic microorganisms (Gaggìa et al. 2010).

Considering the main productive stages in pig farming, which are the reproduction cycle (e.g., pregnancy and lactation), the growing of the pigs (e.g., suckling and nursery) and growing-finishing periods, probiotics have been predominantly used during such phases (Kritas 2018).

In gestating sows, the administration of probiotics was reported to increase the sows' weight development and feed intake during lactation, besides helping to strengthen their immune system and likely reducing the neonatal piglets' mortalities (Böhmer et al. 2006). Recently, another study with gestating sows demonstrated a potential benefit of providing a combination of *Bacillus subtilis* and prebiotics during sows' gestation and lactation achieving an increased lactation feed intake (Menegat et al. 2018). Using the same *Bacillus* species, Krita et al. (2015) reported successful results when applied in sows' and piglets' diets, improving the sows' body condition during pregnancy, increasing feed consumption, reducing weight loss during lactation, decreasing the weaning interval, and achieving higher body weight of piglets at weaning.

In fact, *Bacillus* species plays an important role in swine production due to its ability to produce spores surviving to the low pH of the gastric barrier, which represents an advantage over other probiotic microorganisms (Mingmongkolchai and Panbangred 2018). Treatment of sows during early pregnancy, as well as treatment of their litters with *Bacillus cereus* var. *toyoi*, increased intestinal IgA secretion both in sows and piglets, reduced the incidence of diarrhea, and decreased incidence of *Salmonella typhimurium* infection in piglets (Schierack et al. 2007, Scharek-Tedin et al. 2013). Hu et al. (2018) reported an increased growth performance of piglets when *Bacillus amyloliquefaciens* was administrated. Such result was associated with the positive influence of the strain with the tract digestibility and intestinal integrity of piglets. Pan et al. (2017) have shown that a dietary supplementation associating *Bacillus licheniformis* and *Saccharomyces cerevisiae*, protected the pigs against *E. coli* K88, one of the major challenge for post-weaning pigs. In this study, the researchers demonstrated that the weaned pigs which received the diet containing the probiotic strains were protected of *E. coli* K88 infection by activating the immune response, attenuating intestinal damage and improving the performance and nutrient digestibility. Additionally, *Bacillus* species also proved to be effective on the health status, performance, and carcass quality of grower and finisher pigs (Alexopoulos et al. 2004).

Regarding the growing and finishing period, pigs tend to become more resilient to adverse conditions and less susceptible to diseases. However, the proper growth of pigs after weaning is greatly influenced by pathogens that often profit from the reduced immune response of the animal in each stage and cause clinical or subclinical disease. For that reason, some farmers continue using the AGPs permitted by legislation (Kritas 2018). Considering the previously discussed concerns about antibiotics on animals, researchers are being motivated to evaluate the beneficial effect of probiotics on swine growing and finishing period cycle, allowing the farmer to maintain the performance of their pigs while reserving antibiotics only for therapy or prevention of more serious health conditions (Kritas 2018). Davis et al. (2008) demonstrated that a feed supplementation composed with selected *Bacillus* strains improved feed efficiency and weight gain and decreased mortality of pigs during the growing-finishing period. Dowarah et al. (2017) showed a positive growth response in grower-finisher pigs after feeding them with probiotic LAB. In this same study, an increase of LAB in fecal samples was observed while there was a reduced count of *E. coli* and clostridia, known as the principle causes of diarrheal diseases in pigs. Furthermore, the results suggest that the use of microbes isolated from the host as probiotic is more appropriate technology for maintaining health status and performance of the animals by ameliorating weaning stress. Despite all aforementioned benefits with grower and finisher pigs, in general, supplementing swine diets with probiotics have given more positive and consistent effects in weaned piglets which may be due to their inducing better digestibility of feed, improved immunity, and increased resistance to intestinal disorders than young pigs (Liu et al. 2015).

Other than that, probiotics have been widely administrated to swines in order to improve the meat quality (Tufarelli et al. 2017, Balasubramanian et al. 2018, Chang et al. 2018), and also to decrease the faecal noxious gas emission, resulting in a viable approach to reduce animal excreta pollution (Jeong et al. 2015, Tufarelli et al. 2017). Additionally, probiotic supplementation proved to strengthen the immune system of unvaccinated sows infected with porcine epidemic diarrhea virus (Tsukahara et al. 2018).

Horses

Many probiotics are marketed for application in horses, the main rationale being the prevention or treatment of gastrointestinal illnesses, for instance, as an adjunct therapy in the management of chronic or acute diarrhea or prophylactic administration for prevention of colic related with disorders of gastrointestinal microbiota, i.e., for animals fed a high cereal ration (Geor et al. 2007, Kritas 2018, Lauková et al. 2018). Despite the high availability and employment, scientific data behind commercial formulations of probiotics in equines are still relatively limited. In addition, the quality control of products that are marketed without prescription is not strictly regulated. While promising *in vitro* results have been attained, *in vivo* health benefits have still been one of the key challenges to be overcome. Ambiguous data are caused by strain selection, dosage or true lack of efficacy, and still needs to be better understood. Even with the limitations mentioned above, probiotics are increasingly being employed due to the ease of their administration and application, lack of serious adverse effects, and low costs (Schoster et al. 2014, Kritas 2018).

More specifically, advances in the horses' treatment have risen in last years. In spite of new advances in intensive care of horses, those with severe enterocolitis may have a guarded prognosis and life-threatening complications may develop even if they survive the primary illness. The described survival rate for horses with diarrhea is around 70%; however, the prognosis is much worse for these animals with different kinds of colitis, involving antimicrobial-associated diarrhea and necrotizing enterocolitis (Mair et al. 1990, Stewart et al. 1995, Saville et al. 1996, Cohen and Woods 1999). The diagnosis of colitis is usually not performed in a timely manner; this is an important factor, since the cause of this inflammatory disease is determined in approximately 35% of horses (Mair et al. 1990). Thus, treatment is often supportive, mainly focused on electrolyte and fluids replacement, antidiarrheal agents, colloids, among others medications. Some antimicrobials, such as oxytetracycline and metronidazole have been employed for specific treatment. Other treatments such as live-culture yogurt, probiotic paste or fecal suspensions from a healthy donor have also been used to reestablish normal gastrointestinal microbiota. In the case of biotherapeutic agents for the prevention or treatment of diarrhea in these animals related to their effectiveness, safety, and application is still little known (Parraga et al. 1997, Weese 2002, Desrochers et al. 2005).

Among the several agents being investigated as antibiotic substitute in horses, some probiotics microorganisms have been employed for treatment or prevention of diarrhea. Here we show and discuss the current evidence for probiotic use in equine medicine. In this sense, Landes et al. (2008) discuss that the intake of dirt and sand by horses can cause weight loss, chronic diarrhea, and colic caused by irritation and obstruction of the GIT. For this reason, these researchers hypothesized to increase fecal sand output in clinically normal horses in a natural environment by nutritional supplement mixing probiotics, prebiotics, and psyllium husk (i.e., a type of dietary fiber). They demonstrated that daily supplementation with probiotics and prebiotics in combination with psyllium enhanced fecal sand clearance in horses. According to results obtained, this approach may be an effective prophylactic treatment for sand colic and sand enteropathy in which administration alone of probiotic and prebiotic are not sufficient to prevent sand accumulation in the GIT.

As mentioned before, a diversity of microorganisms, usually LAB, such as *Bifidobacteria, Enterococci,* and *Lactobacilli* have been assessed as potential probiotics. Although *Bifidobacterium* and *Lactobacillus* spp. are regarded as beneficial to the general health of the host, not all LAB fall in the same group (Ouwehand et al. 2002). *L. delbrueckii*, *L. fermentum*, *L. mucosae*, *L. reuteri*, and *L. salivarius* are known to decarboxylate amino acids and have been associated in equine laminitis, i.e., inflammation of the lamina of the hoof (Mungall et al. 2001, Bailey et al. 2003). *L. pentosus* WE7, isolated from the intestinal tract of horses, failed to prevent neonatal diarrhea in foals, and stimulated the development of additional clinical abnormalities (Weese and Rousseau 2005). Moreover, Yuyama et al. (2004) have demonstrated that care of horses with *L. crispatus*, *L. equi*, *L. johnsonii*, and *L. reuteri*, which are originated from horses, improved growth and prevented diarrhea (Endo et al. 2009).

On the other hand, Parraga et al. (1997) investigated the probiotic administration for prevention of *Salmonella* shedding in the postoperative period in horses with colic. Results from this study indicate that the chosen probiotic formulations (a combination of various microorganisms) administered at the recommended manufacturers' dose did not have any effect on fecal shedding of *Salmonella* spp., neither reduced the likelihood of postoperative diarrhea or shorten hospitalization or antimicrobial therapy.

Weese et al. (2003) investigated whether *L. rhamnosus* strain GG can colonize the intestines of adult horses and foals. Fecal levels of this microorganism in horses were low and variable in the 2 lower dose groups. Even in the high dose group, colonization was relatively low. Instead, more consistent intestinal colonization was present in foals, and colonization persisted for up to 9 d after the end of the administration. According to the data, no animal had any adverse effects.

Weese et al. (2004) have screened intestinal microflora of horses for microorganisms, and initially identified *L. pentosus* (WE7) as a possible equine probiotic. However, in a randomized controlled clinical trial of 153 neonatal foals (24 to 48 h old), the administration of this strain (2×10^{11} CFU administered orally) for 7 d and clinically monitored by 14 d was significantly related with development of several signs (i.e., anorexia, colic, and diarrhea) in comparison with the placebo treatment (Weese and Rousseau 2005, Geor et al. 2007). It is important to mention that *L. pentosus* WE7, an organism of equine origin, has been described to inhibit the growth of *C. perfringens* and *E. coli in vitro* (Weese et al. 2004, Ishizaka et al. 2014), but unfortunately in the case reported above the data showed *in vivo* side effects with the treatment of this strain.

The results obtained by Ishizaka et al. (2014), who evaluated if probiotics improves the intestinal disorders in horses concluded that oral administration of probiotics (a combination of *L. acidophilus* and yeast species, including *S. cerevisiae* and *S. boulardii*) may have the ability to improve the intestinal environment microbiologically (with significant reduction for *E. coli* and *C. perfringens*) and biochemically, with no risks of adverse effects.

Although Kim et al. (2001), Parraga et al. (1997), and Weese and Rousseau (2005) reported that the application of probiotics did not prevent diarrhea in horses, different studies with oral administration of commercial probiotics have proven to improve different gut disorders (such as fecal conditions, improves digestion, and symptoms of enterocolitis) without inducing adverse effects (Glade 1991, Desrochers et al. 2005,

Jouany et al. 2008, Swyers et al. 2008). This is important since the digestive disease is a typical problem and it requires long-term treatment, and that can lead the animal to death (Traub-Dargatz et al. 1991), therefore any benefit due to the use of probiotics can have a significant value in equine health (Ishizaka et al. 2014).

Other than the aforementioned application, probiotics have been also used to improve the milk quantity and quality of mares, to decreased risk of lactic acidosis and to limit stress factors generally caused by transportation, races, etc. (Chaucheyras-Durand et al. 2010).

Pets

Some factors such as stress and dietary changes are conditions that can affect the intestinal microbiota of pets and the use of probiotics may be a beneficial approach to improve their health. Important changes in microbiota can occur at weaning, and events during this period of life can have a strong effect on the overall health of the animal throughout its life, especially in the development of its immune system. For this reason, during the first steps of life, the gut microbiota has an important role in various physiological functions of the host, which include maturation of the gut-associated lymphoid tissues. Hence, the addition of probiotics in feed, particularly for puppies, appears to be an advantage and a promising area (Benyacoub et al. 2003, Suchodolski 2011, Lee and Hase 2014).

Dogs and cats have been cohabiting with us for thousands of years; they are the major human companions and are most common pets in urban areas. For this reason, and considering the supposed relationship between pets' microbiota and owners, our focus here will be the probiotics application in these animals (Cohen 2002).

Dogs and cats have diets rich in carbohydrates and face challenges similar to those of humans, so the health of pets, as well as their owners, depends on the microbes present in the intestine. Therefore, providing a nutritionally balanced diet and proper care to pets is recognized as part of our responsibility to maintain their health and well-being. However, as microbiota alterations may facilitate exposure to pathogens and harmful environmental effects, it is wise to explore for novel tools to protect dogs and cats from unwanted strains. Some disorders such as inflammatory bowel diseases, acute gastroenteritis, and allergy can already be treated and prevented in pets by the use of probiotic supplementations. New obstacles for probiotic purposes include maintenance of overweight/obesity, *Helicobacter* gastritis, as well as parasitic infections (Suchodolski 2011, Grześkowiak et al. 2015).

Oral administration of probiotic LAB modulates the immune system of some laboratory animals. In this sense, Benyacoub et al. (2003) studied the capacity of *Ec. faecium* (SF68) to stimulate immune functions in puppies when added to dry food during the growth period. According to the results, fecal immunoglobulin A (IgA) and canine distemper virus vaccine—specific circulating IgG and IgA—were higher in the group receiving the probiotic in comparison with the controls. There were no differences in the measurements of $CD4^+$ and $CD8^+$ T cells between the groups, but the amount of mature B cells ($CD21^+$/major histocompatibility complex class II^+) was superior in those fed the probiotic. Such data show that a dietary probiotic LAB enhance specific immune functions in young dogs, thus offering new prospects for the employment of probiotics in canine diet.

Bybee et al. (2011) studied the effect of the same probiotic (*Ec. faecium* SF68) on presence of diarrhea in cats and dogs. Statistical differences between groups of dogs were not observed, but diarrhea was uncommon in both groups of dogs during the study. For the cats, the results showed that the percentage of them with diarrhea $\geq$ 2 d was significantly inferior in the probiotic group (7.4%) when compared with the placebo group (20.7%). These authors concluded that cats fed *Ec. faecium* SF68 had fewer periods of diarrhea when compared with controls, they also highlighted that the probiotic may have beneficial effects on the GIT.

The microorganisms present in the large intestine plays an important role in maintaining the state of health of the host and the GIT. The employment of specific dietary supplements such as symbiotics (probiotic strains and prebiotics) might positively influence the composition and metabolism of the intestinal microbial population. Several studies have been performed on the application of prebiotics in association with probiotics in dogs; in this case, most of these studies have aimed to assess whether using prebiotics brings about an improvement in the canine intestinal microbiota (Swanson and Fahey 2006, Hernot et al. 2008, Biagi et al. 2010, Pinna and Biagi 2014).

It is worth emphasizing that probiotics of human origin emerge to be among the new promising tools for the maintenance of pets' health, including that the host-derived microorganisms might be the most interesting probiotic source. The current evidence suggests that specific symbiotic strains and/or their defined mixtures may be valuable especially in the canine and feline food and therapy. At the same time the potential to decrease safety risk related to new probiotic strains in animal health and welfare should be considered. More controlled assays are important to both describe and characterize new probiotic formulations with potential application on general health in dogs and cats.

Application of Probiotic in Aquatic Animals

For aquaculture, the concept of probiotics is relatively new. These microorganisms intended for aquatic use have certain influencing factors that are fundamentally different from probiotics used in terrestrial systems (Denev et al. 2009). Fish have an important interaction with the aquatic environment, which made necessary to use specific probiotics for this type of host, since there is no separation line between the microbial community inside and outside the host, allowing a constant interaction with the ecosystem.

According to Fečkaninová et al. (2017) and Jahangiri and Esteban (2018), probiotics applied to aquaculture are a living microbial complement that has a beneficial effect on the host, modifying the environmental microbial community, ensuring a better use of food and nutritional value, interaction with the environment and immune response to diseases.

The colonization of the gastrointestinal tract in fish occurs immediately after hatching, and is completed in a few hours to modulate the expression of genes in the digestive tract, thus creating a favorable habitat and preventing the invasion of other bacteria introduced later in the ecosystem (Carvalho et al. 2012). The gastrointestinal microbiota depends on the external environment, and most of the bacteria present are transient in the intestine due to the constant flow of water and food along with the microorganisms present in them.

Several studies have reported the presence of pathogenic microorganisms in fish, such as *Salmonella* spp., *Listeria* spp. and *E. coli*; there are also reports of probiotic bacteria and other microorganisms, which include Gram-positive bacteria such as *Bacillus*, *Carnobacterium*, *Enterococcus* and various species of *Lactobacillus*, that play a modulatory role in the nutritional and immune balance in aquatic animals. However, the need for sustainable aquaculture promoted research on the use of probiotics in aquatic organisms (Fečkaninová et al. 2017, Wanka et al. 2018). Initial interest centered on the use of growth promoters and pathogenic inhibitors in fish. However, new areas of application, such as stress tolerance and effect on reproductive parameters, have not been fully studied in fish.

The mechanisms of action of probiotics are often difficult to analyze conclusively because of the wide range of possible modes of action and the multifactorial synergistic relationships between them (Merrifield et al. 2010a). However, some possible benefits associated with the administration of probiotics in the fish diet were studied as: production of inhibitory compounds (Vine et al. 2006, Panidyan 2013, Sopková et al. 2016), competition for chemicals or available energy (Mohapatra et al. 2013), inhibition of the expression of virulence genes or the interruption of quorum detection (Fuente et al. 2015), improvement of water quality tolerance (Cruz et al. 2012), optimization of the immune response (Nayak 2010a,b, Magnadottir 2010), sources of macro and/or micronutrients and digestive enzymatic contribution (Tuan et al. 2013, Saranu et al. 2014, Wanka et al. 2018).

The probiotics that are commonly used in aquaculture belong to the genera *Saccharomyces*, *Lactobacillus*, *Bacillus*, *Enterococcus*, *Shewanella*, *Leuconostoc*, *Lactococcus*, *Carnobacterium*, *Aeromonas* and several other species (Kim et al. 2010, Jahangiri and Esteban 2018). Different studies have shown that probiotics can promote growth and immunomodulatory activity (Merrifield et al. 2010b, Nayak 2010a,b, Dimitroglou et al. 2011).

The genus *Bacillus* has been widely used as a feed additive in aquaculture due to resistance to high temperatures and pressures (Ochoa-Solano and Olmos-Soto 2006, Rengpipat et al. 2000). The benefits of using *Bacillus* to improve growth rate, survival, immunity and disease resistance in aquatic animals have been reported (Nayak et al. 2007). Thus, allowing a substantial optimization of the feed conversion factor, specific growth rate and the protein efficiency ratio in the diets supplemented with *Bacillus* spp. for rainbow trout (*Onchorhynchus mykiss*) (Merrifield et al. 2010a). In addition, in gilthead seabream (*Sparus aurata* L.), tilapia (*Oreochromis niloticus*) and carp (*Labeorohita*) the administration of diets with *Bacillus subtilis* significantly increase their growth and resistance to diseases (Salinas et al. 2005, Aly et al. 2008, Kumar et al. 2008).

LAB have been also used as a dietary supplement to protect fish from various infectious diseases (Merrifield et al. 2010b, Nayak 2010a,b, Dimitroglou et al. 2011). In particular, the dietary administration of *Lactobacillus* spp. increased the growth rate, innate immune response, and resistance to certain diseases in *Epinephelus coioides* (Son et al. 2009), *Epinephelus bruneus* (Harikrishnan et al. 2012), and Nile tilapia (Ngamkala et al. 2010).

Salinas et al. (2005) indicated that the combination of two bacteria (*L. delbrueckii lactis* and *Bacillus subtilis* activated or inactivated) may be more effective and more consistent than a single probiotic. In Nile tilapia, when administering a diet containing *Bacillus subtilis* and *L. acidophilus*, a significant increase in immune responses and resistance to certain diseases was observed (Aly et al. 2008). In gilthead seabream (*Sparus aurata* L.), the administration of two inactivated bacteria of the family *Vibrionaceae* had a positive synergistic effect. In addition, a combination of four bacteria was reported in the diet of rainbow trout (*Oncorhynchus mykiss*) (Irianto and Austin 2002). Therefore, mixed cultures may contain bacteria that complement each other and that can modulate differently the immune response of the host (Dimitroglou et al. 2011).

Currently, there are commercial probiotic products prepared from several species of bacteria such as *Bacillus* spp., *Lactobacillus* spp., *Enterococcus* spp., *Carnobacterium* spp., and *Saccharomyces cerevisiae*, among others (Cruz et al. 2012). In particular, only one probiotic was authorized for use in aquaculture in the European Union; *Pediococcus acidilactici* CNCM MA 18/5 M, acid-lactic bacteria (Ramos et al. 2013) Bactocell® Aquaculture (Lallemand Inc., Canada) is internationally certified according to quality and safety standards (QPS status), effectively improving the production processes of some fish and salmonids (EFSA 2012).

Likewise, the use of a preparation of *Bacillus subtilis* (C-3102) (DSM15544) as an additive in ornamental fish feed, known commercially as Calsporin®, which is a preparation containing viable spores from a single strain of *Bacillus subtilis* intended to be used as a zootechnical additive (intestinal microbiota stabilizer) in fish, poultry and pig feed. The minimum recommended dose for use in ornamental fish feed is 1×10^{10} CFU/kg of whole feed (EFSA 2015).

Other important aspects within aquaculture is reproduction, which presents important nutritional requirements in fish, so that the reproductive capacity depends on appropriate concentrations of lipids, proteins, fatty acids, vitamins C, E, and carotenoids. In addition, the relationship of these components influences the reproduction in several processes, such as rate of fertilization, fertility, embryo survival and quality of the larvae produced (Izquierdo et al. 2001).

Currently, for most species of farmed fish, there are commercially available "broodstock diets", but these are not effective. In practice, many fish farms improve the nutrition of their breeders, feeding them only with byproducts of fresh fish or in combination with commercial diets. The use of these unprocessed fish products generally does not provide the adequate levels of nutrients necessary for breeding fish, which increases the risk of transmission of pathogens, parasites, bacteria and viruses. Therefore, it is necessary to incorporate probiotics in foods to prevent infections and exploit their effects on the reproduction of fish of interest for aquaculture (Carnevali et al. 2017, Haygood et al. 2018).

The pioneering study on the effect of probiotic supplementation on the reproductive performance of fish was carried out by Ghosh et al. (2007), using a strain of *Bacillus subtilis* isolated from the intestine of *Cirrhinusmrigala*, incorporated in different concentrations for four species of ornamental fish: *Poecilia reticulata*, *P. sphenops*, *Xiphophorus helleri* and *X. maculastes*. The results showed that the use of *Bacillus subtilis* concentrations of 10^6–10^8 g^{-1} cells increased the gonadomatic index, fertility, survival and larval production of the four species studied. In addition, these researchers proposed that vitamin B complexes synthesized by the probiotic, especially thiamine (vitamin B_1) and vitamin B_{12}, contributed to reduce the number of dead or deformed young. Abasali and Mohamad (2010) conducted similar studies with *X. helleri*, using a commercial probiotic containing *L. acidophilus*, *L. casei*, *Bifidobacterium* (*B.*) *thermophilum* and *Ec. faecium.* In this study, total larval production per female and relative fecundity were evaluated. The results showed significant differences between the control and the groups treated with probiotics, which had a higher rate of relative fertilization and greater production of fingerlings. Similar effects were observed in enriching diets with probiotics (*L. rhamnosus* and *Pediococcus acidilactici*) on reproductive performance and growth of larvae produced by *Ompokpabda* and *Carassius auratus* (Mehdinejad et al. 2018).

Conclusion

The demand for quality and safe animal products has significantly augmented. Controlled research assays display that probiotic LAB can positively stabilize gastrointestinal microbiota, and thereby improve animal production and health. Nonetheless, care must be considered in the way that the probiotic candidate-strains are chosen. Better comprehension of the mode of action, structure and activities of the gut microbiota, functional interactions between gut bacteria and interrelationships between microorganisms and host cells indicate a relevant aspect of future probiotic investigation.

In this sense, the future perspectives could be focused on basic research to characterize and identify existing probiotic LAB, establish optimal doses needed for certain strains, as well as evaluate their stability through digestion and processing. For the probiotics to demonstrate a real and effective alternative to pharmaceuticals and antibiotics it is necessary to guarantee their consistently high efficacy. It is fundamental to look for ways to potentiate the efficiency of probiotic bacteria in all portions of the GIT. The efficacy of them may be boosted by the evaluation or selection of more efficient strains of microorganisms, blend of probiotics and synergistically acting components, the blend of several strains, as well as gene manipulations.

More specifically, genetic engineering procedures can be employed to insert one or more antigens from a pathogen into probiotic strains with good colonizing ability for use in immunotherapeutic approaches (i.e., delivery of immunoregulatory compounds or vaccination). The future target is to increment the genomic information on microbiota and probiotic activities to advance the understanding of the interfaces with particular intestinal illnesses. Thus, the aim is to employ the know-how of GIT health normal microbiota composition in contrast with microbiota existing during illness to select the best probiotic or the best combinations of them. All of these new approaches will be very helpful to better comprehend the special effects of probiotics on the balance of the gastrointestinal microbiota. It will be possible, for instance, to define more powerful or targeted strains on a scientific basis and follow their behavior in the host in order to improve animal health and nutrition.

References

Abdelrahman, W., M. Mohnl, K. Teichmann, B. Doupovec, G. Schatzmayr, B. Lumpkins and G. Mathis. 2014. Comparative evaluation of probiotic and salinomycin effects on performance and coccidiosis control in broiler chickens. Poult. Sci. 93: 3002–3008.

Afsharmanesh, M. and B. Sadaghi. 2014. Effects of dietary alternatives (probiotic, green tea powder, and Kombucha tea) as antimicrobial growth promoters on growth, ileal nutrient digestibility, blood parameters, and immune response of broiler chickens. Comp. Clin. Pathol. 23: 717–724.

Ahasan, A.S.M.L., A. Asml, A. Agazzi, G. Invernizzi, V. Bontempo and G. Savoini. 2015. The beneficial role of probiotics in monogastric animal nutrition and health. J. Dairy Vet. Anim. Res. 2(4): 116–132.

Alagawany, M., M.E.A. El-hack, M.R. Farag, S. Sachan and K. Karthik. 2018. The use of probiotics as eco-friendly alternatives for antibiotics in poultry nutrition. Environ. Sci. Pollut. Res. 25: 10611–10618.

Alexopoulos, C., I.E. Georgoulakis, A. Tzivara, C.S. Kyriakis, A. Govaris and S.C. Kyriakis. 2004. Field evaluation of the effect of a probiotic-containing *Bacillus licheniformis* and *Bacillus subtilis* spores on the health status, performance, and carcass quality of grower and finisher pigs. J. Vet. Med. A 51(6): 306–312.

Aly, S.M., Y.A.G. Ahmed, A.A.A. Ghareeb and M.F. Mohamed. 2008. Studies on *Bacillus subtilis* and *Lactobacillus acidophilus*, as potential probiotics, on the immune response and resistance of Tilapia nilotica (*Oreochromis niloticus*) to challenge infections. Fish Shellfish Immun. 25: 128–136.

Amat, S., S. Subramanian, E. Timsit and T.W. Alexander. 2017. Probiotic bacteria inhibit the bovine respiratory pathogen *Mannheimiahaemolytica* serotype 1 *in vitro*. Lett. Appl. Microbiol. 64: 343–349.

Andrejčáková, Z., D. Sopková, R. Vlčková, L. Kulichová, S. Gancarčíková, V. Almášiová, K. Holovská, V. Petrilla and L. Krešáková. 2015. Synbiotics suppress the release of lactate dehydrogenase, promote non-specific immunity and integrity of jejunum mucosa in piglets. Anim. Sci. J. 87: 1157–1166.

Angelakis, E. 2017. Weight gain by gut microbiota manipulation in productive animals. Microb. Pathog. 106: 162–170.

Apás, A.L., J. Dupraz, R. Ross, S.N. González and M.E. Arena. 2010. Probiotic administration effect on fecal mutagenicity and microflora in the goat's gut. J. Biosci. Bioeng. 110(5): 537–540.

Arikan, O.A., W. Mulbry and C. Rice. 2009. Management of antibiotic residues from agricultural sources: use of composting to reduce chlortetracycline residues in beef manure from treated animals. J. Hazard Mater. 164: 483–489.

Baffoni, L., F. Gaggìa, D.D. Gioia, C. Santini, L. Mogna and B. Biavati. 2012. A *Bifidobacterium*-based synbiotic product to reduce the transmission of *C. jejuni* along the poultry food chain. Int. J. Food Microbiol. 157: 156–161.

Bailey, S.R., M.L. Baillon, A.N. Rycroft, P.A. Harris and J. Elliott. 2003. Identification of equine cecal bacteria producing amines in an *in vitro* model of carbohydrate overload. Appl. Environ. Microbiol. 69: 2087–2093.

Balasubramanian, B., S.I. Lee and I.H. Kim. 2018. Inclusion of dietary multi-species probiotic on growth performance, nutrient digestibility, meat quality traits, faecal microbiota and diarrhoea score in growing–finishing pigs. Ital. J. Anim. Sci. 17(1): 100–106.

Benyacoub, J., G.L. Czarnecki-Maulden, C. Cavadini, T. Sauthier, R.E. Anderson, E.J. Schiffrin and T. von der Weid. 2003. Supplementation of food with *Enterococcus faecium* (SF68) stimulates immune functions in young dogs. J. Nutr. 133(4): 1158–62.

Berruga, M.I., A. Molina, R.L. Althaus and M.P. Molina. 2016. Control and prevention of antibiotic residues and contaminants in sheep and goat's milk. Small Ruminant Res. 142: 38–43.

Bezerra, W.G.A., R.H. Horn, I.N.G. Silva, R.S.C. Teixeira, E.S. Lopes, Á.H. Albuquerque and W.C. Cardoso. 2017. Antibióticos no setor avícola: uma revisão sobre a resistência microbiana. Arch. Zootec. 66(254): 301–307.

Biagi, G., I. Cipollini, M. Grandi and G. Zaghini. 2010. Influence of some potential prebiotics and fibre-rich foodstuffs on composition and activity of canine intestinal microbiota. Anim. Feed Sci. Tech. 159: 50–58.

Böhmer, B.M., W. Kramer and D.A. Roth-Maier. 2006. Dietary probiotic supplementation and resulting effects on performance, health status, and microbial characteristics of primiparous sows. J. Anim. Physiol. Anim. Nutr. 90(7-8): 309–315.

Bouchard, D.S., B. Seridan, T. Saraoui, L. Rault, J. Nicoli, Y. LeLoir and S. Even. 2015. Lactic acid bacteria isolated from bovine mammary microbiota: Potential allies against bovine mastitis. Plos One 10(12): 1–18.

Brashears, M.M., M.L. Galyean, G.H. Loneragan and J.E. Mann. 2003. Prevalence of *Escherichia coli* O157: H7 and performance by beef feedlot cattle given lactobacillus direct-fed microbials. J. Food Prot. 66(5): 748–754.

Broom, L.J. 2018. Gut barrier function: Effects of (antibiotic) growth promoters on key barrier components and associations with growth performance. Poult. Sci. 0: 1–7.

Bybee, S.N., A.V. Scorza and M.R. Lappin. 2011. Effect of the probiotic *Enterococcus faecium* SF68 on presence of diarrhea in cats and dogs housed in an animal shelter. J. Vet. Inter. Med. 25(4): 856–60.

Carter, A., M. Adams, M. La Regine and M.J. Woodward. 2017. Colonisation of poultry by *Salmonella* Enteritidis S1400 is reduced by combined administration of *Lactobacillus salivarius* 59 and *Enterococcus faecium* PXN-33. Vet. Microbiol. 199: 100–107.

Carvalho, E., G.S. David and R.J. Silva. 2012. Health and Environment in Aquaculture. Intech Open, London, UK.

Casewell, M., C. Friis, E. Marco, P. McMullin and I. Phillips. 2003. The European ban on growth-promoting antibiotics and emerging consequences for human and animal health. J. Antimicrob. Chemother. 52(2): 159–161.

Castanon, J.I.R. 2007. History of the use of antibiotic as growth promoters in European poultry feeds. Poult. Sci. 86: 2466–2471.

Cavazzoni, V., A. Adami and C. Castrovilli. 1998. Performance of chickens supplemented with *Bacillus coagulans* as probiotic. Br. Poult. Sci. 39: 526–529.

Chang, S.Y., S.A. Belal, D.R. Kang, Y.II Choi, Y.H. Kim, H.S. Choe, J.Y. Heo and K.S. Shim. 2018. Influence of probiotics-friendly pig production on meat quality and physicochemical characteristics. Korean J. Food Sci. An. 38(2): 403.

Chaucheyras-Durand, F. and H. Durand. 2009. Probiotics in animal nutrition and health. Benef. Microbes 1(1): 3–9.

Chiquette, J., M.J. Allison and M. Rasmussen. 2012. Use of *Prevotellabryantii* 25A and a commercial probiotic during subacute acidosis challenge in mid lactation dairy cows. J. Dairy Sci. 95: 5985–5995.

Choct, M. 2009. Managing gut health through nutrition. Br. Poul. Sci. 50(1): 9–15.

Cohen, N.D. and A.M. Woods. 1992. Characteristics and risk factors for failure of horses with acute diarrhea to survive: 122 cases (1990–1996). J. Am. Vet. Med. Assoc. 214: 382–390.

Cohen, S.P. 2002. Can pets function as family members? West. J. Nurs. Res. 24(6): 621–638.

Cox, C.M. and R.A. Dalloul. 2014. Immunomodulatory role of probiotics in poultry and potential *in ovo* application. Benef. Microbes 6(1): 45–52.

Crispie, F., M. Alonso-go, C.O. Loughlin, K. Klostermann, W. Meaney, R.P. Ross, C. Hill and J. Flynn. 2008. Intramammary infusion of a live culture for treatment of bovine mastitis: effect of live lactococci on the mammary immune response. J. Dairy Res. 75: 374–384.

Cruz, P.M., A.L. Ibáñez, O.A.M. Hermosillo and H.C.R. Saad. 2012. Use of probiotics in aquaculture. ISRN Microbiol. 2012: 1–13.

[DANMAP] Danish Integrated Antimicrobial Resistance Monitoring and Research Programme. 2005. Use of antimicrobial agents and occurrence of antimicrobial resistance in bacteria from food animals, foods and humans in Denmark.

Davis, M.E., T. Parrott, D.C. Brown, B.Z. de Rodas, Z.B. Johnson, C.V. Maxwell and T. Rehberger. 2008. Effect of a *Bacillus*-based direct-fed microbial feed supplement on growth performance and pen cleaning characteristics of growing-finishing pigs. J. Anim. Sci. 86(6): 1459–1467.

Defoirdt, T., N. Boon, P. Bossier and W. Verstraete. 2004. Disruption of bacterial quorum sensing: An unexplored strategy to fight infections in aquaculture. Aquaculture 240: 69–88.

Denev, S., Y. Staykov, R. Moutafchieva and G. Beev. 2009. Microbial ecology of the gastrointestinal tract of fish and the potential application of probiotics and prebiotics in finfish aquaculture. Int. Aquatic Res. 1: 1–29.

Deng, Q., J.F. Odhiambo, U. Farooq, T. Lam, S.M. Dunn, M.G. Gänzle and B.N. Ametaj. 2015. Intravaginally administered lactic acid bacteria expedited uterine involution and modulated hormonal profiles of transition dairy cows. J. Dairy Sci. 98: 1–11.

Deng, Q., J.F. Odhiambo, U. Farooq, T. Lam, S.M. Dunn and B.N. Ametaj. 2016. Intravaginal probiotics modulated metabolic status and improved milk production and composition of transition dairy cows. J. Anim. Sci. 94: 760–770.

Desrochers, A.M., B.A. Dolente, M.F. Roy, R. Boston and S. Carlisle. 2005. Efficacy of *Saccharomyces boulardii* for treatment of horses with acute enterocolitis. J. Am. Vet. Med. Assoc. 227(6): 954–9.

Dibner, J.J. and J.D. Richards. 2005. Antibiotic growth promoters in agriculture: History and mode of action. Poult. Sci. 84: 634–643.

Dimitroglou, A., D.L. Merrifield, O. Carnevali, S. Picchietti, M. Avella, C.L. Daniels, D.Güroy and S.J. Davies. 2011. Microbial manipulations to improve fish health and production a Mediterranean perspective. Fish Shellfish Immun. 30: 1–16.

Dowarah, R., A.K. Verma, N. Agarwal, B.H.M. Patel and P. Singh. 2017. Effect of swine based probiotic on performance, diarrhoea scores, intestinal microbiota and gut health of grower-finisher crossbred pigs. Livest. Sci. 195: 74–79.

Duff, G.C. and M.L. Galyean. 2002. Board-invited review: Recent advances in management of highly stressed, newly received feedlot cattle. J. Anim. Sci. 85: 823–840.

[EC] Economic and Social Committee of the European Union. 2003. Commission Regulation No. 1831/2003. Official Journal of European Union L 268, 29–43 pp.

Economou, V. and P. Gousia. 2015. Agriculture and food animals as a source of antimicrobial-resistant bacteria. Infect. Drug Resist. 8: 49–61.

[EFSA] European Food Safety Authority. Panel on Additives and Products or Substances used in Animal Feed (FEEDAP). 2012. Scientific Opinion on the efficacy of Bactocell (*Pediococcus acidilactici*) when used as a feed additive for fish. EFSA Journal 10(9): 2886.

[EFSA] European Food Safety Authority. Panel on Additives and Products or Substances used in Animal Feed (FEEDAP). 2015. Scientific Opinion on the safety and efficacy of Calsporin® (*Bacillus subtilis* DSM 15544) as a feed additive for ornamental fish. EFSA Journal 13(11): 4274.

Endo, A., Y. Futagawa-Endo and L.M. Dicks. 2009. *Lactobacillus* and *Bifidobacterium* diversity in horse feces, revealed by PCR-DGGE. Curre. Microbiol. 59(6): 651–655.

[FAO] Food and Agriculture Organization of United Nations. 2016. Probiotics in animal nutrition—Production, impact and regulation FAO Animal Production and Health Paper No. 179. Rome. http://www.fao.org/3/a-i5933e.pdf.

[FAO] Food and Agriculture Organization of United Nations. 2018. Meat market review. Rome. http://www.fao.org/3/CA2129EN/ca2129en.pdf. Accessed April 2019.

[FDA] Food and Drug Administration. 2018. Center for Medicine Veterinary. Update: FDA's Proposed withdrawal of approval of poultry flouroquinolones.

Farrell, D. 2013. The role of poultry in human nutrition. Poultry Development Review 1–2.

Fečkaninová, A., J. Koščová, D. Mudroňová, P. Popelka and J. Toropilova. 2017. The use of probiotic bacteria against Aeromonas infections in salmonid aquaculture. Aquaculture 469: 1–8.

Ferreira, A., S.S. Kunh, P.-A. Cremonez, J. Dieter, J.G. Teleken, J.S.C. Sampaio and P.D. Kunh. 2018. Brazilian poultry activity waste: Destinations and energetic potential. Renew. Sust. Energ. Rev. 81: 3081–3089.

Frizzo, L.S., M.V. Zbrun, L.P. Soto and M.L. Signorini. 2011. Effect of probiotics on the growth performance in young calves: a meta-analysis of randomized controlled trials. Anim. Feed Sci. Technol. 169: 147–156.

Frizzo, L.S., M.L. Signorini and M.R. Rosmini. 2018. Probiotics and prebiotics for the health of cattle. pp. 155–174. *In:* Gioia, D.D. and B. Biavati (eds.). Probiotics and Prebiotics in Animal Health and Food Safety. Springer. Cham, Switzerland.

Fuente, M.d.L., C.D. Miranda, P. Jopia, G. González-Rocha, N. Guiliani, K. Sossa and H. Urrutia. 2015. Growth inhibition of bacterial fish pathogens and quorum-sensing blocking by bacteria recovered from Chilean salmonid farms. J. Aquat. Anim. Health 27: 112–122.

Fuller, R. 1989. Probiotics in man and animals. J. Appl. Bacteriol. 66: 365–378.

Gadde, U., W.H. Kim, S.T. Oh and H.S. Lillehoj. 2017. Alternatives to antibiotics for maximizing growth performance and feed efficiency in poultry: A review. Anim. Health Res. Rev. 18: 26–45.

Gaggìa, F., P. Mattarelli and B. Biavati. 2010. Probiotics and prebiotics in animal feeding for safe food production. Int. J. Food Microbiol. 141: S15–S28.

Genís, S., A. Bach, F. Fàbregas and A. Arís. 2016. Potential of lactic acid bacteria at regulating *Escherichia coli* infection and inflammation of bovine endometrium. Theriogenology 85: 625–37.

Genís, S., À. Bach and A. Arís. 2017a. Effects of intravaginal lactic acid bacteria on bovine endometrium: implications in uterine health. Vet. Microbiol. 204: 174–179.

Genís, S., A. Sánchez-Chardi, A. Bach, F. Fàbregas and A. Arís. 2017b. A combination of lactic acid bacteria regulates *Escherichia coli* infection and inflammation of the bovine endometrium. J. Dairy Sci. 100: 479–92.

Genís, S., R.L.A. Cerri, A. Bach, B.F. Silper and M. Baylão. 2018. Pre-calving intravaginal administration of lactic acid bacteria reduces metritis prevalence and regulates blood neutrophil gene expression after calving in dairy cattle. Front. Vet. Sci. 5: 1–10.

Geor, R.J. and P.A. Harris. 2007. How to minimize gastrointestinal disease associated with carbohydrate nutrition in horses. P. Annu. Conv. Am. Equin. 53: 175–185.

Ghorbani, G.R., D.P. Morgavi, K.A. Beauchemin and J.A.Z. Leedle. 2002. Effects of bacterial direct-fed microbials on ruminal fermentation, blood variables, and the microbial populations of feedlot cattle. J. Anim. Sci. 80: 1977–1986.

Ghosh, S., A. Sinha and C. Sahu. 2007. Effect of probiotic on reproductive performance in female livebearing ornamental fish. Aquac. Res. 38(5): 518–526.

Gilliland, S.E. 2011. Probiotics. pp. 923–925. *In*: Pond, W.G., D.E. Ullrey and C.K. Baer (eds.). Encyclopedia of Animal Science (Second Edition). Taylor and Francis, New York, USA.

Glade, M.J. 1991. Dietary yeast culture supplementation of mares during late gestation and early lactation. Effects on dietary nutrient digestibilities and fecal nitrogen partitioning. J. Equine Vet. Sci. 11: 10–16.

Grześkowiak, Ł., A. Endo, S. Beasley and S. Salminen. 2015. Microbiota and probiotics in canine and feline welfare. Anaerobe 34: 14–23.

Harikrishnan, R., C. Balasundaram and M.S. Heo. 2010. *Lactobacillus sakei* BK19 enriched diet enhances the immunity status and disease resistance to streptococcosis infection in kelp grouper, Epinephelusbruneus. Fish Shellfish Immun. 29: 1037–1043.

Harikrishnan, R., J.-S. Kim, C. Balasundaram and M.-S. Heo. 2012. Immunomodulatory effects of chitin and chitosan enriched diets in Epinephelusbruneus against Vibrio alginolyticus infection. Aquaculture 329: 46–52.

Hasman, H. and F.M. Aarestrup. 2005. Relationship between copper, glycopeptide, and macrolide resistance among *Enterococcus faecium* strains isolated from pigs in Denmark between 1997 and 2003. Antimicrob. Agents Chemother. 49(1): 2003–2006.

Haygood, A.M. and R. Jha. 2018. Strategies to modulate the intestinal microbiota of tilapia (*Oreochromis* sp.) in aquaculture: A review. Rev. Aquacult. 10(2): 320–333.

Hemaiswarya, S., R. Raja, R. Ravikumar and I.S. Carvalho. 2013. Mechanism of action of probiotics. Braz. Arch. Biol. Technol. 1: 113–119.

Hernot, D., E. Ogué, G. Fahey and R.A. Rastall. 2008. Prebiotics and synbiotics in companion animal science. pp. 357–370. *In*: Versalovic, J. and M. Wilson (eds.). Therapeutic Microbiology: Probiotics and Related Strategies. ASM Press, Washington, USA.

Hill, C., F. Guarner, G. Reid, G.R. Gibson, D.J. Merenstein, B. Pot, L. Morelli, R.B. Canani, H.J. Flint, S. Salminen, P.C. Calder and M.E. Sanders. 2014. Expert consensus document: The International Scientific Association for Probiotics and Prebiotics consensus statement on the scope and appropriate use of the term probiotic. Nat. Rev. Gastroenterol. Hepatol. 11(8): 506–514.

Hofmann, R.R. 1989. Evolutionary steps of ecophysiological adaptation and diversification of ruminants: a comparative view of their digestive system. Oecologia 78: 443–457.

Hu, S., X. Cao, Y. Wu, X. Mei, H. Xu, Y. Wang, X. Zhang, L. Gong and W. Li. 2018. Effects of probiotic *Bacillus* as an alternative of antibiotics on digestive enzymes activity and intestinal integrity of piglets. Front. Microbiol. 9(2427): 1–9.
Huber, J.T. 1997. Probiotics in cattle. pp. 162–186. *In*: Fuller, R. (ed.). Probiotics 2—Application and Practical Aspects. Springer, Dordrecht, Holand.
Huff, G.R., W.E. Huff, N.C. Rath, F.A. El-Gohary, Z.Y. Zhou and S. Shini. 2015. Efficacy of a novel prebiotic and a commercial probiotic in reducing mortality and production losses due to cold stress and *Escherichia coli* challenge of broiler chicks. Poult. Sci. 94(5): 918–926.
Idowu, F., K. Junaid, A. Paul, O. Gabriel, A. Paul, N. Sati et al. 2010. Antimicrobial screening of commercial eggs and determination of tetracycline residue using two microbiological methods. Int. J. Poult. Sci. 9(10): 959–962.
Irianto, A. and B. Austin. 2002. Use of probiotics to control furunculosis in rainbow trout, *Oncorhynchus mykiss* (Walbaum). J. Fish Dis. 25(6): 333–342.
Ishizaka, S., A. Matsuda, Y. Amagai, K. Oida, H. Jang, Y. Ueda, M. Takai, A. Tanaka and H. Matsuda. 2014. Oral administration of fermented probiotics improves the condition of feces in adult horses. J. EquineSci. 25(4): 65–72.
Izquierdo, M.S., H. Fernandez-Palacios and A.G.J. Tacon. 2001. Effect of broodstock nutrition on reproductive performance of fish. Aquaculture 197(1-4): 25–42.
Jahangiri, L. and M. Esteban. 2018. Administration of probiotics in the water in finfish aquaculture systems: A review. Fishes 3(3): 33.
Jami, E., B.A. White and I. Mizrahi. 2014. Potential role of the bovine rumen microbiome in modulating milk composition and feed efficiency. PLoS One 9(1): e85423.
Jayabal, T., R. Muralidharan, P.T. Gnanaraj and M. Murugan. 2008. Growth performance of stall-fed goats under probiotic supplementation. Tamilnadu. J. Vet. Anim. Sci. 4(5): 179–184.
Jayaraman, S., G. Thangavel, H. Kurian, R. Mani, R. Mukkalil and H. Chirakkal. 2005. *Bacillus subtilis* PB6 improves intestinal health of broiler chickens challenged with *Clostridium perfringens*-induced necrotic enteritis. Poult. Sci. 92: 370–374.
Jeong, J., J. Kim, S. Lee and I. Kim. 2015. Evaluation of *Bacillus subtilis* and *Lactobacillus acidophilus* probiotic supplementation on reproductive performance and noxious gas emission in sows. Ann. Anim. Sci. 15(3): 699–710.
Jinturkar, A.S., B.V. Gujar, D.S. Chauhan and R.A. Patil. 2009. Effect of feeding probiotics on the growth performance and feed conversion efficiency in Goat. Indian J. Anim. Res. 43(1): 49–52.
Jouany, J.P., J. Gobert, B. Medina, G. Bertin and V. Julliand. 2008. Effect of live yeast culture supplementation on apparent digestibility and rate of passage in horses fed a high-fiber or high-starch diet. J. Anim. Sci. 86: 339–347.
Kabir, S.M.L. 2009. The role of probiotics in the poultry industry. Int. J. Mol. Sci. 10: 3531–3546.
Kalavathy, R., N. Abdullah, S. Jalaludin and Y.W. Ho. 2003. Effects of *Lactobacillus* cultures on growth performance, abdominal fat deposition, serum lipids and weight of organs of broiler chickens. Br. Poult. Sci. 44: 139–144.
Khaksefidi, A. and T. Ghoorchi. 2006. Effect of probiotic on performance and immunocompetence in broiler chicken. J. Poult. Sci. 43: 296–300.
Kim, J.S., R. Harikrishnan, M.C. Kim, C. Balasundaram and M.S. Heo. 2010. Dietary administration of Zooshikellasp enhance the innate immune response and disease resistance of Paralichthysolivaceus against *Sreptococcusiniae*. Fish Shellfish Immun. 29: 104–110.
Kim, L.M., P.S. Morley, J.L. Traub-Dargatz, M.D. Salman and C. Gentry-Weeks. 2001. Factors associated with *Salmonella* shedding among equine colic patients at a veterinary teaching hospital. J. Am. Vet. Med. Assoc. 218: 740–748.
Klostermann, K., F. Crispie, J. Flynn, R.P. Ross, C. Hill and W. Meaney. 2018. Intramammary infusion of a live culture of *Lactococcus lactis* for treatment of bovine mastitis: comparison with antibiotic treatment in field trials. J. Dairy Res. 75: 365–373.
Koike, S., R. Mackie and R. Aminov. 2017. Agricultural use of antibiotics and antibiotic resistance. pp. 217–250. *In*: Mirete, S. and M.L. Pérez (eds.). Antibiotic Resistance Genes in Natural Environments and Long-Term Effects. Nova Science Publishers, New York, USA.
Krehbiel, C.R., S.R. Rust, G. Zhang and S.E. Gilliland. 2002. Bacterial direct-fed microbials in ruminant diets: Performance response and mode of action. J. Anim. Sci. 81(E. Suppl. 2): E120–E132.

Kritas, S.K., A. Govaris, G. Christodoulopoulos and A.R. Burriel. 2006. Effect of *Bacillus licheniformis* and *Bacillus subtilis* supplementation of ewe's feed on sheep milk production and young lamb mortality. J. Vet. Med. A. 53(4): 170–173.

Kritas, S.K., T. Marubashi, G. Filioussis, E. Petridou, G. Christodoulopoulos, A.R. Burriel, A. Tzivara, A. Theodoridis and M. Pískoriková. 2015. Reproductive performance of sows was improved by administration of a sporing bacillary probiotic (*Bacillus subtilis* C-3102). J. Anim. Sci. 93: 405–413.

Kritas, S.K. 2018. Probiotics and prebiotics for the health of pigs and horses. pp. 119–126. *In*: Gioia, D.D. and B. Biavati (eds.). Probiotics and Prebiotics in Animal Health and Food Safety. Springer. Cham, Switzerland.

Kumar, R., S.C. Mukherjee, R. Ranjan and S.K. Nayak. 2008. Enhanced innate immune parameters in Labeorohita (Ham.) following oral administration of *Bacillus subtilis*. Fish Shellfish Immun. 24: 168–172.

Lamy, E., S. van Harten, E. Sales-Baptista, M.M.M. Guerra and A.M. Almeida. 2012. Factors influencing livestock productivity. pp. 19–51. *In*: Sejian, V., S.M.K. Naqvi, T. Ezeji, J. Lakritz and R. Lal (eds.). Environmental Stress and Amelioration in Livestock Production. Springer, Berlin, Germany.

Landers, T.F., B. Cohen, T.E. Wittum and E.L. Larson. 2012. A review of antibiotic use in food animals: perspective, policy, and potential. Public. Health Rep. 127: 4–22.

Landes, A.D., D.M. Hassel, J.D. Funk and A. Hill. 2008. Fecal sand clearance is enhanced with a product combining probiotics, prebiotics, and psyllium in clinically normal horses. Journal of Equine Veterinary Science 28(2): 79–84.

Lauková, A., E. Styková, I. Kubašová, S. Gancarčíková, I. Plachá, D. Mudroňová, A. Kandričáková, R. Miltko, G. Belzecki, I. Valocký and V. Strompfová. 2018. Enterocin M and its beneficial effects in horses—a pilot experiment. Probiotics and Antimicrob. Proteins. 1–7.

Leblanc, S.J. 2008. Postpartum uterine disease and dairy herd reproductive performance: A review. Vet. J. 176: 102–114.

Lee, S.H., H.S. Lillehoj, R.A. Dalloul, D.W. Park, Y.H. Hong and J.J. Lin. 2007. Influence of *Pediococcus*-based probiotic on coccidiosis in broiler chickens. Poult. Sci. 83: 63–66.

Lee, W.J. and K. Hase. 2014. Gut microbiota-generated metabolites in animal health and disease. Nat. Chem. Biol. 10: 416–424.

Liu, H., H.F. Ji, D.Y. Zhang, S.X. Wang, J. Wang, D.C. Shan and Y.M. Wang. 2015. Effects of *Lactobacillus brevis* preparation on growth performance, fecal microflora and serum profile in weaned pigs. Livest. Sci. 178: 251–254.

Magnadottir, B. 2010. Immunological control of fish diseases. Mar. Biotechnol. 12: 361–379.

Mair, T.S., L.V. de Westerlaken, P.J. Cripps and S. Love. 1990. Diarrhoea in adult horses: a survey of clinical cases and an assessment of some prognostic indices. Vet. Rec. 126: 479–481.

Maragkoudakis, P.A., K.C. Mountzouris, C. Rosu, G. Zoumpopoulou, K. Papadimitriou, E. Dalaka, A. Hadjipetrou, G. Theofanous, G.P. Strozzi, N. Carlini, G. Zervas and E. Tsakalidou. 2010. Feed supplementation of *Lactobacillus plantarum* PCA 236 modulates gut microbiota and milk fatty acid composition in dairy goats—a preliminary study. Int. J. Food Microbiol. 141: S109–S116.

Martin, W.A., M.W.A. Verstegen and B.A. Williams. 2002. Alternatives to the use of antibiotics as growth promoters for monogastric animals. Anim. Biotechnol. 13(1): 113–127.

Mazanko, M.S., I.F. Gorlov, E.V. Prazdnova, M.S. Makarenko, A.V. Usatov, A.B. Bren, V.A.Chistyakov, A.V. Tutelyan, Z.B. Komarova, N.I. Mosolova, D.N. Pilipenko, O.E. Krotova, A.N. Struk, A. Lin and M.L. Chikindas. 2018. *Bacillus* probiotic supplementations improve laying performance, egg quality, hatching of laying hens, and sperm quality of roosters. Probiotic Antimicrob. Proteins 10: 367–373.

Mehdinejad, N., M.R. Imanpour and V. Jafari. 2018. Combined or individual effects of dietary probiotic, *Pediococcus acidilactici* and nucleotide on reproductive performance in goldfish (Carassius auratus). Probiotics Antimicrob. Proteins 1–6.

Menegat, M.B., K.M. Gourley, M.B. Braun, J.M. DeRouchey, J.C. Woodworth, J. Bryte, M.D. Tokach, S.S. Dritz and R.D. Goodband. 2018. Effects of a *Bacillus*-based probiotic on sow performance and on progeny growth performance, fecal consistency, and fecal microflora. Kans AES Res. Rep. 4(9): 1–23.

Merrifield, D.L., A. Dimitroglou, G. Bradley, R.T.M. Baker and S.J. Davies. 2010a Probiotic applications for rainbow trout (*Oncorhynchus mykiss* Walbaum) I. Effects on growth performance, feed utilization, intestinal microbiota and related health criteria. Aquac. Nutr. 16: 504–510.

Merrifield, D.L., A. Dimitroglou, A. Foey, S.J. Davies, R.T.M. Baker, J. Bøgwald, M. Castex and E. Ringø. 2010b. The current status and future focus of probiotic and prebiotic applications for salmonids. Aquaculture 302: 1–18.

Mingmongkolchai, S. and W. Panbangred. 2018. *Bacillus* probiotics: An alternative to antibiotics for livestock production. J. Appl. Microbiol. 124(6): 1334–1346.

Mirza, R.A. 2018. Probiotics and prebiotics for the health of poultry. pp. 127–154. *In*: Gioia, D.D. and B. Biavati (eds.). Probiotics and Prebiotics in Animal Health and Food Safety. Springer. Cham, Switzerland.

Modesto, M., M.R.D. Aimmo, I. Stefanini, P. Trevisi, S. Filippi, L. De Casini, M. Mazzoni, P. Bosi and B. Biavati. 2009. A novel strategy to select *Bifidobacterium* strains and prebiotics as natural growth promoters in newly weaned pigs. Livest. Sci. 122: 248–258.

Mohapatra, S., T. Chakraborty, V. Kumar, G. Deboeck and K.N. Mohanta. 2013. Aquaculture and stress management: A review of probiotic intervention. J. Anim. Physiol. Anim. Nutr. 97: 405–430.

Moore, P.R., A. Everson, T.D. Luckey, E. McCoy, C.A. Elvehjem and E.B. Hart. 1946. Use of sulfasuxidine, streptothricin, and streptomycin in nutritional studies with the chick. J. Biol. Chem. 2: 437–441.

Mountzouris, K.C., P. Tsirtsikos, E. Kalamara, S. Nitsch, G. Schatzmayr and K. Fegeros. 2007. Evaluation of the efficacy of a probiotic containing *Lactobacillus, Bifidobacterium, Enterococcus,* and *Pediococcus* strains in promoting broiler performance and modulating cecal microflora composition and metabolic activities. Poult. Sci. 86(2): 309–317.

Mungall, B.A., M. Kyaw-Tanner and C.C. Pollitt. 2001. *In vitro* evidence for a bacterial pathogenesis of equine laminitis. Vet. Microbiol. 79: 209–223.

Musa, H.H., S.L. Wu, C.H. Zhu, H.I. Seri and G.Q. Zhu. 2009. The potential benefits of probiotics in animal production and health. J. Anim. Vet. Adv. 8(2): 313–321.

Nader-Macías, M.E.F. and M.S.J. Tomás. 2015. Profiles and technological requirements of urogenital probiotics. Adv. Drug Deliver. Rev. 92: 84–104.

Nayak, S.K., P. Swain and S.C. Mukherjee. 2007. Effect of dietary supplementation of probiotic and vitamin C on the immune response of Indian major carp, Labeorohita (Ham.). Fish Shellfish Immun. 23: 892–896.

Nayak, S.K. 2010a. Role of gastrointestinal microbiota in fish. Aquac. Res. 41: 1553–1573.

Nayak, S.K. 2010b. Probiotics and immunity: a fish perspective. Fish Shellfish Immun. 29: 2–14.

Ngamkala, S., K. Futami, M. Endo, M. Maita and T. Katagiri. 2010. Immunological effects of glucan and *Lactobacillus rhamnosus* GG, a probiotic bacterium, on Nile tilapia Oreochromis niloticus intestine with oral Aeromonas challenges. Fisheries Sci. 76: 833–840.

Nurmi, E. and M. Rantala. 1973. New aspects of *Salmonella* infection in broiler production. Nature 241: 210–211.

O'hara, A.M. and F. Shanahan. 2007. Gut microbiota: mining for therapeutic potential. Clin. Gastroenterol. Hepatol. 5(3): 274–284.

Ochoa-Solano, J.L. and J. Olmos-Soto. 2006. The functional property of *Bacillus* for shrimp feeds. Food Microbiol. 23: 519–525.

Oliveira, J.E., E. van der Hoeven-Hangoor, I.B. van der Linde, R.C. Van De, J.M.B.C. Montijn and M. van der Vossen. 2014. *In ovo* inoculation of chicken embryos with probiotic bacteria and its effect on posthatch *Salmonella* susceptibility. Poult. Sci. 93: 818–829.

Ostermann, A., J. Siemens, G. Welp, Q. Xue, X. Lin, X. Liu and W. Amelung. 2013. Leaching of veterinary antibiotics in calcareous Chinese croplands. Chemosphere 91(7): 928–934.

Ouwehand, A.C., S. Salminen and E. Isolauri. 2002. Probiotics: An overview of beneficial effects. pp. 279–289. *In*: Siezen, R.J., J. Kok, T. Abee and G. Schasfsma (eds.). Lactic Acid Bacteria: Genetics, Metabolism and Applications. Springer, Dordrecht, Holand.

Pan, L., P.F. Zhao, X.K. Ma, Q.H. Shang, Y.T. Xu, S.F. Long, Y. Wu, F.M. Yuan and X.S. Piao. 2017. Probiotic supplementation protects weaned pigs against enterotoxigenic *Escherichia coli* K88 challenge and improves performance similar to antibiotics. J. Anim. Sci. 95(6): 2627–2639.

Panidyan, P. 2013. Probiotics in aquaculture. Drug Invent. Today 5: 55–59.

Parker, R.B. 1974. Probiotics, the other half of the antibiotics story. Anim. Nutr. Health 29: 4–8.

Parraga, M.E., S.J. Spier, M. Thurmond and D. Hirsh. 1997. A clinical trial of probiotic administration for prevention of *Salmonella* shedding in the postoperative period in horses with colic. J. Vet. Intern. Med. 11: 36–41.

Pellegrino, M.S., I.D. Frola, B. Natanael, D. Gobelli, M.E.F. Nader-Macias and C.I. Bogni. 2018. *In vitro* characterization of lactic acid bacteria isolated from bovine milk as potential probiotic strains to prevent bovine mastitis. Probiotics Antimicrob. Proteins 1–11.

Pinna, C. and G. Biagi. 2014. The utilisation of prebiotics and synbiotics in dogs. Ital. J. Anim. Sci. 13(3107): 169–178.

Popova, M., P. Molimard, S. Courau, J. Crociani, C. Dufour, F. Le Vacon and T. Carton. 2012. Beneficial effects of probiotics in upper respiratory tract infections and their mechanical actions to antagonize pathogens. J. Appl. Microbiol. 113: 1305–1318.

Prado-rebolledo, O.F., J.J. Delgado-Machuca, R.J. Macedo-Barragan, L.J. Garcia-márquez, Morales-Barrera, J.D. Latorre, X. Hernandez-Velasco and G. Tellez. 2017. Evaluation of a selected lactic acid bacteria-based probiotic on *Salmonella enterica* serovar Enteritidis colonization and intestinal permeability in broiler chickens. Avian Pathol. 46(1): 90–94.

Pruszynska-Oszmalek, E., P.A. Kolodziejski, K. Stadnicka, M. Sassek, D. Chalupka, B. Kuston, L.P. Nogowski, G.J. Maiorano and M. Bednarczyk. 2015. *In ovo* injection of prebiotics and synbiotics affects the digestive potency of the pancreas in growing chickens. Poult. Sci. 94(8): 1909–1916.

Qadis, A.Q., S. Goya, K. Ikuta, M. Yatsu and A. Kimura. 2014. Effects of a bacteria-based probiotic on ruminal pH, volatile fatty acids and bacterial flora of holstein calves. J. Vet. Med. Sci. 76(6): 877–885.

Rahman, M.L., S. Akhter, M.K.M. Mallik and I. Rashid. 2018. Probiotic enrich dietary effect on the reproduction of butter catfish, Ompokpabda (Hamilton, 1872). IJCRLS 7(02): 866–873.

Ramos, C.L., L. Thorsen, R.F. Schwan and L. Jespersen. 2013. Strain-specific probiotics properties of *Lactobacillus fermentum, Lactobacillus plantarum* and *Lactobacillus brevis* isolates from Brazilian food products. Food Microbiol. 36(1): 22–29.

Rengpipat, S., S. Rukpratanporn, S. Piyatiratitivorakul and P. Menasaveta. 2000. Immunity enhancement in black tiger shrimp (*Penaeus monodon*) by a probiont bacterium (Bacillus S11). Aquaculture 191: 271–288.

Rigobelo, E.E.C., N. Karapetkov, S.A. Maestá, F.A.D. Ávila and D. McIntosh. 2014. Use of probiotics to reduce faecal shedding of Shiga toxin-producing *Escherichia coli* in sheep. Benef. Microbes 6(1): 53–60.

Ronquillo, M.G. and J.C.A. Hernandez. 2017. Antibiotic and synthetic growth promoters in animal diets: Review of impact and analytical methods. Food Control 72: 255–267.

Roto, S.M., Y.M. Kwon and S.C. Ricke. 2016. Applications of *in ovo* technique for the optimal development of the gastrointestinal tract and the potential influence on the establishment of its microbiome in poultry. Front. Vet. Sci. 3: 63.

Saint-Cyr, M.J., N. Haddad, B. Taminiau, T. Poezevara, S. Quesne, M. Amelot, G. Daube, M. Chemaly, X. Dousset and M. Guyard-Nicodème. 2017. Use of the potential probiotic strain *Lactobacillus salivarius* SMXD51 to control *Campylobacter jejuni* in broilers. Int. J. Food Microbiol. 247: 9–17.

Salinas, I., A. Cuesta, M.A. Esteban and J. Meseguer. 2005. Dietary administration of *Lactobacillus delbrueckii* and *Bacillus subtilis*, single or combined, on gilthead seabream cellular innate immune responses. Fish Shellfish Immun. 19: 67–77.

Sanders, M.E. 2010. International scientific association for probiotics and prebiotics 2010 Meeting Report. Functional Food Reviews 2(4): 131–140.

Saranu, L., J. Madhavi, J. Srilakshmi and M.V. Raghavendra Rao. 2014. Probiotics—overview and significance in enhancing crustacean immune system. Int. J. Ins. Pharm. Life Sci. 4: 115–134.

Saville, W.J., K.W. Hinchcliff, B.R. Moore, C.W. Kohn, S.M. Reed, L.A. Mitten and L.J. Rivas. 1996. Necrotizing enterocolitis in horses: a retrospective study. J. Vet. Intern. Med. 10(4): 265–270.

Scharek-Tedin, L., R. Pieper, W. Vahjen, K. Tedin, K. Neumann and J. Zentek. 2013. *Bacillus cereus* var. *toyoi* modulates the immune reaction and reduces the occurrence of diarrhea in piglets challenged with *Salmonella typhimurium* DT104. J. Anim. Sci. 91(12): 5696–5704.

Schierack, P., L.H. Wieler, D. Taras, V. Herwig, B. Tachu, A. Hlinak, M.F.G. Schmidt and L. Scharek. 2007. *Bacillus cereus* var. *toyoi* enhanced systemic immune response in piglets. Vet. Immunol. Immunop. 118(1-2): 1–11.

Schoster, A., J.S. Weese and L. Guardabassi. 2014. Probiotic use in horses–what is the evidence for their clinical efficacy? J. Vet. Intern. Med. 28(6): 1640–52.

Signorini, M.L., L.P. Soto, M.V. Zbrun, G.J. Sequeira, M.R. Rosmini and L.S. Frizzo. 2012. Impact of probiotic administration on the health and fecal microbiota of young calves: A meta-analysis of randomized controlled trials of lactic acid bacteria. Res. Vet. Sci. 93: 250–258.

Smialek, M., S. Burchardt and A. Koncicki. 2018. Research in Veterinary Science The influence of probiotic supplementation in broiler chickens on population and carcass contamination with *Campylobacter* spp. - Field study. Res. Vet. Sci. 118: 312–316.

Sobczak, A. and K. Kozłowsk. 2015. The effect of a probiotic preparation containing *Bacillus subtilis* ATCC PTA-6737 on egg production and physiological parameters of laying hens. Ann. Anim. Sci. 15(3): 711–723.

Son, V.M., C.C. Chang, M.C. Wu, Y.K. Guu, C.H. Chiu and W. Cheng. 2009. Dietary administration of the probiotic, *Lactobacillus plantarum*, enhanced the growth, innate immune responses, and disease resistance of the grouper Epinepheluscoioides. Fish Shellfish Immun. 26(5): 691–698.

Sopková, D., Z. Hertelyová, Z. Andrejčáková, R. Vlčková, S. Gancarčíková, V. Petrilla, S. Ondrašovičová and L. Krešáková. 2016. The application of probiotics and flaxseed promotes metabolism of n-3 polyunsaturated fatty acids in pigs. J. Appl. Anim. Res. 45: 93–98.

Stenkamp-Strahm, C., C. McConnel, S. Magzamen, Z. Abdo and S. Reynolds. 2017. Associations between *Escherichia coli* O157 shedding and the faecal microbiota of dairy cows. J. Appl. Microbiol. 124: 881–898.

Stewart, M.C., J.L. Hodgson, H. Kim, D.R. Hutchins and D.R. Hodgson. 1995. Acute febrile diarrhea in horses: 86 cases (1986–1991). Aust. Vet. J. 72: 41–44.

Suchodolski, J.S. 2011. Intestinal microbiota of dogs and cats: a bigger world than we thought. Vet. Clin. North Am. Small Anim. Pract. 41: 261–272.

Swanson, K.S. and G.C. Fahey. 2006. Prebiotic impacts on companion animals. pp. 217–236. *In*: Gibson, G.R. and R.A. Rastall (eds.). Prebiotics Development and Application. John Wiley and Sons, Chichester, UK.

Swyers, K.L., A.O. Burk, T.G. Hartsock, E.M. Ungerfeld and J.L. Shelton. 2008. Effects of direct-fed microbial supplementation on digestibility and fermentation end products in horses fed low- and high-starch concentrates. J. Anim. Sci. 86: 2596–2608.

Tang, S.G.H., C.C. Sieo, R. Kalavathy, W.Z. Saad, S.T. Yong, SH.K. Wong and Y.W. Ho. 2015. Chemical composition of egg yolk and egg quality of laying hens fed prebiotic, probiotic, and synbiotic diets. 80(8): C1686–C1695.

Tarabees, R., K.M. Gafar, M.S. El-sayed, A.A. Shehata and M. Ahmed. 2018. Effects of dietary supplementation of probiotic mix and prebiotic on growth performance, cecal microbiota composition, and protection against *Escherichia coli* O78 in broiler chickens. Probiotics Antimicrob. Proteins 1–9.

Traub-Dargatz, J.L., M.D. Salman and J.L. Voss. 1991. Medical problems of adult horses, as ranked by equine practitioners. J. Am. Vet. Med. Assoc. 198: 1745–1747.

Tsukahara, T., T. Inatomi, K. Otomaru, M. Amatatsu, G.A. Romero-Pérez and R. Inoue. 2018. Probiotic supplementation improves reproductive performance of unvaccinated farmed sows infected with porcine epidemic diarrhea virus. Anim. Sci. J. 89: 1144–1151.

Tuan, T.N., P.M. Duc and K. Hatai. 2013. Overview of the use of probiotics in aquaculture. Int. J. Res. Fish. Aquac. 3: 89–97.

Tufarelli, V., A.M. Crovace, G. Rossi and V. Laudadio. 2017. Effect of a dietary probiotic blend on performance, blood characteristics, meat quality and faecal microbial shedding in growing-finishing pigs. S. Afr. J. of Anim. Sci. 47(6): 875–882.

Twomey, D.P., A.I. Wheelock, J. Flynn, W.J. Meaney, C. Hill and R.P. Ross. 2000. Protection against *Staphylococcus aureus* mastitis in dairy cows using a bismuth-based teat seal containing the bacteriocin, lacticin 3147. J. Dairy Sci. 83: 1981–1988.

Utz, E.M., A.L. Apás, M.A. Díaz, S.N. González and M.E. Arena. 2018. Goat milk mutagenesis is influenced by probiotic administration. Small Ruminant Res. 161: 24–27.

Uyeno, Y., S. Shigemori and T. Shimosato. 2015. Effect of probiotics/prebiotics on cattle health and productivity. Microbes Environ. 30(2): 126–132.

Villena, J., H. Aso, V.P.M.G. Rutten, H. Takahashi, W. van Eden and H. Kitazawa. 2018. Immunobiotics for the bovine host: their interaction with intestinal epithelial cells and their effect on antiviral immunity. Front. Immunol. 9: 1–0.

Vine, N.G., W.D. Leukes and H. Kaiser. 2006. Probiotics in marine larviculture. FEMS Microbiol. Rev. 30: 404–427.

Wang, H., X. Ni, L. Liu, D. Zeng, J. Lai, X. Qing, G. Guangyao Li, K. Pan and B. Jing. 2017. Controlling of growth performance, lipid deposits and fatty acid composition of chicken meat through a probiotic, *Lactobacillus johnsonii* during subclinical *Clostridium perfringens* infection. Lipids Health Dis. 16: 1–10.

Wanka, K.M., T. Damerau, B. Costas, A. Krueger, C. Schulz and S. Wuertz. 2018. Isolation and characterization of native probiotics for fish farming. BMC Microbiol. 18(1): 119.

Weese, J.S. 2002. Microbiologic evaluation of commercial probiotic. J. Am. Vet. Med. Assoc. 220: 794–797.

Weese, J.S., M.E. Anderson, A. Lowe and G.J. Monteith. 2003. Preliminary investigation of the probiotic potential of *Lactobacillus rhamnosus* strain GG in horses: fecal recovery following oral administration and safety. Can. Vet. J. 44(4): 299–302.

Weese, J.S., M.E.C. Anderson, A. Lowe, R. Penno, T.M. Da Costa, L. Button and K.C. Goth. 2004. Screening of the equine intestinal microflora for potential probiotic organisms. Equine Vet. J. 36(4): 351–355.

Weese, J.S. and J. Rousseau. 2005. Evaluation of *Lactobacillus pentosus* WE7 for prevention of diarrhea in neonatal foals. J. Am. Vet. Med. Assoc. 226: 2031–2034.

Wei, X., X. Liao, J. Cai, Z. Zheng, L. Zhang, T. Shang, Y. Fu, C. Hu, L. Ma and R. Zhang. 2017. Effects of *Bacillus amyloliquefaciens* LFB112 in the diet on growth of broilers and on the quality and fatty acid composition of broiler meat. Anim. Prod. Sci. 57(9): 1899–1905.

Wessels, S., L. Axelsson, E.B. Hansen, L. De Vuyst, S. Laulund, L. Lähteenmäki, S. Lingren, B. Mollet, S. Salminen and A. von Wright. 2004. The lactic acid bacteria, the food chain, and their regulation. Trends Food Sci. Technol. 15(10): 498–505.

[WHO] World Health Organization. 1997. The medical impact of the use of antimicrobials in food animals: Report of a WHO meeting. Berlin, Germany.

[WHO] World Health Organization. 2004. Second joint FAO/OIE/WHO expert workshop on non-human antimicrobial usage and antimicrobial resistance: Management options. Oslo, Norway.

[WHO] World Health Organization. 2017. WHO guidelines on use of medically important antimicrobials in food-producing animals. Geneva, Italy.

Xu, H., W. Huang, Q. Hou, L. Kwok, Z. Sun, H. Ma, F. Zhao, Y. Lee and H. Zhang. 2017. The effects of probiotics administration on the milk production, milk components and fecal bacteria microbiota of dairy cows. Sci. Bull. 62: 767–774.

Yamawaki, R.A., E.L. Milbradt, M.P. Coppola, J.C.Z. Rodrigues, R.L.A. Filho, C.R. Padovani and A.S. Okamoto. 2013. Effect of immersion and inoculation *in ovo* of *Lactobacillus* spp. in embryonated chicken eggs in the prevention of *Salmonella* Enteritidis after hatch. Poult. Sci. 92: 1560–1563.

Yirga, H. 2015. The use of probiotics in animal nutrition. J. Prob. Health 3(2): 1–10.

Yoon, I.K. and M.D. Stern. 1995. Influence of direct-fed microbials on ruminant microbial fermentation and performance of ruminants: A review. Asian-Austr. J. Anim. Sci. 8: 533–555.

Yuyama, T., S. Takai, S. Tsubaki, Y. Kado and M. Morotomi. 2004. Evaluation of a host-specific *Lactobacillus* probiotics in training horses and neonatal foals. J. Intest. Microbiol. 18: 101–106.

Zhang, L., L. Zhang, X. Zeng, L. Zhou, G. Cao and C. Yang. 2016. Effects of dietary supplementation of probiotic, *Clostridium butyricum*, on growth performance, immune response, intestinal barrier function, and digestive enzyme activity in broiler chickens challenged with *Escherichia coli* K88. J. Anim. Sci. Biotechno. 7(1): 3.

Zhao, T., M.P. Doyle, B.G. Harmon, C.A. Brown, E.P.O. Mueller and A.H. Parks. 1998. Reduction of carriage of enterohemorrhagic *Escherichia coli* O157:H7 in cattle by inoculation with probiotic bacteria. J. Clin. Microbiol. 36(3): 641–647.

Zimmermann, J.A., M.L. Fusari, E. Rossler, J.E. Blajman, D.M. Astesana, C.R. Olivero, A.P. Berisvil, M.L. Signorinib, M.V. Zbruna, L.S. Frizzo and L.P. Soto. 2016. Effects of probiotics in swines growth performance: A meta-analysis of randomised controlled trials. Anim. Feed Sci. Tech. 219: 280–293.

Zoumpopoulou, G., M. Kazou, V. Alexandraki, A. Angelopoulou, K. Papadimitriou, B. Pot and E. Tsakalidou. 2018. Probiotics and prebiotics: an overview on recent trends. pp. 1–34. *In*: Gioia, D.D. and B. Biavati (eds.). Probiotics and Prebiotics in Animal Health and Food Safety. Springer. Cham, Switzerland.

Index

P

Q

R

S

T

V

W